NMR of Paramagnetic Molecules in Biological Systems

PHYSICAL BIOINORGANIC CHEMISTRY SERIES

A. B. P. Lever and Harry B. Gray, Series Editors

NUMBER

1. A. B. P. Lever and Harry B. Gray (eds.), *Iron Porphyrins, Part I*, 1982
2. A. B. P. Lever and Harry B. Gray (eds.), *Iron Porphyrins, Part II*, 1982
3. Ivano Bertini and Claudio Luchinat, *NMR of Paramagnetic Molecules in Biological Systems*, 1986

   Other numbers in preparation

NMR of Paramagnetic Molecules in Biological Systems

Ivano Bertini

Claudio Luchinat

Department of Chemistry
University of Florence, Italy

The Benjamin/Cummings Publishing Company, Inc.

Menlo Park, California • Reading, Massachusetts
Don Mills, Ontario • Wokingham, U.K. • Amsterdam
Sydney • Singapore • Tokyo • Mexico City • Bogota
Santiago • San Juan

Sponsoring Editor Diane L. Bowen
Production Editor Greg Hubit
Copy Editor Debra Myson-Etherington

Library of Congress Cataloging-in-Publication Data

Bertini, Ivano.
NMR of paramagnetic molecules in biological systems.

Includes bibliographies and index.
1. Nuclear magnetic resonance. 2. Paramagnetism.
3. Biological chemistry – Methodology. I. Luchinat,
C. (Claudio), 1952- . II. Title. [DNLM:
1. Biochemistry – methods. 2. Nuclear Magnetic Resonance.
QU 25 B544n]
QH324.9.N8B47 1986 574.19′2′01538362 85-26715
ISBN 0-8053-0780-X

ABCDEFGHIJ – MA – 89876

The Benjamin/Cummings Publishing Company, Inc.
2727 Sand Hill Road
Menlo Park, California 94025

To
Luigi Sacconi
who taught us
to love Science

Contents

3 Unpaired Electron-Nuclear Coupling Detected by NMR: Nuclear Relaxation 47

4 Chemical Exchange and Chemical Equilibria 85

5 Experimental Techniques 103

6 Copper 137

7 Iron 165

8 Manganese 231

9 Other Transition Metal Ions 253

10 Lanthanides 273

Appendices

Series Editors' Foreword

In recent years there has been explosive growth in the study of metal ions in living systems, a field commonly referred to as "bioinorganic chemistry." This field crosses the boundaries of inorganic, organic, and physical chemistry, chemical physics, biophysics, biochemistry, and medicine. It demands of its practitioners a broadly based knowledge covering many of these areas. The field is vast, but there have been few textbooks other than some introductory texts and some early monographs. Most "bioinorganickers" were trained in inorganic chemistry or biochemistry. Surprisingly few have come from physics and related areas. One of the goals of the Physical Bioinorganic Chemistry Series is to provide monographs that will stimulate the entrance of more chemical physicists into the field.

The common theme of the series is: "What can we learn about the geometric structures, the electronic structures, and the mechanisms of metalloproteins and other bioinorganic molecules through the use of modern physical techniques?"

In each volume, background information on the chemical physics of simple inorganic systems is provided to form a basis for discussion of the more complex biological environments. In our exchanges with potential authors, we have emphasized the need to include only hard, factual information and accepted theory and to avoid inclusion of speculative materials so that the series will remain current and valid for many years.

Nuclear magnetic resonance (NMR) spectroscopy occupies a special place in the investigations of the structures and mechanisms of biological molecules. An important part of the field, that having to do with the NMR of species containing paramagnetic metal ions, is dealt with thoroughly and expertly in Volume 3 by Professor Ivano Bertini and Dr. Claudio Luchinat of the University of Florence.

A. B. P. Lever
Harry B. Gray

Preface

Nuclear magnetic resonance spectroscopy of paramagnetic systems has never gained wide popularity outside a few groups of specialists. We feel, however, that the theory necessary for the interpretation of NMR parameters of paramagnetic molecules is culturally qualifying by itself, and important within the framework of the general theory of electron-nucleus coupling. We also feel that recent developments in NMR instrumentation have allowed a more profound exploitation of the theory toward an understanding of the structure and reactivity of both chemical and biochemical systems, the relative share of investigations on the latter becoming larger and larger.

The book is, therefore, meant to be an attempt to disseminate this aspect of the NMR technique among biochemists, bioinorganic chemists, biophysicists, and biophysical chemists. With this aim in mind we have dedicated large sections of the text to theoretical and methodological aspects on one hand, and to a survey of investigative approaches for paramagnetic metalloproteins and model compounds on the other.

This book is intended for graduate students and researchers in the fields of chemistry and biochemistry. Since the content spans from the theory of the chemical bond in transition metal compounds to systematic biochemistry – through the theory of electron-nucleus interactions – some knowledge of these areas would be desirable, although not necessary, as well as a fundamental understanding of the NMR experiment.

Requisites in the rapidly expanding field of NMR imaging have recently led to the use of paramagnetic metal ions as contrast agents owing to their nuclear-relaxing properties; we hope that this book can provide some help in choosing the proper metal ion and experimental conditions.

Although perhaps annoying for those already familiar with the theory treated here, we have chosen to use SI units throughout. Radians have been always explicitly used as units to avoid possible computational errors due to the magical appearance or disappearance of 2π.

A researcher might find it difficult to read the entire book because of the complexity of the subject and the breadth to which it is treated. However, we hope the book will be a useful tool in interpreting NMR experiments and to appreciate the level of the research in this area.

We would like to express our deepest gratitude to our friends Harry B. Gray and A. B. P. Lever for giving us the opportunity to write the book and making us feel always confident of writing it.

We would also like to thank W. DeW. Horrocks who introduced us to NMR of paramagnetic molecules and to S. H. Koenig for introducing us to NMRD and letting us use his variable field instrument. They also read, corrected, and discussed several parts of the book.

The following scientists have read and made suggestions on various chapters: R. D. Brown, E. Cedergren-Zeppezauer, R. S. Drago, P. Fantucci, R. D. Fischer, J. H. Freed, L. Latos-Grazynski, A. S. Mildvan, G. Navon, A. V. Xavier, and M. Zeppezauer. We are thankful to them.

Discussions with J. Kowalewski, P. W. Atkins, H. L. Friedman, and H. G. Hertz are also acknowledged, as well as correspondence with R. J. P. Williams.

Finally, the importance of coffee break discussions with friends and students of the Departments of Chemistry and Physics of the University of Florence is stressed. Continuous discussions with D. Gatteschi have always been stimulating and enlightening, and careful suggestions and criticisms by L. Banci, A. Bencini, C. Benelli, M. Mancini, G. Martini, A. Dei, A. Scozzafava, and G. Spina are profoundly appreciated. Likewise, we acknowledge suggestions and criticisms by R. Basosi and G. Valensin of the University of Siena.

The work has been followed through its development by G. Lanini, L. Messori, and R. Monnanni, who discussed the scientific content and helped with the various aspects of organization. Thanks are expressed to S. Papaleo for preparing several figures and to L. Crociani, P. Giacalone, and G. Zanobini for help in typing the manuscript. We take this opportunity to also acknowledge the precious help of V. Piccinotti, an expert in NMR instrumentation, for allowing us to cultivate the NMR technique for a wide range of applications, and for his competent comments on the technical section of the book.

Ivano Bertini
Claudio Luchinat

CHAPTER 1

Introduction

1.1 WHY THE TITLE: *NMR OF PARAMAGNETIC MOLECULES IN BIOLOGICAL SYSTEMS*?

Nuclear magnetic resonance spectroscopy is one of the most fertile physicochemical tools of investigation. Its obvious appeal in biochemistry, as well as in chemistry, relies on the possibility of directly monitoring nuclei of atoms that are located in strategic positions or are directly involved in biological processes.

A substantial fraction of the extensive literature reported on NMR deals with experiments on paramagnetic molecules. The theory of NMR of paramagnetic systems has little in common with the general theory of NMR of diamagnetic systems. In fact, NMR of paramagnetic systems represents a well-shaped physicoinorganic research field in its own right, and is particularly well suited for coverage in a separate treatise.

There are at least two fundamentally important publications dealing with this subject. The first is *NMR of Paramagnetic Molecules* (Academic Press, 1973) edited by G. N. La Mar, W. DeW. Horrocks, Jr., and R. H. Holm. This book outlines the theory of electron-nucleus interactions, the most important of which will be reviewed here, and discusses the applications to inorganic systems. The chapter dedicated to biological applications, although interesting, only covers relevant examples in the literature up to 1971. It is worthy to recall that the first Fourier transform instrument became commercially available at the beginning of 1970 and that, as will be pointed out later, the Fourier transformation technique permitted a giant step in the field.

The second principal publication is *NMR in Biochemistry* (Clarendon Press, 1973) by R. A. Dwek, which represents a very successful attempt at revealing the potential of the NMR technique to researchers in biochemical fields. Within the general scheme of NMR theory and applications, the principles of NMR in paramagnetic systems are also clearly presented.

We, the authors, now feel that there is room and a need for an entire volume committed to NMR of paramagnetic molecules in biological systems. With this in mind we have devoted two chapters to the theoretical aspects of the NMR detectable part of the interaction between nuclei and unpaired electrons: Chapter 2 deals with the origin of the chemical shift and Chapter 3 with nuclear relaxation—we feel that relaxation studies should become more routinely familiar to every researcher in the field. Chapter 4 is dedicated to the treatment of chemical exchange processes, and Chapter 5 provides the theoretical background for the experimental methods that have been specifically developed for the investigation of paramagnetic molecules.

Having presented the principles of the discipline, the book progresses in Chapters 6 to 8 to a discussion of metal ions relevant in the study of biological systems: copper(II), iron(II) and iron(III), manganese(II) and manganese(III). Each chapter deals with the electronic properties of the paramagnetic metal ion according to the symmetry properties of the ground state. The main effects on the NMR parameters are then predicted and compared with typical examples of NMR measurements in biological systems. The examples chosen represent, in our view, some of the most meaningful studies of biological topics. Metal ions that play a minor role in biological systems, such as nickel(II), cobalt(II), chromium(III), vanadium(III), and oxovanadium(IV) are all treated in Chapter 9, which also has a short section dedicated to spin labels. Finally, the essential features of lanthanides and their applications as shift reagents are presented in Chapter 10.

This book does not aim to be an exhaustive review, but rather a critical presentation of the state of the art. Excellent reviews and book chapters that cover specific topics in the area of NMR of paramagnetic molecules in biological systems are available and are referenced at the end of the appropriate chapter. Book series on metalloproteins, metal-protein interactions, and relative models, which contain all biologically relevant results including those of paramagnetic systems, are also available. Among these, and in addition to the series that hosts this book, are: *Inorganic Biochemistry* (specialist periodical reports, The Chemical Society), *Nuclear Magnetic Resonance* (specialist periodical reports, The Chemical Society), *Biological Magnetic Resonance* (L. J. Berliner and J. Reuben, eds., Plenum Press, New York), *Metal Ions in Biological Systems* (H. Sigel, ed., Marcel Dekker, New York), *Magnetic Resonance in Biology* (J. S. Cohen, John Wiley & Sons, New York), *Inorganic Biochemistry and Advances in Inorganic Biochemistry*, (G. N. Eichhorn and L. G. Marzilli, eds., Elsevier North Holland, New York), *The Porphyrins* (D. Dolphin, ed., Academic Press, New York), *Metal Ions in Biology* (T. G. Spiro, ed., John Wiley & Sons, New York).

1.2 INTRODUCTORY REMARKS ON THE ELECTRONIC STRUCTURE OF METAL IONS

The introductory comments in this section are intended to: (a) help the reader who is not familiar with the chemistry of transition metal elements to understand the relationship between the electronic ground state of the metal and the NMR properties of the interacting nuclei; and (b) to provide a uniform vocabulary for the phenomena associated with the electronic properties of metal ions. The present discussion is restricted to $3d$ metal ions, since most of the book deals with these ions; however, extension to $4d$ and $5d$ elements does not require, in principle, any new concept. Extension to f elements will be treated in Chapter 10.

Electronic $3d^n$ configurations give rise to several free ion terms depending on the occupancy of the five d orbitals. Such terms arise because of all the possible combinations, C, of the m_l orbital quantum number and the m_s spin quantum number for each electron. C can be expressed by the general formula

$$C = \frac{10!}{n!(10-n)!} \tag{1.1}$$

where 10 is the number of possibilities for a single d electron to occupy five d orbitals; that is, 10 spin orbitals.

Not all of the C microstates have different energies: they are grouped into degenerate sets of microstates, which define the free ion terms. These sets are indicated as ${}^{2S+1}L$, where S is the total multielectron spin quantum number and L is the total orbital quantum number. The number of degenerate states for each free ion term is $(2S + 1) \times (2L + 1)$, where $2S + 1$ is the spin multiplicity and $2L + 1$ is the orbital multiplicity. Whereas $2S + 1$ is expressed by the appropriate number, L is expressed by a capital letter according to the following convention:

$L = 0$ S

$L = 1$ P

$L = 2$ D

$L = 3$ F

$L = 4$ G

$L = 5$ H

$L = 6$ I

Some examples of terms that will be encountered, therefore, are: ${}^2D$, ${}^3F$, ${}^6S$.

Of the free ion terms, the one of most interest is the ground term. This term can be easily identified through a simple rule: Represent the five d orbitals as boxes, as shown;

d orbitals	□	□	□	□	□
m_l	2	1	0	−1	−2

then find the ground term by placing the electrons in the orbital boxes in such a way as to obtain the maximum value of $\sum m_s$ and $\sum m_l$, which are S and L, respectively.

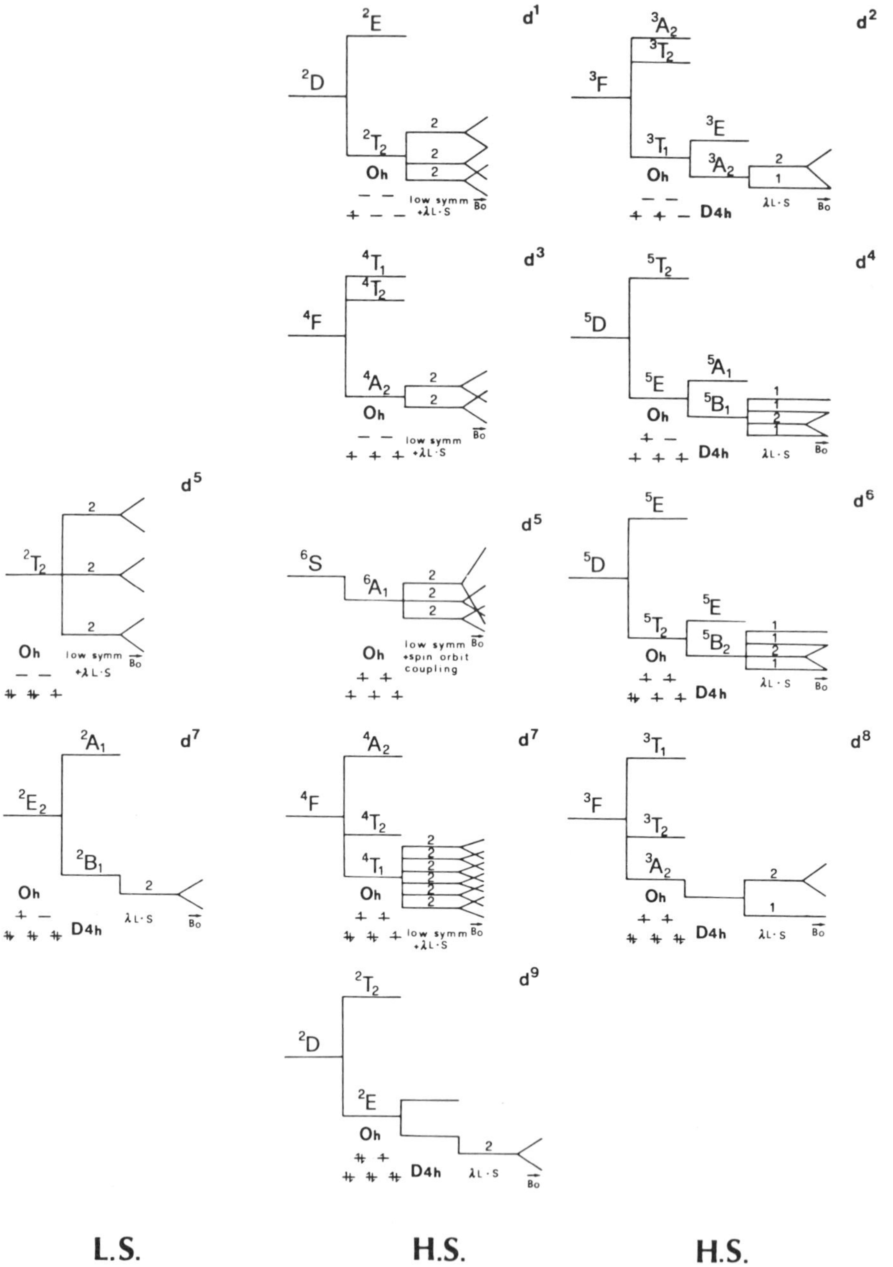

Figure 1.1 Lowest energy levels under successive perturbations (cubic ligand field, low-symmetry components, spin-orbit coupling, external magnetic field) for d^1 to d^9 metal ions. The numbers indicate the overall degeneracy of the levels; λ is the spin-orbit coupling constant of the

For a d^1 configuration (Ti^{3+}, VO^{2+}) the ground term ^2D ($S = \frac{1}{2}$, $L = 2$) is derived by placing the single electron in the orbital with the highest m_l value. The multiplicity of this term is 10, i.e. $(2S + 1) \cdot (2L + 1)$, which corresponds to the states $(2, \frac{1}{2})$, $(1, \frac{1}{2})$, $(0, \frac{1}{2})$, $(-1, \frac{1}{2})$, $(-2, \frac{1}{2})$, $(2, -\frac{1}{2})$, $(1, -\frac{1}{2})$, $(0, -\frac{1}{2})$, $(-1, -\frac{1}{2})$, $(-2, -\frac{1}{2})$, where the numbers ± 2, ± 1, and 0 are the m_l values and $\pm\frac{1}{2}$ are the m_s values. Application of Equation (1.1) shows that the ^2D term is the only free ion term of a d^1 configuration. The same term is also obtained for a d^9 configuration, which can be viewed as a hole in the closed-shell d^{10} configuration. Therefore, Cu^{2+} and Ag^{2+} have a ^2D free ion term. Configurations d^2 and d^8 (V^{3+} and Ni^{2+}, respectively) have a ground term ^3F corresponding to $S = 1$ (two unpaired electrons) and $L = 3$; that is, L is equal to the maximum value of M_L which is $(m_l)_1 + (m_l)_2 = 2 + 1$. Configurations d^3 (Cr^{3+}, V^{2+}) and d^7 (Co^{2+}) have a ground term ^4F with $S = \frac{3}{2}$ and $L = 3$; the latter value can be again determined from the maximum value of M_L which is $(m_l)_1 + (m_l)_2 + (m_l)_3 = 2 + 1 + 0$. Similarly, ^5D ($S = 2$, $L = 2$) is the free ion ground term for the d^4 (Mn^{3+}, Cr^{2+}) and d^6 (Fe^{2+}) configurations. The ground term for the d^5 configuration (Mn^{2+}, Fe^{3+}) is ^6S ($S = \frac{5}{2}$, $L = 0$).

The free ion terms refer to isolated atoms or metal ions. When the metal ions are bound in coordination compounds and metalloproteins, the five d orbitals are no longer degenerate, nor are all the states of a given term. The splitting of the free ion ground term in octahedral symmetry for the various configurations is shown in Figure 1.1 in order to illustrate the effect of coordination on the degeneracy of the terms as well as the formalism that will be used in this book. Although we have referred here to octahedral symmetry (O_h symmetry), it should be noted that such symmetry is so rare that in our discussion it represents an idealized symmetry for more or less distorted six-coordinated chromophores. This is a very important point to keep in mind as it also holds for other idealized symmetries, such as tetrahedral, trigonal bipyramidal, etc. The splitting of a term due to a given symmetry is such that the spin degeneracy remains unchanged; this is at variance with the orbital degeneracy which is reduced. The symbols A or B mean

free ion ground term; the subscripts g to the labelling of the symmetry species are omitted.

For d^1, d^3, low and high spin d^5, and high spin d^7 configurations, low-symmetry components are introduced simultaneously with spin-orbit coupling effects. Low-symmetry components are such that the overall symmetry is rhombic or less. In this way only the double Kramer's degeneracy is left before applying the external magnetic field. In the case of low spin d^7 and d^9 configurations, owing to Jahn-Teller effects, large separations from the split components of the E level are observed as a result of large static distortions. Generally, zero field splitting separations are largest when the level is a T level and smallest when spin-orbit coupling occurs only between the ground level and levels arising from excited free ion terms (high spin d^5). For d^n with even n D_{4h} distortions have been postulated in such a way so the ground state is orbitally nondegenerate. The same symmetry may provide nondegenerate spin-orbit ground states. The relative energies of the levels in the figures are arbitrary and not to scale.

orbitally nondegenerate, E twice degenerate, and T threefold degenerate.* The maximum orbital degeneracy left upon coordination is three in cubic symmetries, such as octahedral or tetrahedral. However, it is expected on theoretical grounds, and is actually found experimentally, that the chromophores are always distorted from the idealized symmetry in such a way as to remove all the orbital degeneracies. Furthermore, the coupling between the magnetic moment operator associated to the electron spin angular momentum ($\hat{\mathbf{S}}$) and that associated to the orbital angular momentum ($\hat{\mathbf{L}}$)—that is, the spin-orbit coupling operator—accounts for the splitting of the S manifold. Theory also expects that the geometry of the chromophore is such that every spin degeneracy of the ground state is removed, except the double degeneracy when S is equal to a half-integer (Jahn-Teller theorem). The double degeneracy is a consequence of the symmetry properties of angular momenta (Kramers doublets). The combined effects of small low-symmetry components and spin-orbit coupling on the ground level originating in the chosen idealized symmetry give rise to a set of levels, relatively close in energy, whose separation is called zero field splitting (ZFS) because this is the splitting at zero magnetic field—that is, before a magnetic field is applied; see Section 1.3. The energy separation between one level and the next within zero field split levels may range from zero up to several hundreds of wave numbers (cm^{-1}).

Sometimes the departure from an idealized geometry is so large that the orbitally nondegenerate ground level is separated by thousands of wave numbers from the first excited state. In such a case zero field splitting exists only if the spin degeneracy is ≥ 1. For example, elongated tetragonal copper(II) complexes have a $^2B_{1g}$ ground level separated by several thousands of wave numbers (cm^{-1}) from the $^2A_{1g}$ level arising from the splitting of a degenerate E_g level in octahedral symmetry. In a case of this kind the energy separation is not referred to as zero field splitting; the ground state is doubly spin degenerate.

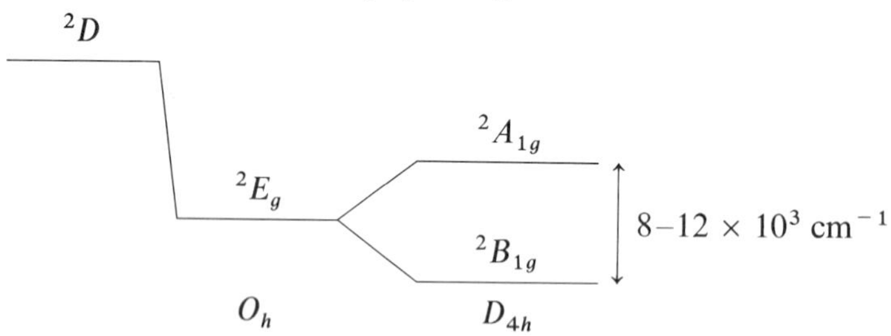

When the system is orbitally nondegenerate even in the reference idealized symmetry, whichever it is, the splitting at zero magnetic field of the S manifold ($S \geq 1$) ranges from zero to few tens of wave numbers (cm^{-1}).

The application of a magnetic field affects the energy of all the levels and splits the residual Kramers degeneracies. Transitions between states with an M_S separation of one ($\Delta M_S = 1$) are the type allowed in EPR spectroscopy; they are also of importance to the topic of the present book (see Section 1.3).

* These symbols can be written as either uppercase or lowercase letters. The latter are used to label orbitals according to their symmetry properties. Subscripts 1, 2, g, or u (for example, A_{1g} and a_{2u}) may also be necessary to specify the symmetry properties of a level or of an orbital.

1.3 PARAMAGNETIC MOLECULES

If a substance is placed in a homogeneous magnetic field, **H**, the field in the space occupied by the substance will normally differ from the field in free space. Such a field, expressed as the ratio between the magnetic induction, **B**, in the sample and the permeability of the vacuum, μ_0, is given by

$$\frac{\mathbf{B}}{\mu_0} = \mathbf{H} + \mathbf{M}$$

where **M** is the magnetization of the substance. The magnetic induction, **B**, which is given by

$$\mathbf{B} = \mu_0(\mathbf{H} + \mathbf{M}) = \mathbf{B}_0 + \mu_0\mathbf{M} \tag{1.2}$$

can be either slightly smaller or somewhat larger than $\mathbf{B}_0$—the induction produced in a vacuum by the same field.* If it is smaller the substance is diamagnetic; if it is bigger the substance is paramagnetic. Paramagnetic substances consist of molecules with unpaired electrons. All substances, even paramagnetic ones, have some intrinsic diamagnetism due to the interaction of paired electrons with the magnetic field; because of this a correction must be applied to magnetic measurements. Paramagnetism generally arises from unpaired electrons in metal ions; diamagnetic contributions are very large in metalloproteins owing to the large diamagnetic protein constituent.

The molar susceptibility, χ_M, of a substance is related to **M** by the equation

$$\chi_M = \frac{\mathrm{MW}}{\rho}\frac{\mu_0\mathbf{M}}{\mathbf{B}_0} = V_M\frac{\mu_0\mathbf{M}}{\mathbf{B}_0} \tag{1.3}$$

where MW is the molar mass, ρ is the density, and V_M is the molar volume. The SI units for χ_M are $\mathrm{m^3\,mol^{-1}}$. To convert to SI units values of χ_M given in the CGS emu units, as is generally found in the literature, the following relationship should be used:

$$\chi_M^{(\mathrm{SI})} = (4\pi \times 10^{-6})\chi_M^{(\mathrm{CGS\ emu})}$$

The molar susceptibility, corrected for the diamagnetic contribution, is the quantity of interest in this book and, for magnetically dilute systems, can be derived by calculating the magnetization, **M**, by a purely quantomechanical treatment. To do this, the magnetization can be expressed in terms of the magnetic moment: the magnetic moment, μ_i, of an atom in a given state, of energy E_i, is defined as

$$\mu_i = -\frac{\delta E_i}{\delta B_0}$$

The variation of energy with magnetic field is now evaluated over all the states belonging to S manifolds populated at a given temperature T. Under certain

* In this book the magnetic induction, $\mathbf{B}_0$, will always be used instead of **H**, although it will currently be called magnetic field. The SI unit for **B** is the tesla (T).

conditions, the energy of each state, i, in a magnetic field can be expressed as

$$E_i = E_i^0 + E_i' B_0 + E_i'' B_0^2$$

where E_i^0 is the energy of the ith state at zero field, and E_i' and E_i'' are called the first- and second-order Zeeman coefficients. Therefore μ_i is given by

$$\mu_i = -E_i' - 2E_i'' B_0$$

The magnetization M of a sample containing N_A atoms is defined as

$$M = \frac{N_A}{V_M} \langle \mu \rangle$$

where N_A is Avogadro's constant and $\langle \mu \rangle$ is obtained from μ_i by taking into account the Boltzmann distribution over the populated i levels. Therefore

$$M = \frac{N_A}{V_M} \langle \mu \rangle = \frac{N_A \sum_i (-E_i' - 2E_i'' B_0) \exp(-E_i/kT)}{V_M \sum_i \exp(-E_i/kT)} \tag{1.4}$$

where k is the Boltzmann constant and T is the absolute temperature. E_i' and E_i'' can be calculated through the magnetic moment operator

$$-\mu_B(\hat{\mathbf{L}} + g_e \hat{\mathbf{S}})$$

where $\mu_B = e\hbar/2m_e$ and $g_e = 2.0023$.* Here $\hbar = h/2\pi$ is the Planck constant, e is the absolute value of the charge of the electron, and m_e is the rest mass of the electron. For an external magnetic field along the z direction

$$E_i' = \mu_B \langle i | \hat{L}_z + g_e \hat{S}_z | i \rangle$$

$$E_i'' = \sum_{j \neq i} \frac{\mu_B^2 |\langle i | \hat{L}_z + g_e \hat{S}_z | j \rangle|^2}{E_i^0 - E_j^0}$$

In these expressions i refers to the ground multiplet state and j to all the excited states.

For atoms in low-symmetry environments, the magnetization and therefore the magnetic susceptibility are anisotropic. In such cases, these quantities become tensors for which the principal directions are defined along the molecular axes. If $E_i' B_0 \ll kT$ then, using Equations (1.3) and (1.4) and rearranging, the χ_M tensor is given by

$$\chi_{M_{\alpha\alpha}} = \mu_0 N_A \mu_B^2 \frac{\sum_i \left\{ \frac{|\langle i | \hat{L}_\alpha + g_e \hat{S}_\alpha | i \rangle|^2}{kT} - 2 \sum_{j \neq i} \frac{|\langle i | \hat{L}_\alpha + g_e \hat{S}_\alpha | j \rangle|^2}{E_i^0 - E_j^0} \right\} \exp(-E_i^0/kT)}{\sum_i \exp(-E_i^0/kT)} \tag{1.5}$$

which is the Van Vleck equation. The latter is of fundamental importance for

* In this book, g_e is taken as positive, and the operators containing g_e are explicitly written in such a way to contain positive g_e.

this book. Reference will often be made to the molecular χ values, which are obtained by dividing χ_M by Avogadro's constant, and are expressed in m^3.

It is implicit in the foregoing treatment that the energy of interaction between a magnetic moment, $\boldsymbol{\mu}$, and a magnetic field, $\mathbf{B}_0$, from a classical point of view is

$$E = -\boldsymbol{\mu} \cdot \mathbf{B}_0 \tag{1.6}$$

and that the related Hamiltonian is

$$\hat{\mathscr{H}} = \mu_B(\hat{\mathbf{L}} + g_e\hat{\mathbf{S}}) \cdot \mathbf{B}_0 = \mu_B(\hat{L}_z + g_e\hat{S}_z)B_{0z} + (\hat{L}_x + g_e\hat{S}_x)B_{0_x} + (\hat{L}_y + g_e\hat{S}_y)B_{0_y} \tag{1.7}$$

where B_{0_α} ($\alpha = x, y, z$) are the components of $\mathbf{B}_0$ along the molecular axes.

In order to eliminate $\hat{\mathbf{L}}$ from the Hamiltonian, it is convenient to write a spin-Hamiltonian of the form

$$\hat{\mathscr{H}} = \mu_B(g_{zz}\hat{S}_zB_{0_z} + g_{xx}\hat{S}_xB_{0_x} + g_{yy}\hat{S}_yB_{0_y}) = \mu_B\mathbf{B}_0 \cdot \mathbf{g} \cdot \hat{\mathbf{S}} \tag{1.8}$$

where g_{zz}, g_{xx}, g_{yy} are the g values along the molecular axes; they are also called the principal values of the **g** tensor. Under the conditions in which the second-order term in Equation (1.5) is negligible such a Hamiltonian is equivalent to Equation (1.7).

The principal g values are different from g_e in that they contain the orbital contribution to the magnetic moment. Since the orbital contribution depends on the spin-orbit coupling constant of the free ion ground term, λ, $g_{\alpha\alpha}$ ($\alpha = x, y, z$) is smaller than g_e if λ is positive (d^1 to d^4 ions), and larger than g_e if λ is negative (d^6 to d^9 ions). For d^5 ions ($S = \frac{5}{2}$) the $g_{\alpha\alpha}$ values are very close to g_e.

Operating with the spin-Hamiltonian of Equation (1.8) on an S manifold, $2S + 1$ different energies are obtained, all proportional to M_S. Without loss of generality, for $\mathbf{B}_0$ parallel to S_z

$$\mu_B B_0 g_{zz}\langle S, M_S|\hat{S}_z|S, M_S\rangle = \mu_B B_0 g_{zz} M_S \tag{1.9}$$

Therefore, the separation between levels differing by $\Delta M_S = 1$ is

$$\Delta E = \mu_B B_0 g_{zz}$$

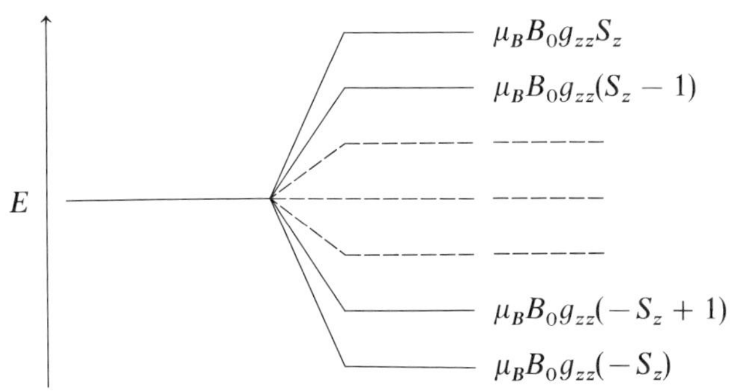

Transitions between these levels may be induced by applying electromagnetic microwave radiation such that

$$h\nu = \hbar\omega_S = \Delta E$$

1.4 THE MAGNETIC MOMENTS OF ELECTRONS AND NUCLEI

It is usual to define an effective magnetic moment, μ_{eff}, related to the experimentally determined value for χ_M through the relationship

$$\chi_M = \mu_0 \frac{N_A \mu_{eff}^2}{3kT} \tag{1.10}$$

Using Equations (1.5) and (1.10), and assuming that there are no contributions from the orbital angular momentum operator $\hat{\mathbf{L}}$—no orbital contributions to the magnetic susceptibility—the magnetic moment μ_{eff} associated with the spin angular momentum operator $\hat{\mathbf{S}}$, would be

$$\mu_{eff}^2 = g_e^2 \mu_B^2 S(S + 1) \tag{1.11}$$

The problem with electrons in compounds, however, is that there are always some orbital contributions so that Equation (1.11) does not rigorously hold. Sometimes this equation is used with a numerical g value different from g_e. With some approximations Equation (1.11) can be written

$$\mu_{eff}^2 = \bar{g}^2 \mu_B^2 S(S + 1) \tag{1.12}$$

where

$$\bar{g}^2 = \tfrac{1}{3}(g_{xx}^2 + g_{yy}^2 + g_{zz}^2) \tag{1.13}$$

For nuclei, the approach is analogous, except that the nuclear constant, g_N, which corresponds to the electronic g_e, for a given nucleus, is not sensitive to the symmetry of the chemical environment. When a nucleus possesses a nonzero total spin quantum number, I, its magnetic moment is given by

$$\mu^2 = g_N^2 \mu_N^2 I(I + 1) \tag{1.14}$$

where $\mu_N = e\hbar/2m_p$, m_p being the proton rest mass.

The Hamiltonian for the interaction of the nuclear magnetic moment with the magnetic field is

$$\hat{\mathscr{H}} = -\mu_N g_N \mathbf{B}_0 \cdot \hat{\mathbf{I}} \tag{1.15}$$

In analogy with electrons, the splitting of the I manifold in a magnetic field, $\mathbf{B}_0$, which determines the z direction, gives $2I + 1$ nuclear states separated by $\Delta E = \mu_N B_0 g_N$.

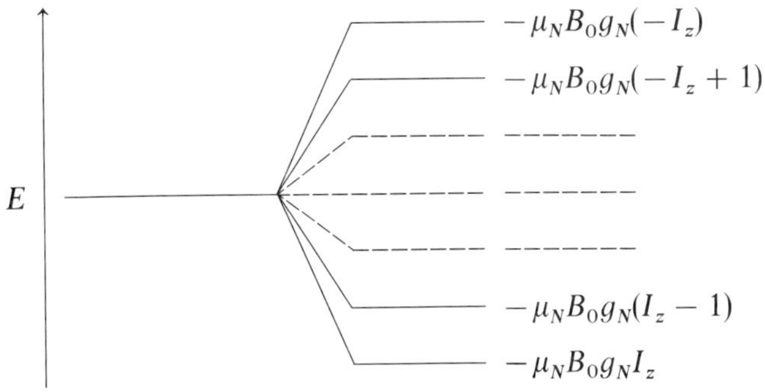

Note that the I manifold splits in the opposite sense to the S manifold previously described.

Again, transitions between nuclear levels may be induced by applying a radio-frequency such that

$$h\nu = \hbar\omega_I = \Delta E$$

The values of g_N for nuclei with $I \neq 0$ are reported in Table 1.1, together with their magnetogyric ratios, γ_N, given by $\gamma_N = g_N\mu_N/\hbar$. The larger $|g_N|$, or $|\gamma_N|$, the larger the magnetic moment and therefore the energy of interaction with the external magnetic field. Note that the NMR sensitivity of a nucleus N with spin I, S_N, relative to that of the ^{1}H nucleus, S_H, depends on the third power of their g_N ratio:

$$\frac{S_N}{S_H} = \frac{4}{3}\left(\frac{|g_N|}{|g_H|}\right)^3 I(I+1) \tag{1.16}$$

When $I > \frac{1}{2}$, the nucleus has a quadrupole moment, Q, which allows the nucleus itself to interact with electric field gradients. The quadrupole moments are also reported in Table 1.1.

1.5 NUCLEI SUITABLE FOR NMR INVESTIGATION IN PARAMAGNETIC SYSTEMS

As will be shown in Chapters 2 and 3, unpaired electrons affect the NMR linewidth and give rise to a paramagnetic shift. This line broadening decreases the quality of the NMR spectra although it contains valuable information. The technique cannot be used when the signal is too broad to be detected or when it is so broad that its resolution is below the threshold of usefulness. As far as the former problem is concerned, the use of nuclei that are more NMR sensitive can be helpful.

A lot of work is available on (in order) ^{1}H, ^{19}F, ^{31}P, ^{13}C, and ^{15}N. Nuclei with quadrupole moments, such as ^{17}O, ^{2}H, and ^{14}N, have in general large linewidths because the quadrupole moments efficiently couple with electric field gradients.

Table 1.1 NMR Properties of Nuclei

Isotope	Spin (I)	Natural Abundance (%)	Relative Sensitivity[a]	g_N	γ_N[b]	NMR Frequency (MHz at 2.3488 T)	Q[c]
^{1}H	$\frac{1}{2}$	99.98	1.00	5.5856912	2.6751987×10^{8}	100.000	—
^{2}H	1	1.56×10^{-2}	9.65×10^{-3}	0.8574376	4.106593×10^{7}	15.351	2.875×10^{-3}
^{3}H	$\frac{1}{2}$	0	1.21	5.957920	2.853473×10^{8}	106.664	—
^{3}He	$\frac{1}{2}$	1.3×10^{-4}	0.442	−4.255248	-2.037999×10^{8}	76.181	—
^{6}Li	1	7.42	8.50×10^{-3}	0.8220467	3.937092×10^{7}	14.717	-6.44×10^{-4}
^{7}Li	$\frac{3}{2}$	92.58	0.294	2.170949	1.039750×10^{8}	38.866	-3.66×10^{-2}
^{9}Be	$\frac{3}{2}$	100	1.39×10^{-2}	−0.7852	-3.7606×10^{7}	14.057	5.3×10^{-2}
^{10}B	3	19.58	1.99×10^{-2}	0.600217	2.87466×10^{7}	10.746	8.472×10^{-3}
^{11}B	$\frac{3}{2}$	80.42	0.165	1.792425	8.58460×10^{7}	32.090	4.065×10^{-2}
^{13}C	$\frac{1}{2}$	1.108	1.59×10^{-2}	1.404822	6.72822×10^{7}	25.150	—
^{14}N	1	99.63	1.01×10^{-3}	0.4037607	1.933763×10^{7}	7.228	1.56×10^{-2}
^{15}N	$\frac{1}{2}$	0.37	1.04×10^{-3}	−0.5663784	-2.712600×10^{7}	10.140	—
^{17}O	$\frac{5}{2}$	3.7×10^{-2}	2.91×10^{-2}	−0.75752	-3.62803×10^{7}	13.562	-2.578×10^{-2}
^{19}F	$\frac{1}{2}$	100	0.834	5.257732	2.518127×10^{8}	94.129	—
^{21}Ne	$\frac{3}{2}$	0.257	2.46×10^{-3}	−0.441197	-2.11306×10^{7}	7.899	1.029×10^{-1}
^{23}Na	$\frac{3}{2}$	100	9.27×10^{-2}	1.478347	7.080361×10^{7}	26.467	1.01×10^{-1}
^{25}Mg	$\frac{5}{2}$	10.13	2.68×10^{-3}	−0.34218	-1.63883×10^{7}	6.126	0.22
^{27}Al	$\frac{5}{2}$	100	0.207	1.456602	6.976216×10^{7}	26.077	0.140
^{29}Si	$\frac{1}{2}$	4.7	7.86×10^{-3}	−1.11058	-5.3190×10^{7}	19.883	—
^{31}P	$\frac{1}{2}$	100	6.65×10^{-2}	2.26320	1.08393×10^{8}	40.518	—
^{33}S	$\frac{3}{2}$	0.76	2.27×10^{-3}	0.429214	2.05567×10^{7}	7.684	-6.4×10^{-2}
^{35}Cl	$\frac{3}{2}$	75.53	4.72×10^{-3}	0.5479157	2.624176×10^{7}	9.809	-8.249×10^{-2}
^{37}Cl	$\frac{3}{2}$	24.47	2.72×10^{-3}	0.4560820	2.184349×10^{7}	8.165	-6.493×10^{-2}
^{39}K	$\frac{3}{2}$	93.1	5.10×10^{-4}	0.2609772	1.249918×10^{7}	4.672	4.9×10^{-2}

^{41}K	$\frac{3}{2}$	6.88	8.43×10^{-5}	0.1432466	6.860621×10^{6}	2.565	6.0×10^{-2}
^{43}Ca	$\frac{7}{2}$	0.145	6.42×10^{-3}	−0.37636	-1.80253×10^{7}	6.738	—
^{45}Sc	$\frac{7}{2}$	100	0.302	1.358995	6.508741×10^{7}	24.330	−0.22
^{47}Ti	$\frac{5}{2}$	7.28	2.10×10^{-3}	−0.31539	-1.51053×10^{7}	5.646	0.29
^{49}Ti	$\frac{7}{2}$	5.51	3.78×10^{-3}	−0.31548	-1.51094×10^{7}	5.648	0.24
^{50}V	6	0.24	5.58×10^{-2}	0.557908	2.67203×10^{7}	9.988	7×10^{-2}
^{51}V	$\frac{7}{2}$	99.76	0.384	1.47183	7.0491×10^{7}	26.350	-5.2×10^{-2}
^{53}Cr	$\frac{3}{2}$	9.55	9.08×10^{-4}	−0.31636	-1.51517×10^{7}	5.664	2.2×10^{-2}
^{55}Mn	$\frac{5}{2}$	100.00	0.176	1.3813	6.6155×10^{7}	24.729	0.40
^{57}Fe	$\frac{1}{2}$	2.19	3.42×10^{-5}	0.1812459	8.680549×10^{6}	3.245	—
^{59}Co	$\frac{7}{2}$	100.00	0.278	1.322	6.332×10^{7}	23.67	0.404
^{61}Ni	$\frac{3}{2}$	1.19	3.59×10^{-3}	−0.50001	-2.39475×10^{7}	8.952	0.162
^{63}Cu	$\frac{3}{2}$	69.09	9.34×10^{-2}	1.4822	7.0988×10^{7}	26.536	−0.209
^{65}Cu	$\frac{3}{2}$	30.91	0.115	1.5878	7.6046×10^{7}	28.426	−0.195
^{67}Zn	$\frac{5}{2}$	4.11	2.87×10^{-3}	0.350191	1.677198×10^{7}	6.269	0.150
^{69}Ga	$\frac{3}{2}$	60.4	6.97×10^{-2}	1.34439	6.43881×10^{7}	24.069	0.168
^{71}Ga	$\frac{3}{2}$	39.60	0.143	1.70818	8.18112×10^{7}	30.581	0.106
^{73}Ge	$\frac{9}{2}$	7.76	1.41×10^{-3}	−0.1954371	-9.360221×10^{6}	3.499	−0.173
^{75}As	$\frac{3}{2}$	100.00	2.54×10^{-2}	0.95965	4.5961×10^{7}	17.180	0.29
^{77}Se	$\frac{1}{2}$	7.58	7.03×10^{-3}	1.07012	5.1252×10^{7}	19.158	—
^{79}Br	$\frac{3}{2}$	50.54	7.94×10^{-2}	1.404266	6.72556×10^{7}	25.140	0.293
^{81}Br	$\frac{3}{2}$	49.46	9.95×10^{-2}	1.513707	7.24971×10^{7}	27.100	0.27
^{83}Kr	$\frac{9}{2}$	11.55	1.90×10^{-3}	−0.215704	-1.033089×10^{7}	3.862	0.270
^{85}Rb	$\frac{5}{2}$	72.15	1.06×10^{-2}	0.54121	2.59207×10^{7}	9.689	0.274
^{87}Rb	$\frac{3}{2}$	27.85	0.177	1.83416	8.78449×10^{7}	32.837	0.132

[a] NMR sensitivity at constant field and equal number of nuclei, relative to ^{1}H nuclei, given by $\frac{4}{3}(|g_N|/|g_H|)^3 I(I+1)$.

[b] γ_N is defined by: $\gamma_N = g_N \mu_N / \hbar$, where the nuclear magneton μ_N is 5.050824×10^{-27} J T^{-1}, and $\hbar$ is $1.0545887 \times 10^{-34}$ J s rad^{-1}.

[c] In units of 10^{-24} cm^2.

Table 1.1 NMR Properties of Nuclei (cont.)

Isotope	Spin (I)	Natural Abundance (%)	Relative Sensitivity[a]	g_N	γ_N[b]	NMR Frequency (MHz at 2.3488 T)	Q[c]
^{87}Sr	$\frac{9}{2}$	7.02	2.71×10^{-3}	−0.242849	-1.16309×10^{7}	4.348	0.15
^{89}Y	$\frac{1}{2}$	100.00	1.19×10^{-4}	−0.2748306	-1.316268×10^{7}	4.920	—
^{91}Zr	$\frac{5}{2}$	11.23	9.49×10^{-3}	−0.52145	-2.49741×10^{7}	9.335	—
^{93}Nb	$\frac{9}{2}$	100.00	0.488	1.37122	6.5673×10^{7}	24.549	−0.36
^{95}Mo	$\frac{5}{2}$	15.72	3.27×10^{-3}	−0.3657	-1.7514×10^{7}	6.547	-1.9×10^{-2}
^{97}Mo	$\frac{5}{2}$	9.46	3.49×10^{-3}	−0.3734	-1.7884×10^{7}	6.785	−0.102
^{99}Ru	$\frac{5}{2}$	12.72	1.13×10^{-3}	−0.2565	-1.2286×10^{7}	4.592	7.6×10^{-2}
^{101}Ru	$\frac{5}{2}$	17.07	1.59×10^{-3}	−0.2875	-1.3770×10^{7}	5.147	0.44
^{103}Rh	$\frac{1}{2}$	100.00	3.17×10^{-5}	−0.17680	-8.468×10^{6}	3.165	—
^{105}Pd	$\frac{5}{2}$	22.23	1.13×10^{-3}	−0.2568	-1.230×10^{7}	4.60	0.8
^{107}Ag	$\frac{1}{2}$	51.82	6.72×10^{-5}	−0.227140	-1.08786×10^{7}	4.066	—
^{109}Ag	$\frac{1}{2}$	48.18	1.02×10^{-4}	−0.2613810	-1.251852×10^{7}	4.679	—
^{111}Cd	$\frac{1}{2}$	12.75	9.66×10^{-3}	−1.1897712	-5.698264×10^{7}	21.300	—
^{113}Cd	$\frac{1}{2}$	12.26	1.11×10^{-2}	−1.2446010	-5.960865×10^{7}	22.282	—
^{113}In	$\frac{9}{2}$	4.28	0.351	1.22864	5.8844×10^{7}	21.996	0.846
^{115}In	$\frac{9}{2}$	95.72	0.353	1.23129	5.8971×10^{7}	2.044	0.861
^{115}Sn	$\frac{1}{2}$	0.35	3.56×10^{-2}	−1.83766	-8.8012×10^{7}	32.899	—
^{117}Sn	$\frac{1}{2}$	7.61	4.60×10^{-2}	−2.00208	-9.5887×10^{7}	35.843	—
^{119}Sn	$\frac{1}{2}$	8.58	5.27×10^{-2}	−2.09456	-1.00316×10^{8}	37.499	—
^{121}Sb	$\frac{5}{2}$	57.25	0.163	1.34536	6.4434×10^{7}	24.086	−0.20
^{123}Sb	$\frac{7}{2}$	42.75	4.66×10^{-2}	0.72851	3.4891×10^{7}	13.043	−0.26
^{123}Te	$\frac{1}{2}$	0.87	1.84×10^{-2}	−1.47358	-7.0575×10^{7}	26.381	—
^{125}Te	$\frac{1}{2}$	6.99	3.22×10^{-2}	−1.77656	-8.5086×10^{7}	31.806	—
^{127}I	$\frac{5}{2}$	100.00	9.54×10^{-2}	1.12531	5.38953×10^{7}	20.146	−0.789

^{129}Xe	$\frac{1}{2}$	26.44	2.16×10^{-2}	−1.555952	-7.45204×10^{7}	27.856	—
^{131}Xe	$\frac{3}{2}$	21.18	2.82×10^{-3}	0.461241	2.209056×10^{7}	8.258	−0.120
^{133}Cs	$\frac{7}{2}$	100.00	4.84×10^{-2}	0.7377209	3.533224×10^{7}	13.207	-3×10^{-3}
^{135}Ba	$\frac{3}{2}$	6.59	5.00×10^{-3}	0.558629	2.675484×10^{7}	10.001	0.18
^{137}Ba	$\frac{3}{2}$	11.32	7.00×10^{-3}	0.624910	2.992930×10^{7}	11.188	0.28
^{138}La	5	0.089	9.41×10^{-2}	0.74278	3.5575×10^{7}	13.298	0.51
^{139}La	$\frac{7}{2}$	99.911	6.06×10^{-2}	0.79520	3.8085×10^{7}	14.236	0.22
^{141}Pr	$\frac{5}{2}$	100.00	0.303	1.6544	7.924×10^{7}	29.62	-5.89×10^{-2}
^{143}Nd	$\frac{7}{2}$	12.17	3.39×10^{-3}	−0.3043	-1.457×10^{7}	5.45	−0.484
^{145}Nd	$\frac{7}{2}$	8.30	7.93×10^{-4}	−0.1874	-8.98×10^{6}	3.36	−0.253
^{147}Sm	$\frac{7}{2}$	14.97	1.52×10^{-3}	−0.23280	-1.1150×10^{7}	4.168	−0.18
^{149}Sm	$\frac{7}{2}$	13.83	8.52×10^{-4}	−0.19191	-9.192×10^{6}	3.436	5.2×10^{-2}
^{151}Eu	$\frac{5}{2}$	47.82	0.179	1.3887	6.6509×10^{7}	24.861	1.15
^{153}Eu	$\frac{5}{2}$	52.18	1.54×10^{-2}	0.6132	2.9368×10^{7}	10.978	2.94
^{155}Gd	$\frac{3}{2}$	14.73	1.48×10^{-4}	−0.1727	-8.273×10^{6}	3.092	1.59
^{157}Gd	$\frac{3}{2}$	15.68	3.34×10^{-4}	−0.2265	-1.0850×10^{7}	4.056	2.03
^{159}Tb	$\frac{3}{2}$	100.00	6.94×10^{-2}	1.343	6.431×10^{7}	24.04	1.18
^{161}Dy	$\frac{5}{2}$	18.88	4.75×10^{-4}	−0.1922	-9.205×10^{6}	3.441	2.44
^{163}Dy	$\frac{5}{2}$	24.97	1.30×10^{-3}	0.2690	1.2885×10^{7}	4.817	2.57
^{165}Ho	$\frac{7}{2}$	100.00	0.204	1.1923	5.710×10^{7}	21.35	2.73
^{167}Er	$\frac{7}{2}$	22.94	5.11×10^{-4}	−0.16186	-7.752×10^{6}	2.898	2.827
^{169}Tm	$\frac{1}{2}$	100.00	5.70×10^{-4}	−0.4632	-2.218×10^{7}	8.293	—
^{171}Yb	$\frac{1}{2}$	14.31	5.52×10^{-3}	0.98734	4.7287×10^{7}	17.676	—
^{173}Yb	$\frac{5}{2}$	16.13	1.35×10^{-3}	−0.271956	-1.30250×10^{7}	4.869	2.8
^{175}Lu	$\frac{7}{2}$	97.41	3.13×10^{-2}	0.63791	3.0552×10^{7}	11.421	5.68

[a] NMR sensitivity at constant field and equal number of nuclei, relative to ^{1}H nuclei, given by $\frac{4}{3}(|g_N|/|g_H|)^3 I(I+1)$.

[b] γ_N is defined by: $\gamma_N = g_N\mu_N/\hbar$, where the nuclear magneton μ_N is 5.050824×10^{-27} J T^{-1}, and $\hbar$ is $1.0545887 \times 10^{-34}$ J s rad^{-1}.

[c] In units of 10^{-24} cm^2.

Table 1.1 NMR Properties of Nuclei (cont.)

Isotope	Spin (I)	Natural Abundance (%)	Relative Sensitivity[a]	g_N	γ_N[b]	NMR Frequency (MHz at 2.3488 T)	Q[c]
^{176}Lu	7	2.59	4.05×10^{-2}	0.456	2.18×10^{7}	8.16	8
^{177}Hf	$\frac{7}{2}$	18.5	1.40×10^{-3}	0.22671	1.0858×10^{7}	4.059	4.5
^{179}Hf	$\frac{9}{2}$	13.75	5.47×10^{-4}	−0.14242	-6.821×10^{6}	2.550	5.1
^{181}Ta	$\frac{7}{2}$	99.99	3.74×10^{-2}	0.6771	3.243×10^{7}	12.12	3.9
^{183}W	$\frac{1}{2}$	14.40	7.50×10^{-5}	0.2355694	1.128231×10^{7}	4.217	—
^{185}Re	$\frac{5}{2}$	37.07	0.139	1.2748	6.1057×10^{7}	22.823	2.36
^{187}Re	$\frac{5}{2}$	62.93	0.143	1.2879	6.1681×10^{7}	23.057	2.24
^{187}Os	$\frac{1}{2}$	1.64	1.24×10^{-5}	0.12930	6.193×10^{6}	2.315	—
^{189}Os	$\frac{3}{2}$	16.10	2.44×10^{-3}	0.439955	2.107112×10^{7}	7.876	0.91
^{191}Ir	$\frac{3}{2}$	37.30	2.65×10^{-5}	0.0974	4.665×10^{6}	1.744	0.78
^{193}Ir	$\frac{3}{2}$	62.70	3.42×10^{-5}	0.1061	5.080×10^{6}	1.899	0.70
^{195}Pt	$\frac{1}{2}$	33.80	1.04×10^{-2}	1.21898	5.8382×10^{7}	21.823	—
^{197}Au	$\frac{3}{2}$	100.00	2.76×10^{-5}	0.098772	4.73056×10^{6}	1.768	0.594
^{199}Hg	$\frac{1}{2}$	16.84	5.94×10^{-3}	1.0117702	4.845750×10^{7}	18.114	—
^{201}Hg	$\frac{3}{2}$	13.22	1.49×10^{-3}	−0.373483	-1.788753×10^{7}	6.686	0.455
^{203}Tl	$\frac{1}{2}$	29.50	0.196	3.244514	1.553920×10^{8}	58.086	—
^{205}Tl	$\frac{1}{2}$	70.50	0.202	3.2764268	1.5692047×10^{8}	58.657	—
^{207}Pb	$\frac{1}{2}$	22.60	9.06×10^{-3}	1.16438	5.57666×10^{7}	20.846	—
^{209}Bi	$\frac{9}{2}$	100.00	0.144	0.91347	4.3749×10^{7}	16.354	−0.46
^{235}U	$\frac{7}{2}$	0.72	1.2×10^{-4}	−0.10	-4.8×10^{6}	1.8	4.55

[a] NMR sensitivity at constant field and equal number of nuclei, relative to ^{1}H nuclei, given by $\frac{4}{3}(|g_N|/|g_H|)^3 I(I+1)$.

[b] γ_N is defined by: $\gamma_N = g_N \mu_N / \hbar$, where the nuclear magneton μ_N is 5.050824×10^{-27} J T^{-1}, and $\hbar$ is $1.0545887 \times 10^{-34}$ J s rad^{-1}.

[c] In units of 10^{-24} cm^2.

Therefore, the effect of the paramagnetic center on the linewidth should be at least of the same order of magnitude of the quadrupolar coupling in order to be detected. This requirement narrows the range of usefulness of paramagnetic line broadening. For example, investigations on ^{33}S in biological systems are scarce owing to its low nuclear sensitivity and quadrupolar moment; in fact, to the authors' knowledge, there are no studies on ^{33}S in paramagnetic systems because the quality of the spectrum would be too low.

Metal ion nuclei, such as ^{113}Cd, ^{205}Tl, and ^{199}Hg, are sometimes investigated using NMR. However, it is not possible to study paramagnetic metal ions with the instrumentation currently available because the coupling between the nucleus and the unpaired electron is too large. These particular systems may be investigated through EPR and ENDOR spectroscopies.

All the considerations so far (and all the examples reported in this book, with the exception of Section 5.7) apply to NMR experiments in solution. As will be discussed in Chapter 2, both the diamagnetic and paramagnetic contributions to the chemical shift contain terms that are orientation dependent. Therefore, in the solid or frozen state the NMR signal of each equivalent set of nuclei will be spread out over such a large chemical shift range—owing to contributions from each particular orientation of the molecule—that both sensitivity and resolution will be dramatically lowered. Single crystal studies, not taking into account technical complications, would reduce this problem owing to the resultant crystalline order, but suitable crystals of large molecules are often impossible to grow. In solution, owing to molecular rotation, the anisotropy of the chemical shift is effectively averaged out. The borderline between a real solution and a medium with characteristics of the frozen state depends on the overall spreading in frequency of a given nuclear signal compared with the rotational time (see Section 3.1.1) of the molecule bearing that nucleus. For a frequency spreading of 10^5 Hz—which is an extremely high value to encounter, even at very high magnetic fields—the rotational time has only to be less than $\sim 10^{-6}$ s to average out the signal. Most of the systems of interest to this book fall well within this limit.

GENERAL REFERENCES

The energy levels in transition metal ions are treated in the following books:

Ballhausen, C. J. *Introduction to Ligand Field Theory*, McGraw-Hill, New York, 1962.
Figgis, B. N. *Introduction to Ligand Fields*, Interscience, New York, 1966.
Golding, R. M. *Applied Wave Mechanics*, D. Van Nostrand, London, 1969.
Ballhausen, C. J., and Gray, H. B. In *Coordination Chemistry*, Vol. I, A. E. Martell (ed.), Van Nostrand-Reinhold, New York, 1971, p. 1–83.
Drago, R. S. *Physical Methods in Chemistry*, Saunders, Philadelphia, 1977.
Ballhausen, C. J., and Gray, H. B. *Molecular Electronic Structures—An Introduction*, Benjamin/Cummings, Reading, MA, 1980.
Lever, A. B. P. *Inorganic Electronic Spectroscopy*, Elsevier, Amsterdam, Second Edition, 1984.

A comprehensive and classic book is:

Griffith, J. S. *The Theory of Transition Metal Ions*, Cambridge University Press, London, 1971.

An easy-to-read introduction to group theory as applied to transition metal ions is:

Cotton, F. A. *Chemical Applications of Group Theory*, Wiley-Interscience, New York, Second Edition, 1971.

The theory of EPR and the **g** *tensor is thoroughly covered in:*

Abragam, A., and Bleaney, B. *Electron Paramagnetic Resonance of Transition Ions*, Clarendon Press, Oxford, 1970.

The fundamentals of EPR are also clearly given in:

Wertz, J. E., and Bolton, J. R. *Electron Spin Resonance—Elementary Theory and Practical Applications*, McGraw-Hill, New York, 1972.
Atherton, N. M. *Electron Spin Resonance—Theory and Applications*, John Wiley & Sons, New York, 1973.

For a recent review of EPR parameters in transition metal ions see:

Bencini, A., and Gatteschi, D. *Trans. Met. Chem.* **8** (1982) 1.

The theory of the NMR experiment is fully developed in:

Abragam, A. *The Principles of Nuclear Magnetism*, Oxford University Press, Oxford, 1961.
Slichter, C. P. *Principles of Magnetic Resonance*, Springer Verlag, Berlin, 1978.

A more simple but clear presentation is:

Pople, J. A., Schneider, W. G., and Bernstein, H. J. *High-Resolution Nuclear Magnetic Resonance*, McGraw-Hill, New York, 1959.

Fourier transform experiments are fully described in the following books:

Farrar, T. C., and Becker, E. D. *Pulse and Fourier Transform Nuclear Magnetic Resonance*, Academic Press, London, 1971.
Shaw, D. *Fourier Transform NMR Spectroscopy*, Elsevier, Amsterdam, 1976.

CHAPTER 2

Unpaired Electron-Nuclear Coupling Detected by NMR: Isotropic Shift

2.1 THE NMR PARAMETERS

The NMR parameters in a classic NMR experiment are the chemical shift, the longitudinal nuclear relaxation time, T_1, and the transverse nuclear relaxation time, T_2. For a given nucleus signal, the chemical shift is generally defined as the difference in frequency, $\Delta\nu$, between the resonating frequency of the nucleus, ν, and that of the same nuclide in a reference compound, ν_0, which is specified in each experiment. The chemical shift can be expressed as a frequency shift in Hz, specifying the intensity of the applied magnetic field, or as the ratio between the frequency shift and the absolute resonance frequency, $(\Delta\nu/\nu_0)$, in ppm. The shift can also be measured at fixed frequency as a field shift, ΔB, or as $\Delta B/B_0$.* Shifts expressed in ppm are independent of the intensity of the applied magnetic field and values of $\Delta\nu/\nu_0$ and $\Delta B/B_0$ are numerically equal although opposite in sign. Since $h\nu = g_N\mu_N B_0$, the larger the resonance frequency at fixed field, the smaller the magnetic field at fixed frequency (Figure 2.1). Signals resonating at higher frequency (or lower field) with respect to the reference signal are said to be downfield shifted, whereas signals resonating at lower frequency (or higher field) are said to be upfield shifted. In the literature, downfield shifts are reported as negative when expressed as $\Delta B/B_0$ and positive when expressed as $\Delta\nu/\nu_0$; the reverse holds for upfield shifts. In this book all the shifts will be expressed as $\Delta\nu/\nu_0$, and therefore the downfield shifts will be positive and the upfield shifts negative.

Two factors contribute to the chemical shift in diamagnetic compounds: the first, called the diamagnetic shift, arises from the shielding of the nucleus from the

* In the literature, $\Delta H/H_0$ is often used instead of $\Delta B/B_0$, where H is the magnetic field strength.

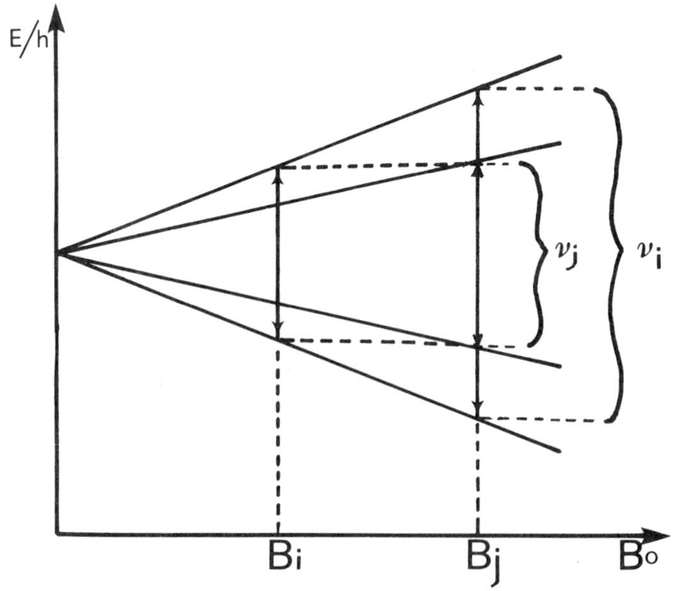

Figure 2.1 The chemical shift of signal i relative to signal j can be expressed as the difference in resonating field, $\Delta B = B_i - B_j$, at fixed frequency, ν_j, or as the difference in resonating frequency, $\Delta\nu = \nu_i - \nu_j$, at fixed field, B_j. If signal j is resonant at frequency ν_j and field B_j, signal i can be brought to resonance either by increasing the exciting frequency up to ν_i or by decreasing the external magnetic field down to B_i. In this example $\Delta\nu$ is positive and thus ΔB is negative.

external magnetic field as a result of the electron density around, and the circulation of electrons near, the nucleus itself; the second, called the paramagnetic shift, is due to excited electronic states. A detailed discussion of the importance of these contributions to shielding can be found in books dealing with classical NMR. As will be shown in Section 2.2, the presence of unpaired electrons in a molecule contribute further to the chemical shift. In this book, the term paramagnetic shift always refers to the presence of unpaired electrons.

The longitudinal relaxation time, T_1, is related to the time necessary for an ensemble of nuclear spins, which are equally distributed among the degenerate spin states in the absence of a magnetic field, to reach Boltzmann equilibrium after the magnetic field, $\mathbf{B}_0$, has been applied. More precisely, if reaching such an equilibrium is regarded as a first-order kinetic process, then T_1 is the time constant for that process. In general, T_1 is unequivocally defined as the time constant for reaching the Boltzmann distribution from any nonequilibrium initial state. At equilibrium, there is a finite component of the nuclear magnetization, $\mathbf{M}$, along the direction of $\mathbf{B}_0$ (z axis). Therefore, T_1 is also a measure of the time constant for the evolution of M_z toward its equilibrium value, justifying the term longitudinal relaxation time (see also Section 5.3).

Since longitudinal relaxation involves a change in the population of the nuclear spin levels and, therefore, a change in the total energy of the system, only those mechanisms that allow a nucleus to switch spin state through energy exchange with the environment are operative in such a process. For this reason, T_1 is also called the spin-lattice relaxation time. The more numerous and efficient the mechanisms that allow the nuclear spin to jump from one nuclear energy level to another, the shorter the longitudinal relaxation time.

T_1 processes, in fact, allow detection of the NMR signal. The resonance phenomenon occurs because of the different populations of the spin levels. The Boltzmann distribution is such that the difference between the populations of the various spin levels is small and, in the absence of mechanisms that allow longitudinal relaxation processes to occur, irradiation at the resonant frequency destroys the population difference, and no steady-state signal is observed. This phenomenon is called saturation. When the longitudinal relaxation mechanisms are effective in maintaining a population difference, the energy of the incident radiation is continuously absorbed, and NMR occurs. Such relaxation mechanisms depend ultimately on the coupling with magnetic fields that are fluctuating at the proper frequency and are generated by all the motions of the molecules in solution. The occurrence of large magnetic moments due to the presence of unpaired electrons, which relax at a rate orders of magnitude greater than that for the nuclei, provides additional, efficient relaxation pathways. As a consequence, T_1 decreases and the rate constant, T_1^{-1}, increases. The contribution of the unpaired electron-nucleus coupling to T_1^{-1} is expressed as T_{1M}^{-1}; note here that only the rates are additive. T_1 is currently determined using Fourier transform NMR instruments by disturbing the spin population equilibrium and measuring the magnetization recovery along the z axis as a function of time.

The transverse relaxation time, T_2, is related to the lifetime of the magnetization component in the xy plane; that is, the component perpendicular to the static magnetic field, $\mathbf{B}_0$. Such a component is zero at equilibrium because the components of the magnetic moment of the individual spins are randomly oriented in the xy plane. However, as a result of a perturbation of the spin system (see Section 5.3.1), a finite magnetization can be created in a general direction perpendicular to $\mathbf{B}_0$. T_2 is thus the time constant for this component to decay to zero.

T_2 is determined by the same mechanisms that determine T_1, that is, coupling with the lattice, and at low magnetic fields these mechanisms are equally effective on T_1 and T_2. At high magnetic fields, however, the coupling with slow magnetic fluctuations in the environment becomes relatively more effective for T_2 than for T_1. An additional mechanism contributing to T_2 is the simultaneous flipping of a pair of nuclear spins, which leaves the spin distribution constant without transferring energy to the lattice. Therefore, in general, $T_2 \leq T_1$ or $T_2^{-1} \geq T_1^{-1}$; the interrelationship between T_1 and T_2 in diamagnetic systems is given in any classical treatment of the NMR experiment. In analogy with T_1^{-1}, the presence of unpaired electrons gives independent contributions to T_2^{-1}, which are indicated as T_{2M}^{-1}.

In order to give an operative definition of T_2, the average lifetime of an ensemble of nuclei in a particular spin state is related to the energy uncertainty of that state through the Uncertainty Principle

$$\Delta t \cdot \Delta E \geq \hbar$$

Experimentally speaking, T_2 is related to the Lorentzian linewidth of the resonance signal, as in all types of spectroscopy, according to the relation

$$T_2^{-1} = \pi \, \Delta \nu$$

where $\Delta \nu$ is the half-height linewidth.

When the external magnetic field is inhomogeneous, inhomogeneous broadening of the line occurs due to the superposition of signals from nuclei resonating at different frequencies. This contribution to the line broadening should be minimized—by experienced operators—or made ineffective by appropriate experimental conditions if the observed linewidth is to be taken as a realistic measure of T_2^{-1}.

2.2 THE ISOTROPIC SHIFT

The nature of the coupling between unpaired electrons and nuclei is rather complex. Consider first a molecular orbital, ψ_1, of a paramagnetic molecule containing a single electron with $m_s = -\frac{1}{2}$. According to the LCAO-MO scheme, the distribution of this unpaired electron over the entire molecule is then given by the squares of the coefficients of each atomic orbital, ϕ, contributing to the spin orbital, ψ_1^- (Figure 2.2). For another molecular orbital, ψ_2, containing two electrons, the total

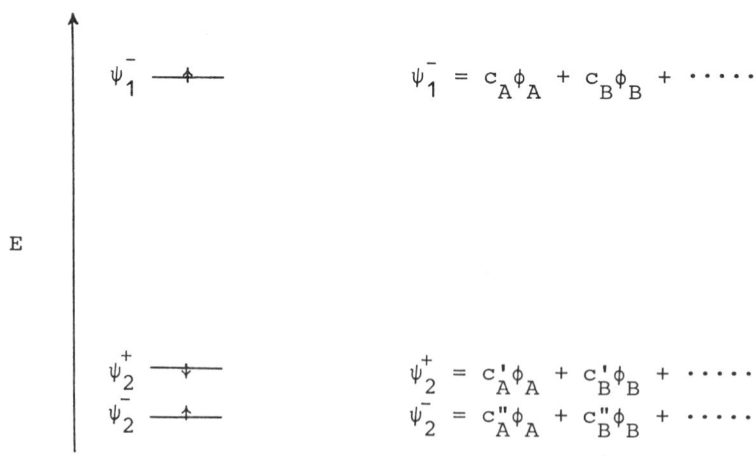

Figure 2.2 The unpaired spin density on the ϕ_A atomic orbital is given by the algebraic sum of a direct contribution (proportional to c_A^2) and a spin polarization effect on ψ_2 ($c'_A \neq c''_A$).

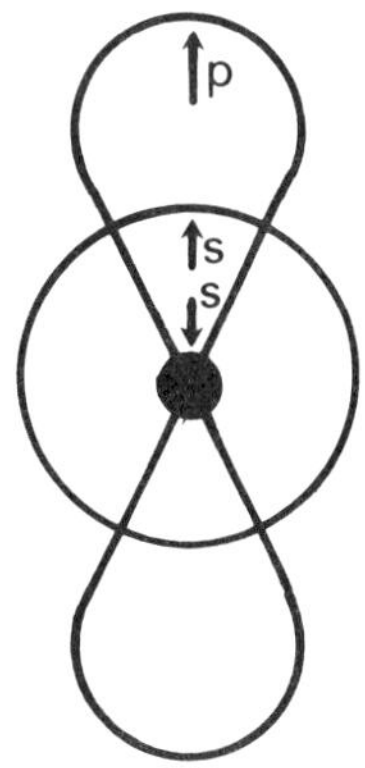

Figure 2.3 Spin polarization induced on a full 2*s* orbital by unpaired spin density in a 2*p* orbital.

spin is zero because one electron must have spin $-\frac{1}{2}$ and the other spin $\frac{1}{2}$. However, if the electrons in ψ_2 interact with the unpaired electron in ψ_1^-, the distribution of the two spins over each atomic orbital, ϕ, of ψ_2 may not be the same, and therefore the two spin orbitals, ψ_2^- and ψ_2^+, may differ slightly in energy. In fact, if ψ_1^- places a large amount of spin $-\frac{1}{2}$ on a given atomic orbital—for example, ϕ_A—spin correlation effects require a larger density of spin $-\frac{1}{2}$ with respect to spin $\frac{1}{2}$ on the same atomic orbital, and therefore ϕ_A will contribute more to ψ_2^- than to ψ_2^+. The electron with spin $\frac{1}{2}$ will thus have a minor density on that atomic orbital. For an atomic orbital that contributes to ψ_1^- to a smaller extent, the electron with spin $\frac{1}{2}$ will have a larger density than the electron with spin $-\frac{1}{2}$. So, for each atomic orbital the electron density is added if it has the same spin ($-\frac{1}{2}$ or $\frac{1}{2}$) and subtracted if it is different. The resultant density is the spin density on the atomic orbital in question.*

Atomic orbitals of the *s* type have nonzero values of ϕ and ϕ^2 at the nucleus. If an *s* orbital of an atom participates in a molecular orbital containing an unpaired electron, unpaired spin density is directly placed on the nucleus. However, this is rarely the only mechanism by which spin density is placed on a nucleus because spin polarization effects of the type just described are also present. Of these, spin polarization effects on σ molecular orbitals due to unpaired electrons delocalized in π molecular orbitals are quite important. π molecular orbitals do not place any spin density directly on nuclei because the nuclei lie in the nodal planes; however, unpaired spin density in p_π orbitals induces spin density on the nucleus through spin polarization of the *s* orbitals (Figure 2.3). In fact, electrons in $2p_\pi$ orbitals have a larger average distance from the nucleus than 2*s* or 1*s* orbitals. Therefore, if the $2p_\pi$ orbital is occupied by an unpaired electron, the *s* electron with the same spin will be farther from the nucleus than the other *s* electron with opposite spin.

Irrespective of the mechanism by which unpaired electrons on a molecule give rise to spin density on nuclei, the resultant effect is a large chemical shift. This

* Spin density arising from an excess of spin $-\frac{1}{2}$ is taken as positive. This is also called α spin density, although the quantomechanical label for the $-\frac{1}{2}$ state is β.

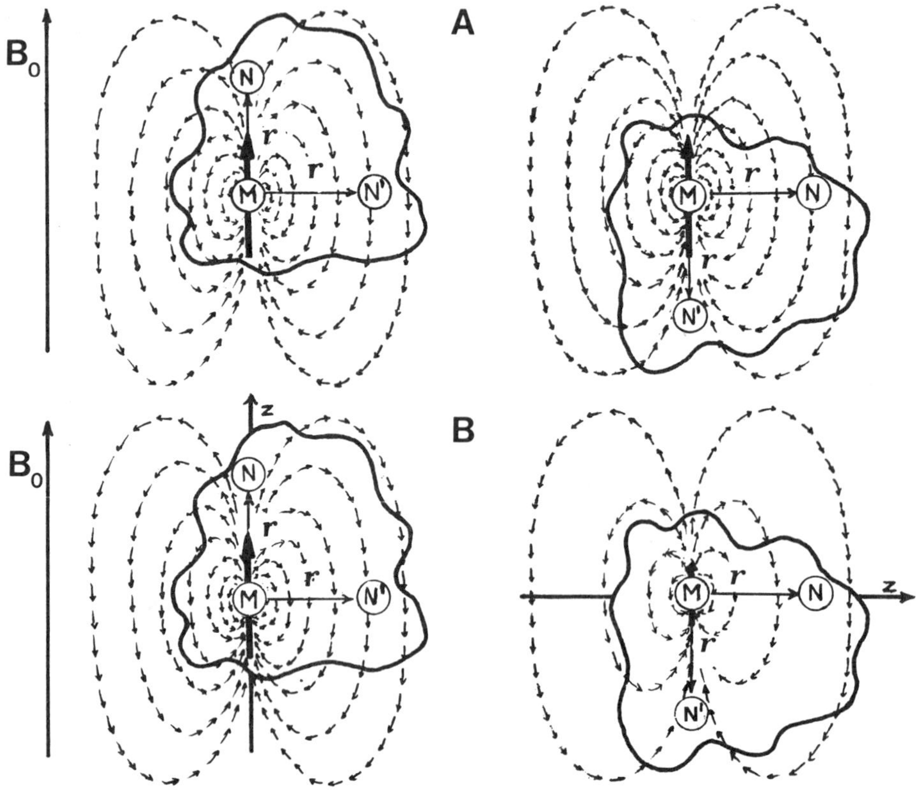

Figure 2.4 Effect on the N and N' nuclei of a magnetic field generated by unpaired electrons at the metal M. The solid line is the contour of a section of a rigid macromolecule containing M, N and N'. The electronic magnetic moment generates a dipolar magnetic field whose intensity in any volume element is proportional to the density of the field strength lines (→→→), and whose direction is given by the arrows. This field vectorially adds to the external magnetic field $\mathbf{B}_0$ which is homogeneous and therefore constant in the entire region. The electronic magnetic moment in the $\mathbf{B}_0$ direction is represented by the large solid arrow on M. If the electronic magnetic moment is isotropic, as illustrated in **A**, it always has the same value in the direction of $\mathbf{B}_0$ independently of the orientation of the molecule; the dipolar magnetic field experienced by the nucleus N at distance r from the metal ion is added to (left) or subtracted from (right) the external magnetic field $\mathbf{B}_0$, depending upon the relative orientation of r with respect to $\mathbf{B}_0$. Its average upon rapid rotation can be shown to be zero. For nucleus N' the dipolar magnetic field is also averaged to zero. When magnetic anisotropy is present, as illustrated in **B**, the electronic magnetic moment in the direction of $\mathbf{B}_0$ changes with the orientation of the molecule. The conditions are chosen so that the electronic magnetic moment is the same as in **A** when the molecular z axis is parallel to $\mathbf{B}_0$, and smaller when the molecular z axis is perpendicular to $\mathbf{B}_0$. Therefore the nuclei N and N' experience the same dipolar field in the left-hand diagrams of **B** and **A**, whereas they experience a smaller absolute value of the dipolar field in the right-hand diagram of **B** than in that of **A**. By keeping in mind that in **A** the dipolar field averaged over all the orientations is zero for both nuclei, it is possible to visualize that the average dipolar field experienced by N is >0 and that experienced by N' is <0.

contribution to the shift is known as the contact shift, or Fermi contact shift because Fermi accounted for the hyperfine splitting in the electronic spectrum of the hydrogen atom, pointing out that the electron has a finite density at the nucleus [1].

Another mechanism contributing to the chemical shift arises from dipolar interactions, that is, through-space interactions between the nuclear and electronic magnetic moments. Unlike the Fermi contact shift, which is isotropic, this mechanism depends on the direction of the magnetic field, with respect to the molecular axes, and, under conditions of rapid rotation, which are always met for the systems discussed here (see Section 1.5), averages to an isotropic contribution known as the pseudocontact or dipolar shift.

If it is assumed that the electron is localized on the metal ion, then magnetic anisotropy is entirely responsible for the dipolar shift. This can be visualized with the aid of Figure 2.4: even upon rapid rotation of the molecule, a nucleus, which is fixed in the molecular frame of reference according to which the χ tensor is defined, experiences an additional field that depends on the molecular coordinates of the nucleus itself. Obviously, a through-space interaction between magnetic dipoles also depends on their separation, r. However, the electron is not entirely localized on the metal ion; it is, at the very least, delocalized within the coordination polyhedron, and this complicates the theoretical treatment.

Finally, unpaired spin density on a p orbital of an atom may cause sizeable ligand-centered dipolar shifts of the atomic nucleus because of the proximity of the electron to the nucleus [2]. For reasons of symmetry, unpaired electrons in s orbitals do not provide dipolar contributions: since protons only have the $1s$ orbital there will be no local contribution to the dipolar coupling, unless the effects of spin density on the neighboring atoms are considered.

The algebraic sum of the Fermi contact shift and that part of the dipolar shift detected in solution is termed *isotropic shift*. Although theoretically well defined, the isotropic shift cannot be rigorously related to an experimental quantity. In fact, it should be defined as the difference between the shift measured in the paramagnetic compound and the shift of the same compound if it were diamagnetic (!). In practice, the experimental isotropic shift is obtained by subtracting from the chemical shift of the paramagnetic species the shift of a similar compound, which is often an analogous compound with a diamagnetic metal ion, or the shift of the ligand itself. For most purposes, the difference in shifts of the diamagnetic references is quite negligible with respect to the absolute value of the isotropic shift, and therefore experimentally defining the isotropic shift is no problem.

Sometimes the isotropic shift may be so large that the signals fall in a range completely different from those of the analogous diamagnetic compound. If the disadvantages of line broadening (see Chapter 3 for more detail) are not too severe, the isotropic shift technique can provide a quite sensitive tool for detecting subtle structural differences in solution. As an example, a spectrum is reported (Figure 2.5). In this case, the presence of a methyl group or a proton on the central nitrogen makes the two sides of the ligand nonequivalent, and therefore almost every proton gives rise to its own signal [3, 4].

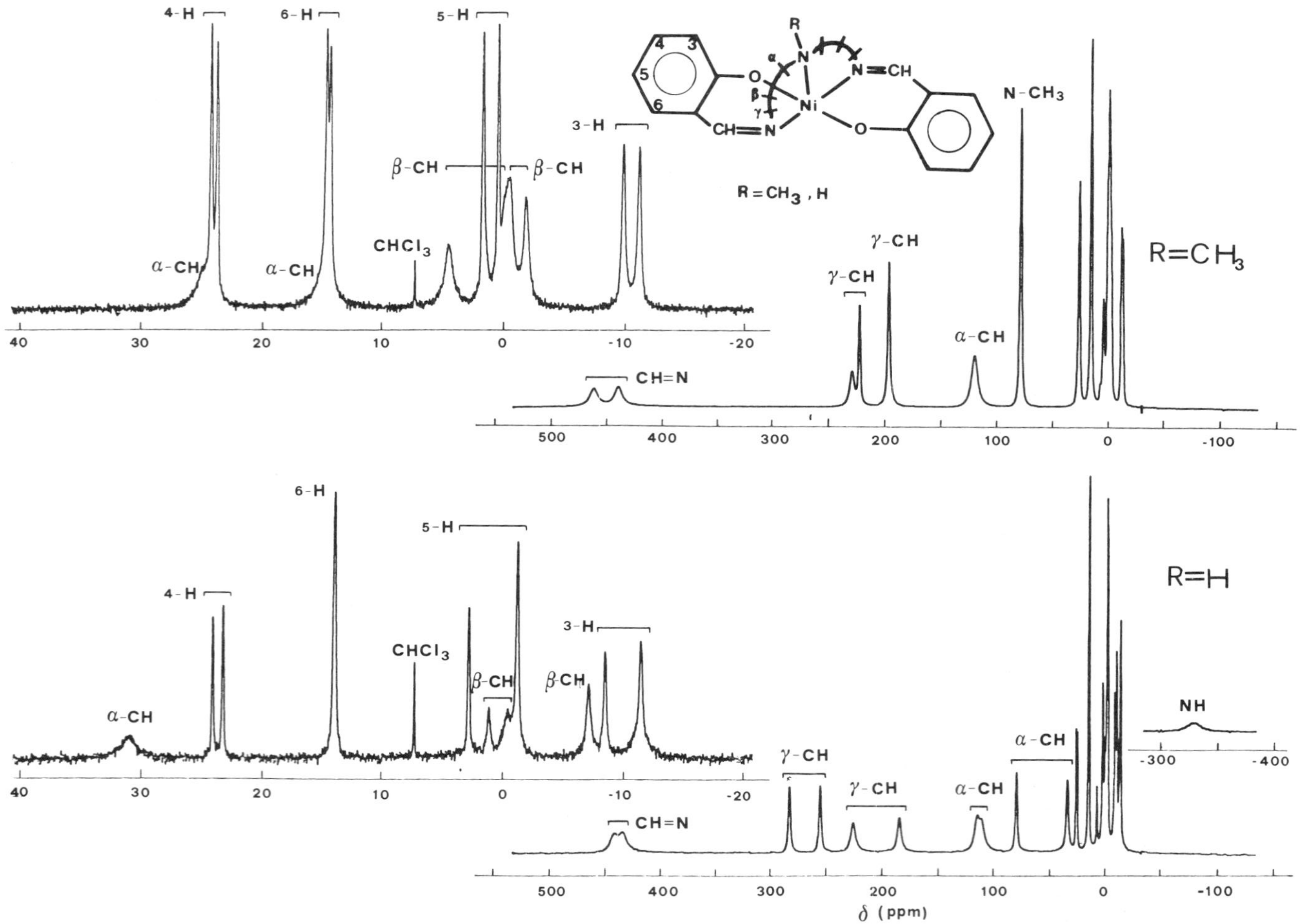

Figure 2.5 Room temperature ^{1}H NMR traces of $CDCl_3$ solutions of the complex shown in the upper part of the figure. The spectra on an expanded scale are recorded at 90 MHz; the full spectra are recorded at 60 MHz. The ^{1}H NMR signals of the corresponding diamagnetic zinc(II) complexes (not shown) lie in the 0–10 ppm range [3].

2.3 THE FERMI CONTACT SHIFT

To study the overall hyperfine interaction between unpaired electrons and a nucleus with spin I, the spin-Hamiltonian can be written as

$$\hat{\mathscr{H}} = A_c \hat{\mathbf{I}} \cdot \hat{\mathbf{S}} + \hat{\mathbf{I}} \cdot \mathbf{A}_{\text{dip}} \cdot \hat{\mathbf{S}}$$

where $\hat{\mathbf{S}}$ is the total electron spin operator of the molecule, A_c is the Fermi contact coupling constant, and $\mathbf{A}_{\text{dip}}$ is the dipolar tensor for the electron-nucleus coupling. Note that whereas A_c is a constant and does not depend on the coordinates of the nucleus with respect to the molecular axes, the contribution of $\mathbf{A}_{\text{dip}}$ to the overall hyperfine coupling depends on the actual coordinates of the resonating nucleus.

The Fermi contact coupling constant, A_c, is defined by

$$A_c = \frac{\mu_0}{3S} \hbar \gamma_N g_e \mu_B \sum_i [|\psi_i^-(0)|^2 - |\psi_i^+(0)|^2] \tag{2.1}$$

where ψ_i are the occupied molecular orbitals, $|\psi_i^-(0)|^2$ and $|\psi_i^+(0)|^2$ represent the $-\frac{1}{2}$ and $\frac{1}{2}$ spin density on the given nucleus due to the ith molecular orbital, S is the total spin of the molecule, and all the other symbols are defined as in Sections 1.3 and 1.4. The Fermi contact shift is then proportional to A_c:

$$\left(\frac{\Delta\nu}{\nu_0}\right)^{\text{con}} = -\left(\frac{\Delta B}{B_0}\right)^{\text{con}} = -\frac{A_c}{\hbar \gamma_N B_0} \langle S_z \rangle \tag{2.2}$$

where $\langle S_z \rangle$ is the average value of S_z for all the spin states weighted according to the Boltzmann distribution. Only the average value of S_z is meaningful with respect to the NMR experiment since electrons relax several orders of magnitude faster than the coupling frequency, $A_c/\hbar$, expressed in rad s^{-1}.

In the simplest case of an orbitally nondegenerate ground state under isotropic conditions with $g = g_e$ and no zero field splitting, $\langle S_z \rangle$ is given by

$$\langle S_z \rangle = -S(S+1) \frac{g_e \mu_B B_0}{3kT} \tag{2.3}$$

and the Fermi contact shift is

$$\left(\frac{\Delta\nu}{\nu_0}\right)^{\text{con}} = -\left(\frac{\Delta B}{B_0}\right)^{\text{con}} = \left(\frac{A_c}{\hbar}\right) \frac{g_e \mu_B S(S+1)}{3\gamma_N kT} \tag{2.4}$$

where k is the Boltzmann constant and T is the absolute temperature [5]. In NMR, the A_c parameter is usually referred to as the hyperfine coupling constant, although the overall hyperfine coupling also contains a dipolar contribution. If μ_B, $\hbar$, k, T, and γ_N are expressed in SI units, then the A_c parameter is in joules:

$$\left(\frac{\Delta\nu}{\nu_0}\right)^{\text{con}} = \left[\frac{A_c(\text{J})}{\hbar(\text{J s rad}^{-1})}\right]\left[\frac{g_e \mu_B(\text{J T}^{-1}) S(S+1)}{3\gamma_N(\text{rad s}^{-1}\,\text{T}^{-1}) k(\text{J K}^{-1}) T(\text{K})}\right]$$

However, it is customary to report the value of the hyperfine coupling constant in terms of $A_c/\hbar$ (rad s^{-1}) or A_c/h (Hz). Furthermore, to make the comparison with values obtained by EPR more meaningful, it may be convenient to express the hyperfine coupling constant in tesla. For this purpose, Equation (2.4) must be multiplied by $g_e\mu_B$; when $g \neq g_e$ the actual g value, see Equation (1.13), should be used and its value explicitly stated (*vide infra*):

$$\left(\frac{\Delta\nu}{\nu_0}\right)^{\mathrm{con}} = \left(\frac{A_c}{\hbar}\right)\frac{\bar{g}^2\mu_B^2 S(S+1)}{3\gamma_N kT} \tag{2.5}$$

A more general expression that takes into account the effect of zero field splitting in an orbitally nondegenerate spin multiplet is

$$\left(\frac{\Delta\nu}{\nu_0}\right)^{\mathrm{con}} = \frac{1}{\mu_0}\frac{A_c}{3\hbar\gamma_N\mu_B}\left[\frac{\chi_{xx}}{g_{xx}} + \frac{\chi_{yy}}{g_{yy}} + \frac{\chi_{zz}}{g_{zz}}\right] \tag{2.6}$$

where χ_{ii} and g_{ii} are the principal components of the molecular susceptibility and **g** tensors. However, this equation is difficult to use as such because the values in square brackets are generally unknown. By assuming negligible zero field splitting and with some approximations, χ and g can be related by the following formula—cf. Equations (1.10) and (1.12):

$$\chi = \frac{\chi_M}{N_A} = \mu_0\frac{\bar{g}^2\mu_B^2 S(S+1)}{3kT} \tag{2.7}$$

The Fermi contact shift then becomes

$$\left(\frac{\Delta\nu}{\nu_0}\right)^{\mathrm{con}} = \left(\frac{A_c}{\hbar}\right)\frac{\bar{g}\mu_B S(S+1)}{3\gamma_N kT} \tag{2.8}$$

This equation differs from Equation (2.4) in that the average g value of the system has been substituted for the free ion, g_e, value. The average g value can be obtained from Equation (2.7) through a single magnetic susceptibility measurement. When there are several zero field split levels occupied to a different extent at room temperature, deviations from Equation (2.8) may be severe.

At this point, it is appropriate to develop a formula expressing the Fermi contact shift for a simple system; that is, $S = \frac{5}{2}$ with tetragonal symmetry and no orbital angular momentum. The zero field split energy levels, for $\mathbf{B}_0$ along z, can be represented as a function of the M_S values as follows:

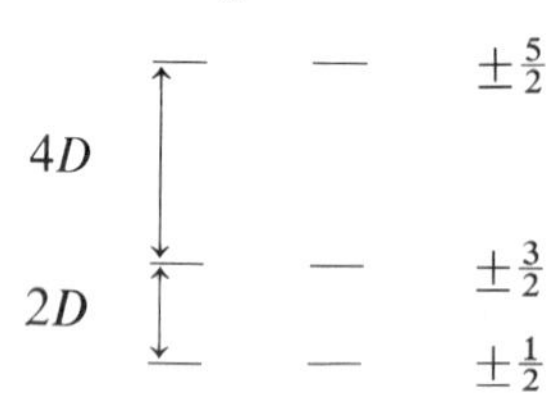

where D is the zero field splitting parameter. The principal susceptibility values can then be calculated through the Van Vleck equation, shown in Equation (1.5), as long as $2D$ is larger than the perturbation due to an external magnetic field:

$$\chi_{zz} = \mu_0 \frac{g_{\|}^2 \mu_B^2}{4kT} \left(\frac{1 + 9e^{-2X} + 25e^{-6X}}{1 + e^{-2X} + e^{-6X}} \right); \qquad X = \frac{D}{kT} \tag{2.9}$$

$$\chi_{xx} = \chi_{yy} = \mu_0 \frac{g_{\perp}^2 \mu_B^2}{4kT} \left(\frac{9 + \frac{8}{X} - \frac{11}{2X} e^{-2X} - \frac{5}{2X} e^{-6X}}{1 + e^{-2X} + e^{-6X}} \right) \tag{2.10}$$

By substituting into Equation (2.6), and taking $e^X = 1 + X$ $(D \ll kT)$, the following expression is obtained [6]:

$$\left(\frac{\Delta \nu}{\nu_0} \right)^{\text{con}} = \left(\frac{A_c}{\hbar} \right) \frac{35 \bar{g} \mu_B}{12 \gamma_N kT} \left[1 - \frac{32(g_{\|} - g_{\perp})D}{45 \bar{g} kT} \right] \tag{2.11}$$

This expression contains one term dependent on T^{-1} and another dependent on T^{-2}—the latter term being a consequence of the Boltzmann population of the zero field split levels. The g values here are not the experimental values obtained by EPR but the molecular g values, which can be obtained through a proper analysis of the experimental values. For $S = \frac{5}{2}$ and excited states far elevated in energy, the molecular g values are close to the free electron, g_e, value and D is relatively small; therefore, the term in T^{-2} can be neglected. But this may not be the case when zero field splitting arises from orbitally nearly degenerate systems; in such cases the treatment is more complicated and a more general expression would be needed [6]. As the use of a general expression requires a very detailed knowledge of the electronic structure, reference is made to either Equation (2.4) or (2.8) for most practical purposes.

The Fermi contact shift is independent of the magnitude of the nuclear magnetogyric ratio, γ_N, since γ_N is also included in the definition of A_c in Equation (2.1), and so $\Delta\nu/\nu_0$ is insensitive to γ_N. The large Fermi contact shifts experienced by nuclei other than ^{1}H and ^{2}H are due to larger spin densities on those nuclei.

2.3.1 The hyperfine coupling constant and spin density

Since the hyperfine coupling constant, A_c, is related to the spin density on the resonating nucleus, it holds some important information about the nature of the interaction between the resonating nucleus and the paramagnetic metal ion.

A_c is the sum of a direct delocalization mechanism (A_d) and a polarization mechanism (A_p). For protons (and deuterons),

$$A_d = \frac{\mu_0}{3S} \hbar \gamma_N g_e \mu_B \sum_i |\psi_i(0)|^2 \tag{2.12}$$

where summation is performed on all the molecular orbitals containing only one electron. Equation (2.12) can also be written as

$$\frac{A_d}{h} = \frac{K\rho}{2S} \tag{2.13}$$

where K is the experimental hyperfine splitting, which in the hydrogen atom is 1.42×10^9 Hz, and $\rho/2S$ is the unpaired electron spin density, normalized to one electron.

For protons (and deuterons), A_p arises from polarization effects between electrons in singly occupied molecular orbitals and those in doubly occupied molecular orbitals. Its evaluation from first principles is a difficult task; however, it may be estimated semiempirically using the spin density due to the A_d mechanism of various atomic orbitals in the molecule. Much is known for the cases in which unpaired electrons are in π orbitals of aromatic moieties. For example, a positive p_π spin density (corresponding to an excess of $-\frac{1}{2}$ spin) on an sp^2 carbon polarizes negative spin density at the nucleus of the attached proton (Figure 2.6A). The so-called McConnell equation [7] relates the π spin density on the carbon atom, ρ_C^π, to the hyperfine coupling constant of the attached proton:

$$\frac{A_p}{h} = \frac{Q\rho_C^\pi}{2S} \tag{2.14}$$

where $\rho_C^\pi/2S$ is the π spin density on the carbon atom, normalized to one electron, and Q is proposed [7] to be a constant equal to $\simeq -70$ MHz (see later). Values

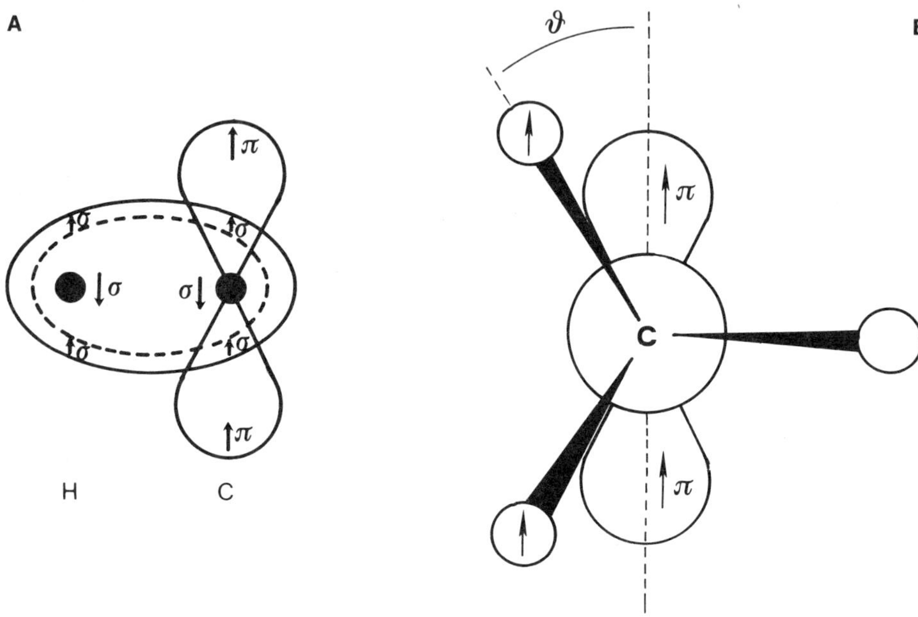

Figure 2.6 **A** Positive p_π spin density on an sp^2 carbon induces unbalancing of the spin density on the doubly occupied C—H bonding σ orbital. The sign of the induced spin density is positive in the outer region and negative in the inner region of the σ orbital. **B** Positive p_π spin density on an sp^2 carbon induces positive spin density on the 1s proton orbitals of an attached CH_3 group. The extent of the overlap depends on the angle θ between the axis of the p_π orbital of the sp^2 carbon and the C—C—H plane.

between −60 and −70 MHz have been proposed for the OH [8] and NH [9] fragments.

A peculiarity of these π spin density delocalizations is that the absolute values of the isotropic shifts are not sensitive to the distance from the paramagnetic center. Furthermore, both upfield and downfield shifts are observed, often alternating between adjacent positions. This is due to unpaired spin density in π atomic orbitals, which alternates as a consequence of spin polarization effects of unpaired electrons on doubly occupied molecular orbitals. In cases where it is reasonable to assume that the isotropic shifts are contact in origin and arise from π spin delocalization, the McConnell equation relates the experimental A_c value to the π spin density. Under these conditions, when a methyl group substitutes a hydrogen attached to an aromatic carbon, the ^{1}H methyl NMR isotropic shift changes its sign with respect to that of the substituted proton. The Q_{CCH_3} value is theoretically predicted to be +75 MHz [10], although it has been found to be quite variable (from 14 to 115 MHz). The positive sign of Q_{CCH_3} can be accounted for on the basis of direct overlap between a p_π orbital of a carbon and the appropriate linear combination of 1s orbitals of the three hydrogens (Figure 2.6B). The value of Q_{CCH_3} is calculated [10] for the case of rapid rotation around the C—C bond; that is, when rotation is faster than A_c/h. Some typical A_c/h values for nickel(II)-containing systems whose dipolar contributions to the isotropic shift have been neglected or factorized out are reported in Table 2.1. Analogous patterns are expected for d^5 to d^9 ions (if the delocalization mechanism is the same), whereas for d^1 to d^4 ions a reverse sign of the shifts may be expected. The ^{2}H contact shifts for the analogous deuterated complexes are expected to be very similar.

For nuclei other than protons and deuterons the electronic environment is considerably more complicated, and a variety of interactions have to be considered. For a nucleus of an element in the second period of the periodic table, the 2s electrons make an important contribution to the hyperfine coupling constant. Another significant contribution comes from the polarization of the 1s electrons. In addition, the neighboring atoms with unpaired spin density have to be considered. As an example, consider a system containing sp^2 hybridized carbon atoms with a single unpaired electron in a π orbital and all other electrons paired in σ atomic orbitals. For a given carbon atom the hyperfine contribution [16] from the 2s orbital spin density, arising from spin polarization in the bond C—X$_i$, is due to spin polarization from π spin density on that atom, ρ_C^π, and from that on neighboring atoms, $\rho_{X_i}^\pi$:

$$\left(\frac{A_p}{h}\right)_{2s} = \sum_{i=1}^{3} Q_{CX_i}^{C}\, \rho_C^\pi + \sum_{i=1}^{3} Q_{X_iC}^{C}\, \rho_{X_i}^\pi \tag{2.15}$$

where X$_i$ are the three atoms bound to the carbon under consideration, and Q^C is a constant with a subscript representing the bond involved. The first letter (C or X) of the subscript designates the atom on which the π electron is located. For example, Q_{CX} represents the hyperfine splitting due to the interaction of the C—X bond with a π electron spin on the atom C. Some Q values are reported in Table 2.2 [16–19]. The hyperfine contribution from the 1s orbital spin density

Table 2.1 Some Indicative A_c/h Values for Protons in Systems with a Predominant π Spin-Delocalization Mechanism

Structure	Proton	A_c/h (MHz)	Reference
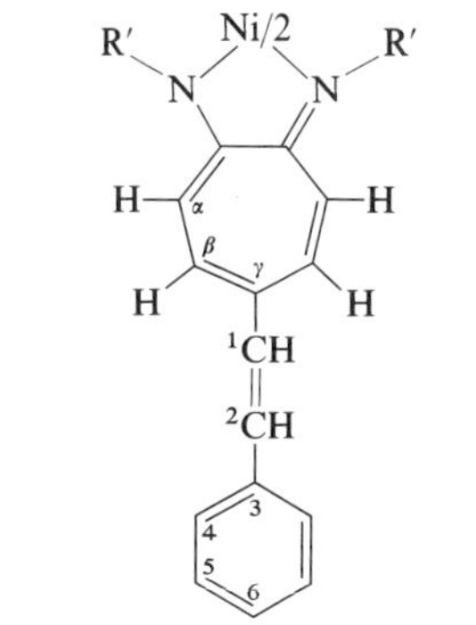	α	-1.29^a	[11]
	β	$+0.66^a$	
	(γ)	$(-1.80)^a$	
	1	$+0.60^a$	[12]
	2	-0.87^a	
	4	-0.19^a	
	5	$+0.05^a$	
	6	-0.21^a	
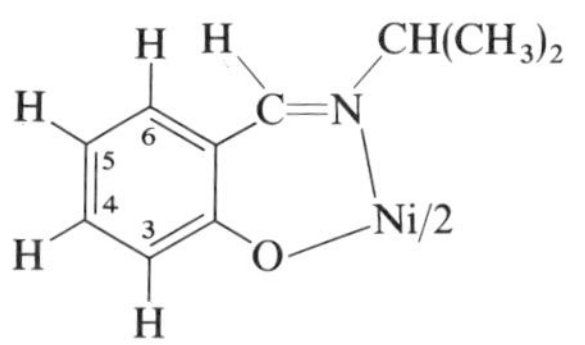	3	-0.30^b	[13]
	4	$+0.30^b$	
	5	-0.29^b	
	6	$+0.14^b$	
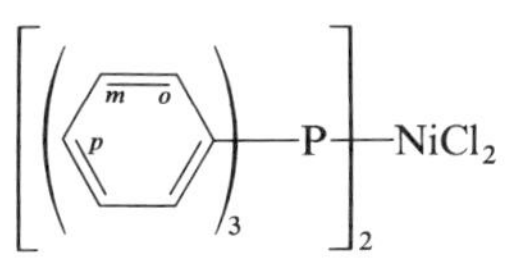	*o*	-0.10^b	[14]
	m	$+0.11^b$	
	p	-0.16^b	
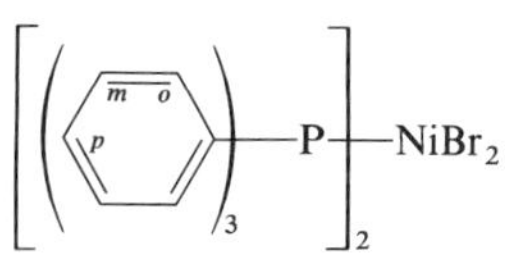	*o*	-0.07^a	[15]
	m	$+0.09^a$	
	p	-0.07^a	
	(p—CH_3)	$(+0.09)^a$	

[a] Dipolar contributions to the isotropic shift have been neglected.
[b] Dipolar contributions have been factorized out.

Table 2.2 Q Values (MHz) for Fragments Involved in π Electron Spin Delocalization Systems

Q^{C}_{CH}	Q^{H}_{CH}	$Q^{C}_{CC'}$	$Q^{C}_{C'C}$	$Q^{H}_{CCH_3}$	Q^{F}_{CF}	Q^{F}_{FC}	Reference
+54.6	−65.8	+40.3	−39.0				[16]
	−75.6	+53.2					[17]
				+14–+115			[18]
					−410	+2370	[19]

of the carbon atom is given by

$$\left(\frac{A_p}{h}\right)_{1s} = S^{\mathrm{C}} \rho_{\mathrm{C}}^{\pi}$$

For CHC_2 fragments, $S^{\mathrm{C}} = -35.5$ MHz, and is relatively insensitive to variations in the bonding because of the localization of the 1*s* orbitals [16].

From the foregoing analysis it appears that there is more than one term making a significant contribution to the hyperfine coupling constant. The polarization of the 1*s* electron produces a somewhat large spin density at the nucleus which is proportional, but opposite in sign, to the π spin density on the atom containing the nucleus. The contribution from polarization of the three σ bonds by π spin density on the same atom has the same sign as ρ_{C}^{π}, because $Q_{\mathrm{CX}}^{\mathrm{C}}$ are positive. Finally, there is the polarization of the three σ bonds by π electron spin density on adjacent atoms, which is again a contribution opposite in sign to ρ_{X}^{π} (Table 2.2). The final value of the hyperfine coupling constant is therefore the result of considerable cancellation among the various contributions. In general, the ρ^{π} values are determined through ^{1}H Fermi contact shifts and Equation (2.14), and the hyperfine splitting for carbon atoms can, in principle, be predicted.

For ^{19}F Fermi contact shifts of fluorine bound to sp^2 carbon atoms, spin density on the nucleus arises from spin polarization by ρ_{C}^{π}, as for analogous CH moieties, and from spin polarization by ρ_{F}^{π}, which occurs via direct delocalization through C—F π bonding (Figure 2.7). The hyperfine coupling is therefore [19]

$$\frac{A_p}{h} = Q_{\mathrm{CF}}^{\mathrm{F}} \rho_{\mathrm{C}}^{\pi} + Q_{\mathrm{FC}}^{\mathrm{F}} \rho_{\mathrm{F}}^{\pi}$$

where $Q_{\mathrm{CF}}^{\mathrm{F}}$ is analogous to $Q_{\mathrm{CH}}^{\mathrm{H}}$ and is negative, and $Q_{\mathrm{FC}}^{\mathrm{F}}$, which is the polarization effect of ρ_{F}^{π} on the C—F σ bond, is positive by analogy to $Q_{\mathrm{CX}}^{\mathrm{C}}$ (see Table 2.2).

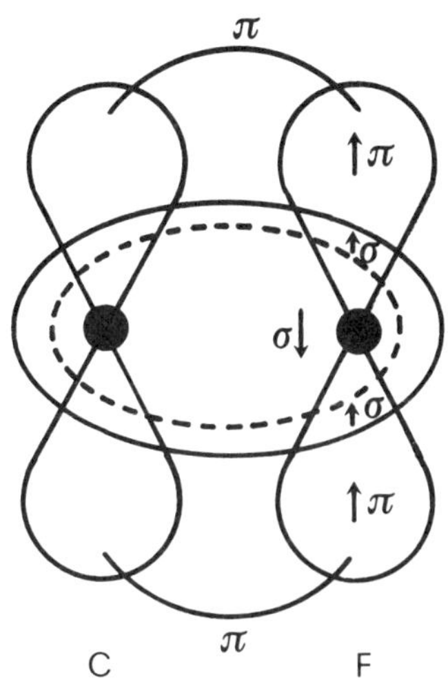

Figure 2.7 Spin polarization (through σ bonding) and direct delocalization (through π bonding) mechanisms for unpaired spin density transfer from an sp^2 carbon to a fluorine nucleus.

Table 2.3 Estimated A_c/h Values (MHz) for ^1H, ^{13}C and ^{14}N Nuclei of *n*-Ethyl Amine Coordinated to bis-Acetylacetonate Nickel(II)[a]

^{14}N	N^1H_2	$^{13}CH_2$	C^1H_2	$^{13}CH_3$	C^1H_3
14.4	−2.64	−0.70	1.02	1.80	0.22

[a] From Reference [23].

In the case of σ spin delocalization, spin polarization effects are more difficult to predict. Along an aliphatic chain, proton isotropic shifts are found positive for nickel complexes and decrease with increasing number of bonds from the paramagnetic center. Exceptions are the N—H proton isotropic shifts of amines [20, 21] and ammonia [22], for which polarization effects prevail, causing large upfield shifts. Similarly, the ^{13}C Fermi contact shift for the α carbon is also upfield (because of predominant spin polarization effects), whereas that for other carbon atoms is downfield and rapidly attenuates with the number of bonds [23, 24]. A sizeable downfield shift is experienced by the ^{14}N nucleus when nitrogen is a donor atom [23] (see Table 2.3). In this case, however, ligand-centered dipolar shifts may be operative; unpaired electrons in non *s* orbitals of heteroatoms produce a further shift, dipolar in origin, which represents another unknown (see Section 2.4).

Substitution of a proton by a CH_3 group, when the shift is due to σ-transferred spin density, attenuates the proton shift. A geometrical relationship has been found in M—N—CH_2 moieties which enables axial and equatorial geminal protons to be distinguished through the following relation [25–28]:

$$\frac{A_d}{h} = (\beta_0 + B_2 \cos^2 \phi)\frac{\rho_C}{2S} \tag{2.16}$$

where ϕ is the torsion angle about the C_1—N bond (Figure 2.8) and β_0 and B_2 are positive constants.

There have been numerous calculations of spin density in *p* orbitals of carbon atoms and at the proton nuclei by using modified simple Hückel theory [29–31],

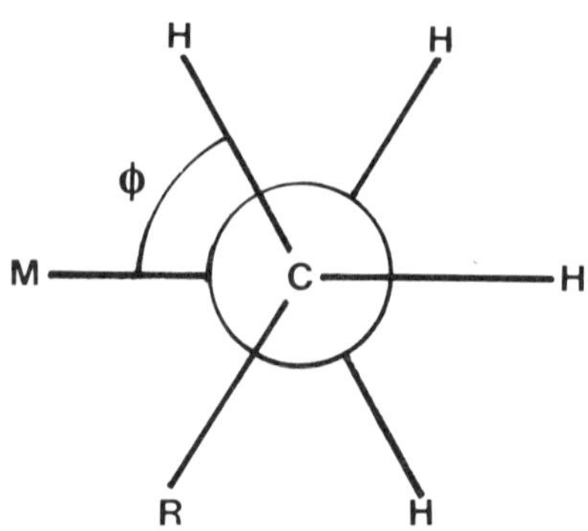

Figure 2.8 Dihedral angle, ϕ, between the H—C—N and the C—N—M planes in metal-coordinated ethylamine. The molecule is viewed along the C—N axis.

extended Hückel [32–34], and INDO or MINDO [35–37] methods. Recently, the X_α method has also proven quite promising [38]. The spin density patterns have been tentatively rationalized for many systems, and often the predominant spin density transfer mechanism has been proposed. When the proton and carbon contact shifts are simultaneously analyzed, fewer degrees of freedom are left to the researcher's imagination.

2.3.2 The hyperfine coupling of water nuclei in aqua complexes

Because of the importance of water in biological systems some comments are in order concerning the mechanism of spin delocalization in bound water. In Table 2.4 the proton A_c/h values for some hexaaqua metal complexes are reported. Spin density has been evaluated through Equation (2.13). The proton isotropic shifts, assumed to be mainly contact in origin as a result of the averaged high symmetry, are downfield and the hyperfine coupling constants are positive. With the possible exception of Mn^{2+}, it appears that the hydrogen spin density, which is also listed in Table 2.4, is proportional to the number of unpaired t_{2g} electrons. The t_{2g} orbitals have the correct symmetry for π bonding. Unpaired electrons in the e_g orbitals, which have the correct symmetry for σ bonding, do not affect the value of ρ, and indeed the experimental proton hyperfine coupling constants in Cu^{2+} and Ni^{2+} complexes are very small [42].

It appears therefore that unpaired π electrons of the metal ions delocalize on to the H_2O moiety. One fully occupied water molecular orbital of σ type is capable of directly overlapping with the t_{2g} metal orbital set [42]. Thus the unpaired

Table 2.4 Hyperfine Coupling Constants of Water Nuclei in Hexaaqua Complexes with Paramagnetic 3*d* Metal Ions. Positive Signs of A_c/h Indicate Downfield Shift.

				A_c/h (MHz)		
		$2S$	n^a	$^1H^b$	^{17}O	Reference
Ti^{3+}	d^1	1	1		4.4^c	[39]
VO^{2+}	d^1	1	1		3.8^c	[40]
V^{2+}	d^3	3	3	2.0 (4.2×10^{-3})		[41]
Cr^{3+}	d^3	3	3	2.0 (4.2×10^{-3})		[41, 42]
Mn^{2+}	d^5	5	3	0.6–1.0 (2.2–3.5×10^{-3})	5.3	[41, 43]
Fe^{3+}	d^5	5	3	1.2 (4.2×10^{-3})		[41]
Fe^{2+}	d^6	4	2		9.4	[43]
Co^{2+}	d^7	3	1	0.4 (1.1×10^{-3})	14.8–15.7	[42–44]
Ni^{2+}	d^8	2	0	0.2 (3×10^{-4})	20.4	[42, 45]
Cu^{2+}	d^9	1	0	0.06–0.19 (4.2×10^{-5}–1.3×10^{-4})	50.0	[42, 46]

[a] Unpaired electrons in t_{2g} orbitals for high-spin complexes with O_h symmetry.

[b] ρ_H values in parentheses, estimated from Equation (2.13).

[c] Absolute values (obtained from relaxation data).

t_{2g} electron is capable of directly producing spin density on the protons. A further π interaction with the metal t_{2g} set arises from a nonbonding p orbital of oxygen, which would place spin density on the protons only through spin polarization. The sign of the observed shifts shows that the former mechanism is dominant [42].

The ^{17}O data (Table 2.4) show large hyperfine coupling constants when unpaired electrons are located in the e_g orbitals, which have the correct symmetry for σ bonding. The data for $Cu(H_2O)_6^{2+}$ and $Ni(H_2O)_6^{2+}$ demonstrate this clearly [45, 46]. It is obvious that direct delocalization through metal-oxygen σ bonding involving oxygen s orbitals gives rise to quite sizeable spin density on the oxygen nucleus.

2.4 THE DIPOLAR SHIFT

The $\hat{\mathbf{I}} \cdot \mathbf{A}_{\text{dip}} \cdot \hat{\mathbf{S}}$ term of the Hamiltonian describes the dipolar part of the electron-nucleus coupling. For a single unpaired electron with spin angular momentum, $\hat{\mathbf{s}}$, this term can be written as

$$\hat{\mathscr{H}}^{\text{dip}} = \frac{\mu_0}{4\pi} \hbar \gamma_N \mu_B \left\{ \frac{2\hat{\ell}_N \cdot \hat{\mathbf{I}}}{r_N^3} + g_e \left[\frac{3(\mathbf{r}_N \cdot \hat{\mathbf{s}})(\mathbf{r}_N \cdot \hat{\mathbf{I}})}{r_N^5} - \frac{\hat{\mathbf{s}} \cdot \hat{\mathbf{I}}}{r_N^3} \right] \right\} \tag{2.17}$$

where $\hat{\ell}_N$ is the electron orbital angular momentum, $\mathbf{r}_N$ is the electron-nucleus distance, and the other symbols have their usual meanings. The part of the Hamiltonian containing $\hat{\ell}_N$ accounts for the magnetic anisotropy contribution to the coupling. By substituting the metal-nucleus distance r for $\mathbf{r}_N$ in Equation (2.17), the metal-centered dipolar shift can be expressed [6, 47] in terms of the magnetic susceptibility anisotropy for any spin system S:

$$\left(\frac{\Delta\nu}{\nu_0}\right)^{\text{dip}} = \frac{1}{4\pi 2r^3} \left[(3\cos^2\theta - 1)\left(\frac{2}{3}\chi_{zz} - \frac{1}{3}\chi_{xx} - \frac{1}{3}\chi_{yy}\right) + \sin^2\theta \cos 2\Omega\,(\chi_{xx} - \chi_{yy}) \right] \tag{2.18}$$

which, making the same assumptions as were used to generate Equation (2.7), is reduced to the following form [48]:

$$\left(\frac{\Delta\nu}{\nu_0}\right)^{\text{dip}} = \frac{\mu_0}{4\pi} \frac{\mu_B^2 S(S+1)}{18kTr^3} \{[2g_{zz}^2 - (g_{xx}^2 + g_{yy}^2)](3\cos^2\theta - 1) + 3(g_{xx}^2 - g_{yy}^2)\sin^2\theta \cos 2\Omega\} \tag{2.19}$$

where θ and Ω are the angles formed by r with the molecular z axis and by the projection of r on the xy plane with the molecular x axis. It should be noted that the magnitude of the dipolar shift does not depend on any magnetic property of the particular nucleus.

The unpaired electron magnetic moment, however, cannot always be considered as localized on the metal and treated as a point dipole. The electron is, at the very least, delocalized within the coordination polyhedron. In the case of $4f$ electrons this does not produce a significant effect [49], whereas a $3d$ electron appears to the nucleus as a magnetic point dipole only at large distances [50]. An estimate of the lower limit of this distance is 700 pm. In order to properly evaluate the pseudocontact term at shorter distances it is necessary to consider the appropriate molecular orbitals and to apply $\hat{\mathscr{H}}^{\text{dip}}$, as given in Equation (2.17), to the relative ψ functions [50]. Attempts at this evaluation assuming a series of point multipoles instead of a simple point dipole [51–54] with the aim of mimicking the electron distribution within the metal atomic orbitals have been partially successful.

Finally, if the electron is considered to be delocalized on the entire molecule, the problem is further complicated. For example, fractions of unpaired electrons delocalized on to nitrogen p orbitals in $Fe(CN)_6^{3-}$ are predicted to affect seriously the overall ^{14}N isotropic shifts [2]. Therefore, the analysis of isotropic shifts for heteronuclei requires ligand-centered dipolar shifts to be taken into account; examples will be given in Chapter 7. Hydrogen nuclei have no accessible p orbitals and hence any ligand-centered effect can only arise from unpaired spin density on p_π orbitals of neighboring atoms. Thus, the effects are expected to be smaller, although not necessarily negligible.

With the above limitations in mind, dipolar shifts will be referred to as metal centered (as is customary in the literature), unless otherwise specified.

In the case where the symmetry of the magnetic susceptibility tensor is axial, Equations (2.18) and (2.19) take the more familiar form:

$$\left(\frac{\Delta\nu}{\nu_0}\right)^{\text{dip}} = \frac{1}{4\pi 3r^3}(3\cos^2\theta - 1)(\chi_{\parallel} - \chi_{\perp}) \tag{2.20}$$

$$\left(\frac{\Delta\nu}{\nu_0}\right)^{\text{dip}} = \frac{\mu_0}{4\pi}\frac{\mu_B^2 S(S+1)}{9kTr^3}(3\cos^2\theta - 1)(g_{\parallel}^2 - g_{\perp}^2) \tag{2.21}$$

Equations (2.18) or (2.19) are used to predict the dipolar shifts. For $S = \frac{1}{2}$ systems, it is of course much more convenient in general to use Equation (2.19), which utilizes g values, rather than Equation (2.18), which involves χ values. Furthermore, no anisotropic susceptibility measurements are available for biomolecules. The difference in using one or the other of these equations can be fully appreciated if reference is made to the Van Vleck equation for magnetic susceptibility, as shown in Equation (1.5).

It should also be borne in mind (Section 1.3) that $\hat{L}_\alpha + g_e\hat{S}_\alpha$ is equivalent to $g_{\alpha\alpha}\hat{S}_\alpha$ when $\hat{L}_\alpha + g_e\hat{S}_\alpha$ operates to first order—i.e., the first term in curly brackets in Equation (1.5). Therefore, Equations 2.18 and 2.19 differ because the former considers only the first-order Zeeman contribution (FOZ). The estimated error arising from the omission of the second-order Zeeman contribution (SOZ) may vary between a few percent to 50% depending on how close the excited levels are to the ground state [55]. It should also be kept in mind that the molecular g values can be obtained from EPR data only for $S = \frac{1}{2}$ systems. For systems with

$S > \frac{1}{2}$, the resonance positions in EPR do not correspond to the molecular g values, although they may be obtained by making several assumptions.

A comment is now necessary concerning the dipolar shift in the case of the $S = \frac{5}{2}$ system. Here, the molecular g values are close to 2 because $L = 0$ for the ^6S free ion term. Nevertheless, the dipolar shifts may be large because of magnetic anisotropy arising from sizeable zero field splitting. By substituting for the magnetic susceptibility in Equation (2.20) with expressions (2.9) and (2.10), and again putting $e^X = 1 + X$, the following equation is obtained [6]:

$$\left(\frac{\Delta\nu}{\nu_0}\right)^{\text{dip}} = \frac{\mu_0}{4\pi}\frac{35\mu_B^2(g_\parallel^2 - g_\perp^2)}{36kT}\frac{3\cos^2\theta - 1}{r^3}\left(1 - \frac{32(g_\parallel^2 + \frac{1}{2}g_\perp^2)}{15(g_\parallel^2 - g_\perp^2)}\frac{D}{kT}\right) \quad (2.22)$$

This equation gives the sum of the magnetic susceptibility contributions of each state, i, weighted by the population according to the Boltzmann distribution. In contrast to the case of the Fermi contact shift, the term depending on T^{-2} cannot be neglected if D/kT is large enough, as is the case for high spin iron(III) porphyrins. The term in T^{-1}, which depends on g anisotropy, is always small.

The preceding equations show that the magnitude of the dipolar shifts strictly depends on the geometrical coordinates of the resonating nuclei with respect to the molecular axes. This is a peculiar property of the dipolar shift which has the potential of providing detailed structural information on molecules in solution.

2.5 SEPARATION OF DIPOLAR AND CONTACT CONTRIBUTIONS TO THE ISOTROPIC SHIFT

As a result of the nature of the dipolar shifts, a knowledge of the magnetic susceptibility tensor provides a method for an independent evaluation of the dipolar contribution to the isotropic shift, if it is assumed that the latter is metal-centered dipolar in origin. In principle, therefore, by performing magnetic susceptibility measurements on single crystals with several orientations of the magnetic field, it is possible to determine the magnetic anisotropy and the principal axes of the magnetic tensor. Knowing the geometrical coordinates, the dipolar shifts can be easily predicted using Equation (2.18). Such a procedure assumes that the magnetic anisotropy and the molecular axes are the same in the solid state and in solution. Furthermore, single crystal magnetic susceptibility data are difficult to obtain and in any case no direct measure is available for metalloproteins. However, in order to give an idea of the magnitude and relative contributions of contact and dipolar shifts, calculated values for various metal ion complexes are reported in Tables 2.5, 2.6, and 2.7.

If single crystal EPR data are available, the magnetic susceptibility tensor components can be indirectly estimated [14, 47, 58] for $S = \frac{1}{2}$ systems such as low spin iron(III). This is done by theoretically reconstructing the principal elements of the $\hat{\mathbf{L}} + g_e\hat{\mathbf{S}}$ operator and the excited states functions and energies from the experimental $g_{\alpha\alpha}$ values, and inserting them in the Van Vleck Equation (1.5).

Table 2.5 Determination of the Metal-Centered Dipolar Contribution to the ^{1}H and ^{13}C Isotropic Shifts (ppm) of the Pyridine Signals in the bis-Pyridine Adduct of Cobalt(II) bis-Acetylacetonate Using Single Crystal Magnetic Susceptibility Data [34, 47]

Atom	$\left(\frac{\Delta\nu}{\nu_0}\right)^{iso}$	$\left(\frac{\Delta\nu}{\nu_0}\right)^{dip}$
α—H	+32.9	−39.5
β—H	+5.0	−18.1
γ—H	−9.4	−15.6
α—C	−199	−92.5
β—C	+229	−35.7
γ—C	−73.8	−28.3

Table 2.6 Factorization into Dipolar and Contact Contributions of the Isotropic Shifts (ppm) of the Aromatic Protons in Nickel(II) bis-Salicylaldiminate Using Single Crystal Magnetic Susceptibility Data [13, 56, 57]

Proton	Structure	$\left(\frac{\Delta\nu}{\nu_0}\right)^{iso}$	$\left(\frac{\Delta\nu}{\nu_0}\right)^{dip}$	$\left(\frac{\Delta\nu}{\nu_0}\right)^{con}$
3	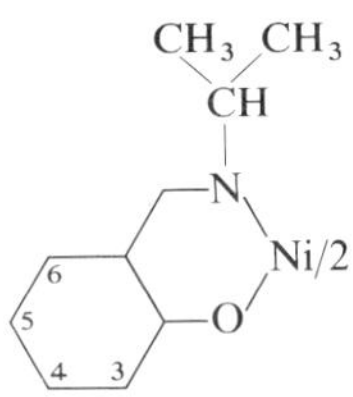	−23.7	+1.4	−25.1
4		+19.0	−2.3	+21.3
5		−23.7	−4.5	−19.2
6		+3.7	−8.4	+12.1

Table 2.7 Factorization into Dipolar and Contact Contributions of the Isotropic Shifts (ppm) of the Phenyl Protons in Dichlorobis(triphenylphosphine)Cobalt(II) and Nickel(II) Using Single Crystal Magnetic Susceptibility Data [14, 58]

Metal/Proton	$\left(\frac{\Delta\nu}{\nu_0}\right)^{iso}$	$\left(\frac{\Delta\nu}{\nu_0}\right)^{dip}$	$\left(\frac{\Delta\nu}{\nu_0}\right)^{con}$
Co *ortho*	−10.35	−2.13	−8.22
Co *meta*	+8.08	−0.54	+8.62
Co *para*	−11.51	−0.53	−10.98
Ni *ortho*	−6.38	+0.56	−6.94
Ni *meta*	+8.03	+0.14	+7.89
Ni *para*	−11.16	+0.14	−11.30

The procedure based on the direct use of the **g** tensor anisotropy and Equation (2.19) is much more common for $S = \frac{1}{2}$ systems since g values from frozen solutions are easily obtainable. In this case, both the SOZ contributions and the effects of temperature on the g values are neglected. Furthermore, the directions of the molecular axes are arbitrarily assumed unless single crystal data are available. The latter is a most unjustified assumption since the principal axes of the χ or **g** tensors, unless fixed by symmetry, are very dependent on small orbital mixing, thus giving rise to a large range of errors in the determination of the geometrical coordinates.

Other semiquantitative methods for evaluating the metal-centered dipolar contributions to the shifts are based on the predicted pattern of the dipolar shifts according to the ratios of the geometric factors. Equation (2.20), which is valid for the axial case, can be rewritten as

$$\left(\frac{\Delta\nu}{\nu_0}\right)^{\text{dip}} = G_{r,\theta} \cdot D_\chi$$

where $G_{r,\theta}$ is the geometric factor and D_χ is the term containing the magnetic susceptibility. Since the latter term is constant for a given molecule, the ratios between the dipolar shifts should be the same as the ratios between the G values calculated assuming axial symmetry. When this equation is obeyed, it can be concluded that the observed isotropic shift is mainly dipolar in origin. In other cases this same property allows the dipolar contribution on all nuclei to be estimated by assuming that at a certain nucleus position the contact contribution is negligible. Horrocks has developed a method, called the ratio method [59], which is based on a comparison between the proton isotropic shift pattern for a molecule where the shifts are only contact in origin and an analogous molecule whose proton shifts have to be factorized. If the spin delocalization mechanism is the same in the two molecules, the contact shift ratios for each proton position of the two complexes should be the same:

$$\left[\frac{(\Delta\nu_i)_{\text{M}}}{(\Delta\nu_i)_{\text{M}'}}\right]^{\text{con}} = \left[\frac{(\Delta\nu_j)_{\text{M}}}{(\Delta\nu_j)_{\text{M}'}}\right]^{\text{con}} \tag{2.23}$$

where M and M′ are two metal ions (most often Ni^{2+} and Co^{2+}) and i and j are two proton positions. This condition may allow the estimation of the dipolar contribution present in one of the complexes:

$$\left[\frac{(\Delta\nu_i)_{\text{Co}}}{(\Delta\nu_j)_{\text{Co}}}\right]^{\text{con}} = \left[\frac{(\Delta\nu_i)_{\text{Ni}}}{(\Delta\nu_j)_{\text{Ni}}}\right]^{\text{con}} = \frac{(\Delta\nu_i)_{\text{Co}}^{\text{iso}} - (\Delta\nu_i)_{\text{Co}}^{\text{dip}}}{(\Delta\nu_j)_{\text{Co}}^{\text{iso}} - (\Delta\nu_j)_{\text{Co}}^{\text{dip}}} \tag{2.24}$$

In the case of iron(III) porphyrins, La Mar *et al.* [60] factorized the shifts by exploiting the different temperature dependence of contact and dipolar shifts. In fact, the observed isotropic shifts were found to follow a relation of the type

$$\left(\frac{\Delta\nu}{\nu_0}\right)^{\text{iso}} = \alpha T^{-1} + \beta T^{-2}$$

where the T^{-1} dependence was attributed to the contact term and the T^{-2} dependence to the metal-centered dipolar term—see Equations (2.11) and (2.22).

2.6 THE NATURE OF THE ISOTROPIC SHIFTS AS RELATED TO THE TYPE OF METAL ATOM

The broadening of an NMR line caused by the presence of a nearby paramagnetic ion, for a given distance and spin density delocalization, varies dramatically from ion to ion and, for the same ion, with oxidation number, spin state, and orbital degeneracy (see Chapter 3). Although under appropriate conditions NMR signals of ligand nuclei have been detected for coordination compounds of almost all the paramagnetic metal ions, applications of the isotropic shift technique to solutions of paramagnetic biomolecules are restricted, mainly because of the limited concentrations attainable, to relatively few metal ions with low nuclear-relaxing capability. These will be specifically covered in Chapters 7 to 10, but it is appropriate to discuss here some general considerations regarding the nature of the isotropic shifts. In principle, the isotropic shift should be always considered as a sum of contact, metal-centered dipolar, and ligand-centered dipolar contributions, and, in fact, there are no cases of practical importance where any of the above contributions is rigorously zero. However, depending on the degree of sophistication required in the interpretation of the experimental data, some contributions may be sufficiently small as to be neglected. As discussed in Section 2.4, this is often the case for the ligand-centered dipolar term on hydrogen nuclei. As far as the metal-centered dipolar term is concerned, this is often assumed to be negligible in six-coordinated nickel(II) (d^8) compounds. This assumption is based on the theoretical consideration that the ground state in O_h symmetry is a $^3A_{2g}$ state; that is, it is orbitally nondegenerate and the first excited level is at a relatively high energy ($\sim 10^4$ cm^{-1}). If there were no excited levels, the magnetic susceptibility would simply be

$$\chi_M = \mu_0 \frac{N_A \mu_{\text{eff}}^2}{3kT}$$

where

$$\mu_{\text{eff}} = g_e \sqrt{S(S+1)}\mu_B = 2.83\ \mu_B$$

and there would be no magnetic anisotropy. The presence of the excited levels raises μ_{eff} to $\sim 3.2\ \mu_B$ and gives rise to a small magnetic anisotropy. A nickel(II) compound in a rigorously tetrahedral environment would have a $^3T_1$ ground state; that is, orbitally triply degenerate with a large orbital contribution. Small deviations from purely cubic symmetry maintain the orbital contribution to χ and give rise to a large magnetic anisotropy. Pseudotetrahedral nickel(II) complexes do have magnetic moments as large as 4.0 μ_B and the dipolar shifts become sizeable (see Tables 2.6 and 2.7). Pseudooctahedral high spin cobalt(II) complexes (ground state $^4T_{1g}$ in O_h symmetry) have particularly large dipolar shifts (see Table 2.5), and this property has been exploited in the study of biological problems (see Chapter 9).

The investigation of synthetic porphyrins has established that a high spin Fe^{3+} ion with a ${}^6A_{1g}$ ground state displays measurable dipolar shifts as a result of relatively large zero field splitting energies ($D = 5$–$20\ cm^{-1}$), which are due to a second-order spin-orbit coupling perturbation (see Chapter 7). This gives rise to a relatively large magnetic anisotropy, although the magnetic moment is very close to that which would arise from "spin only" contributions, and the g values are close to 2. A more forseeable finding is the occurrence of dipolar shifts in low-spin Fe^{3+} porphyrins (${}^2T_{2g}$ ground state in O_h symmetry) owing to large g and χ anisotropies [58].

With the exception of complexes of gadolinium(III), which is an f^7 ion with a ${}^8S$ ground state, lanthanide-containing compounds have shifts that are largely metal-centered dipolar in origin, although sometimes the contact contributions cannot be neglected (see Chapter 10). This is due to large magnetic anisotropies and to the lack of mechanisms that allow the unpaired electrons residing in the inner f orbitals to delocalize on to the ligands. This property has been widely exploited in the area of chemistry and biochemistry to obtain structural information for molecules in solution, and will therefore be discussed in some detail in Chapter 10. Actinide elements have properties similar to the above elements.

2.7 THE ISOTROPIC SHIFT IN EXCHANGE-COUPLED SYSTEMS

Metal ions, when close to each other, can be magnetically coupled; that is, the unpaired electrons interact with each other through dipolar and Coulomb (exchange or superexchange) terms in such a way that the S quantum numbers for each metal ion are no longer good quantum numbers with which to describe the system. In the case of two metal ions, the Hamiltonian for isotropic coupling is given by

$$\hat{\mathscr{H}} = J \cdot \hat{\mathbf{S}}_1 \cdot \hat{\mathbf{S}}_2 \tag{2.25}$$

where the subscripts 1 and 2 refer to the two metal ions and J is a scalar that describes the extent of the magnetic interaction. The coupling can be described in terms of pairs of molecular orbitals each containing a single electron from each metal ion. If there is no interaction between the two molecular orbitals, they will have the same energy and the usual Van Vleck Equation (1.5) for magnetic susceptibility holds. If there is an interaction, the coupling can be further described in terms of bonding and antibonding.

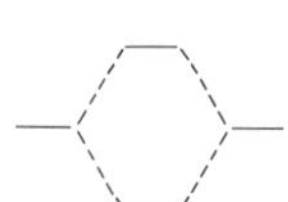

The two electrons in the low-lying orbital have total $S' = 0$, whereas one electron per orbital provides a ground state with total $S' = 1$. The overlap between the two initial molecular orbitals is small (since otherwise there would be a normal chemical bond) and therefore the separation between the resulting molecular orbitals is small with the consequence that the singlet ($S' = 0$) and triplet ($S' = 1$) states are close in energy. The energy separation between the singlet and triplet states depends on the relative magnitude of the energy separation of the one-electron molecular orbitals and the exchange energy between the two electrons.

Exchange effects may be large enough to make the triplet state lower in energy than the singlet state (ferromagnetic coupling); often, however, the singlet state is the ground state and the system is said to be antiferromagnetically coupled. According to Equation (2.25), J is negative in the former case and positive in the latter.

Antiferromagnetic coupling is not rare in biological systems. As far as the isotropic shifts are concerned, the effect of magnetic coupling has been considered only for contact contributions [61]. The Van Vleck equation for the magnetic susceptibility is used for each magnetically coupled level, S'_i, weighted according to its Boltzmann population:

$$\chi = \mu_0 \frac{\mu_B^2}{3kT} \frac{\sum_i g_i^2(2S'_i + 1)S'_i(S'_i + 1)\exp(-E_i^0/kT)}{\sum_i (2S'_i + 1)\exp(-E_i^0/kT)} \tag{2.26}$$

where $(2S'_i + 1)$ is the multiplicity of each S'_i level. The energy of each S'_i level relative to the ground level, for $S_1 \geq S_2$, is

$$E_i^0 = \frac{J}{2}\{S'_i(S'_i + 1) - [S_1(S_1 + 1) - S_2(S_2 + 1)]\} \tag{2.27}$$

For example, for two antiferromagnetically coupled metal ions with $S_1 = S_2 = \frac{5}{2}$, the levels are as shown below.

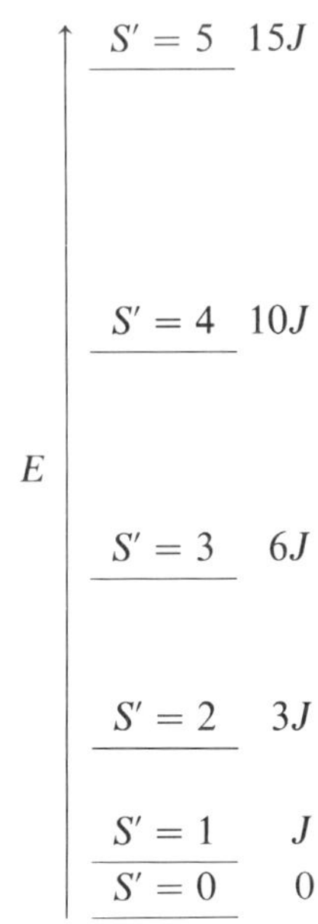

If the two metal ions are equivalent the contact shift is [60]

$$\left(\frac{\Delta\nu}{\nu_0}\right)^{\mathrm{con}} = \frac{\mu_B}{6\hbar\gamma_N kT} \frac{\sum_i g_i A_i(2S'_i + 1)S'_i(S'_i + 1)\exp(-E_i/kT)}{\sum_i (2S'_i + 1)\exp(-E_i/kT)} \tag{2.28}$$

Here it is assumed that each level S'_i contributes to the contact shift with its own g_i and A_i, although in general the g_i values are simply taken as a single average value. If A_i is also taken to be equal for each S'_i level, the temperature dependence of the contact shifts provides the parameter J. Therefore the temperature dependence of contact shifts in antiferromagnetically coupled systems can in principle be a valuable source of information. By increasing temperature, the populations of levels with larger S'_i values and with concomitantly larger susceptibility increase, leading to an increase in the isotropic shifts, despite the opposing temperature dependence given by the term $1/T$.

However, the picture can be complicated if the various A_i constants are different. Furthermore, if the two metals have different S values, each metal ion contributes in a different way to the overall shift. Each contribution will depend on the mixing of the individual atomic orbitals in each S'_i manifold of the coupled system, the mixing being in general different in each S' multiplet [62, 63], and so complicated temperature dependences may arise. This point will be further discussed in Section 7.5.2.

2.8 A FURTHER COMMENT ON THE TEMPERATURE DEPENDENCE OF THE ISOTROPIC SHIFTS

Aside from the case of antiferromagnetically coupled systems, for which the temperature dependence of the isotropic shifts is rather unpredictable, the isotropic shifts invariably decrease with increasing temperature. In the small temperature range in which solutions exist, the isotropic shifts are usually inversely proportional to temperature, as expected from the simplified Equations (2.4) and (2.18). Deviations from the behavior predicted by these formulas cannot usually be detected by the curvature in a plot of isotropic shifts versus T^{-1}, since a small T^{-2} component is masked by the small temperature range investigated. The usual procedure for verifying a non T^{-1} dependence of the shifts is to extrapolate the plot to infinite temperatures. Of course, when Equations (2.4) and (2.18) hold, the isotropic shifts extrapolate to zero.

Whenever there are spin-orbit coupling effects that result in differently populated zero field splitting levels in the temperature range investigated, T^{-2} contributions are expected for both contact and dipolar shifts. These contributions are usually larger for the latter than for the former—compare Equations (2.11) and (2.22).

However, because of the complexity of the mechanisms determining the isotropic shift, both contact and dipolar, the temperature dependence is rarely predictable. For example, the general treatment for dipolar contributions not restricted to metal-centered effects predicts a temperature-independent term that may be sizeable in some cases [2, 64]. In addition, temperature dependences close to T^{-3} have been observed (see Chapter 10) in lanthanide-protein systems [65].

REFERENCES

1. Fermi, E. *Z. Phys.* **60** (1930) 320.
2. Golding, R. M., Pascual, R. O., and Vrbancich, J. *Mol. Phys.* **31** (1976) 731.
3. La Mar, G. N., and Sacconi, L. *J. Am. Chem. Soc.* **89** (1967) 2282.
4. Bertini, I., Sacconi, L., and Speroni, P. *Inorg. Chem.* **11** (1972) 1323.
5. McConnell, H. M., and Chesnut, D. B. *J. Chem. Phys.* **28** (1958) 107.
6. Kurland, R. J., and McGarvey, B. R. *J. Magn. Reson.* **2** (1970) 286.
7. McConnell, H. M. *Proc. Natl. Acad. Sci. U.S.A.* **69** (1972) 335.
8. Rodgers, J. E., Lee, T., Das, T. P., and Ikenberry, D. *Phys. Rev. A* **7** (1973) 51. Rodgers, J. E., and Das, T. P. *Phys. Rev. A* **8** (1973) 2195.
9. Mun, S. K., Mallick, M. K., Mishra, S., Chang, J. C., and Das, T. P. *J. Am. Chem. Soc.* **103** (1981) 5024.
10. Stone, E. W., and Maki, A. H. *J. Chem. Phys.* **37** (1962) 1326.
11. Eaton, D. R., Josey, A. D., Phillips, W. D., and Benson, R. E. *J. Chem. Phys.* **37** (1962) 347.
12. Eaton, D. R., Josey, A. D., Benson, R. E., Phillips, W. D., and Cairns, T. L. *J. Am. Chem. Soc.* **84** (1962) 4100.
13. Benelli, C., Bertini, I., and Gatteschi, D. *J. Chem. Soc., Dalton Trans.* (1972) 661.
14. Horrocks, W. DeW. Jr., and Greenberg, E. S. *Inorg. Chem.* **10** (1971) 2190.
15. La Mar, G. N., Horrocks, W. DeW. Jr., and Allen, L. C. *J. Chem. Phys.* **41** (1964) 2126.
16. Karplus, M., and Fraenkel, G. K. *J. Chem. Phys.* **35** (1961) 1312.
17. Strom, E. T., Underwood, G. R., and Jurkowitz, D. *Mol. Phys.* **24** (1972) 901.
18. Derbishire, W. *Mol. Phys.* **5** (1962) 225.
19. Eaton, D. R., Josey, A. D., Phillips, W. D., and Benson, R. E., *Mol. Phys.* **5** (1962) 407.
20. Fitzgerald, R. J., and Drago, R. S. *J. Am. Chem. Soc.* **89** (1967) 2879.
21. Wayland, B. B., and Rice, W. L. *J. Chem. Phys.* **45** (1966) 3150.
22. Wayland, B. B., and Rice, W. L. *Inorg. Chem.* **6** (1967) 2270.
23. Quaegebeur, J.-P., Chachaty, C., and Yasukawa, T. *Mol. Phys.* **37** (1979) 409.
24. Morishima, I., Okada, K., Yonezawa, T., and Goto, K. *Chem. Commun.* (1970) 1535.
25. Pratt, L., and Smith, B. B. *Trans. Faraday Soc.* **65** (1969) 915.
26. Ho, F. F.-L., and Reilley, C. N. *Anal. Chem.* **41** (1969) 1835.
27. Fitzgerald, R. J., and Drago, R. S. *J. Am. Chem. Soc.* **90** (1968) 2523.
28. Zamarov, K. I., Molin, Y. N., and Skubnevskaya, G. I. *J. Struct. Chem.* **7** (1966) 740.
29. McLachlan, A. D. *Mol. Phys.* **3** (1960) 233.

30. Kluiber, R. W., and Horrocks, W. DeW. Jr. *Inorg. Chem.* **6** (1967) 430.
31. Bertini, I., Johnston, D. L., and Horrocks, W. DeW. Jr. *Inorg. Chem.* **9** (1970) 698.
32. Perry, W. D., Drago, R. S., Herlocker, D. W., Pogenkopf, G. K., and Czworniak, K. *Inorg. Chem.* **10** (1971) 1087.
33. Cramer, R. E., and Drago, R. S. *J. Am. Chem. Soc.* **92** (1970) 66.
34. Doddrell, D., and Roberts, J. D. *J. Am. Chem. Soc.* **92** (1970) 6839.
35. Scarlett, M. J., Casey, A. T., and Craig, R. A. *Aust. J. Chem.* **23** (1970) 1333; *Aust. J. Chem.* **24** (1971) 31.
36. Horrocks, W. DeW. Jr., and Johnston, D. L. *Inorg. Chem.* **10** (1971) 1835.
37. Fantucci, P., and Valenti, V. *J. Am. Chem. Soc.* **98** (1976) 3832. Fantucci, P., Balzarini, P., and Valenti, V. *Inorg. Chim. Acta* **25** (1977) 113.
38. Sontum, S. F., Case, D. A., and Karplus, M. *J. Chem. Phys.* **79** (1983) 2881.
39. Chmelnick, A. M., and Fiat, D. *J. Chem. Phys.* **51** (1969) 4238.
40. Reuben, J., and Fiat, D. *Inorg. Chem.* **6** (1967) 579.
41. Bloembergen, N., and Morgan, L. O. *J. Chem. Phys.* **34** (1961) 842.
42. Luz, Z., and Shulman, R. G. *J. Chem. Phys.* **43** (1965) 3750.
43. Ducommun, Y., Newman, K. E., and Merbach, A. E. *Inorg. Chem.* **19** (1980) 3696.
44. Swift, T. J., and Connick, R. E. *J. Chem. Phys.* **37** (1962) 307; *J. Chem. Phys.* **41** (1964) 2553.
45. Ducommun, Y., Earl, W. L., and Merbach, A. E. *Inorg. Chem.* **18** (1979) 2754.
46. Lewis, W. B., Alei, M. Jr., and Morgan, L. O. *J. Chem. Phys.* **45** (1966) 4003 and references therein.
47. Horrocks, W. DeW. Jr., and Hall, D. D. *Inorg. Chem.* **10** (1971) 2368.
48. McConnell, H. M., and Robertson, R. E. *J. Chem. Phys.* **27** (1958) 1361.
49. Golding, R. M., Pascual, R. O., and McGarvey, B. R. *J. Magn. Reson.* **46** (1982) 30.
50. Golding, R. M., and Stubbs, L. C. *J. Magn. Reson.* **33** (1979) 627.
51. Buckingham, A. D., and Stiles, P. J. *Mol. Phys.* **24** (1972) 99.
52. Stiles, P. J. *Mol. Phys.* **27** (1974) 501. Golding, R. M., Pascual, R. O., and Stubbs, L. C. *Mol. Phys.* **31** (1976) 1933.
53. Stiles, P. J. *Mol. Phys.* **29** (1975) 1271.
54. Riley, J. P., and Raynes, W. T. *Mol. Phys.* **33** (1977) 619.
55. Horrocks, W. DeW. Jr, and Greenberg, E. S. *Biochim. Biophys. Acta* **322** (1973) 38.
56. Gerloch, M., and Slade, R. C. *J. Chem. Soc. A* (1969) 1022.
57. Holm, R. H., Everett, G. W. Jr., and Chakravorty, A. *Prog. Inorg. Chem.* **7** (1966) 83.
58. La Lancette, E. A., and Eaton, D. R. *J. Am. Chem. Soc.* **86** (1964) 5145.
59. Horrocks, W. DeW. Jr. *Inorg. Chem.* **9** (1970) 670.
60. La Mar, G. N., Eaton, G. R., Holm, R. H., and Walker, F. A. *J. Am. Chem. Soc.* **95** (1973) 63.
61. Martin, R. L. In *New Pathways in Inorganic Chemistry*, E. A. V. Ebsworth, A. G. Maddock, and A. G. Sharpe (Eds.), Cambridge University Press, London, 1968, Chapter 9.
62. Dunham, W. R., Palmer, G., Sands, R. H., and Bearden, A. J. *Biochim. Biophys. Acta* **253** (1971) 373.
63. Wicholas, M., Mustacich, R., and Jayne, D. *J. Am. Chem. Soc.* **94** (1972) 4518.
64. Golding, R. M., and Pascual, R. O. *Bull. Magn. Reson.* **5** (1983) 126.
65. Lee, L., and Sykes, B. D. *J. Magn. Reson.* **41** (1980) 512.

CHAPTER 3

Unpaired Electron-Nuclear Coupling Detected by NMR: Nuclear Relaxation

Coupling of nuclear and electronic magnetic moments provides further relaxation pathways for nuclear spins. As in the case of the isotropic shift, the electron-nucleus spin interaction may be described in terms of dipolar and contact contributions. While the contact term is conceptually similar to that described in Section 2.3, the dipolar term is substantially different. In contrast to dipolar shifts, magnetic anisotropy is not important for dipolar relaxation. Indeed, the simplest and most often used dipolar relaxation equations are derived assuming an isotropic **g** tensor.

Relaxation measurements provide a wealth of information both on the extent of the interaction between the resonating nuclei and the paramagnetic center, and on the time dependence of the parameters associated with the interaction. Whereas the dipolar coupling depends on the electron-nucleus distance and therefore contains structural information, the contact contribution is related to the unpaired spin density on the various resonating nuclei. The time-dependent phenomena associated with electron-nucleus interactions are related to the relaxation processes of the electrons themselves, to the tumbling of the molecular system, and to the lifetimes of different chemical situations—if they are available for the resonating nucleus. Obtaining either structural or dynamic information, however, is only possible if an in-depth analysis of a series of experimental results provides sufficient data to characterize the system within the theoretical framework discussed in this chapter.

3.1 DIPOLAR RELAXATION

From the classic description of the coupling between two point dipoles with magnetic moments $\boldsymbol{\mu}_1$ and $\boldsymbol{\mu}_2$, the energy of the interaction is proportional to their scalar product and to the inverse of the third power of their distance—cf Equation (2.17):

$$E^{\text{dip}} \propto \frac{\boldsymbol{\mu}_1 \cdot \boldsymbol{\mu}_2}{r^3} - \frac{3(\boldsymbol{\mu}_1 \cdot \mathbf{r})(\boldsymbol{\mu}_2 \cdot \mathbf{r})}{r^5}$$

If the relative orientation of the two point dipoles changes with time, the fluctuating magnetic field produced by one at the location of the other may induce spin transitions on the other center, and thus contribute a further relaxation pathway. The interest here is limited to nuclear spin transitions induced by fluctuating magnetic fields originating from unpaired electron spins. It should be remembered that the electronic magnetic moment is orders of magnitude larger than that of nuclei; for example, the magnetic moment of the free electron is 658 times that of the proton. [1]

In dilute solutions, there are three general mechanisms that cause fluctuating magnetic fields. One is provided by electron spin relaxation itself, which occurs at rates orders of magnitude faster than those of the nuclear spins. A second arises from molecular tumbling. This can be visualized by assuming that both the nucleus and electron are within the same framework and that the associated magnetic moments are always oriented along the direction of the external magnetic field. Molecular tumbling then allows the electronic magnetic moment to change its orientation with respect to the nuclear magnetic moment (Figure 3.1). The third mechanism occurs when a nucleus belongs to an exchanging moiety and chemical exchange takes place. When the moiety is attached to the species containing the unpaired electron, the effect of the electronic magnetic moment on the nucleus is different to when the nucleus is free in the bulk solvent. A limiting case of chemical exchange is outer-sphere relaxation, which will be treated in Section 4.5.

The enhancement of the nuclear longitudinal relaxation rate, T_{1M}^{-1}, due to coupling with unpaired electrons is proportional to the square of the interaction energy, and to a function of both the frequency of the spin transitions, ω,* and the correlation time, τ_c, for the motion that modulates the interaction:

$$T_{1M}^{-1} \propto \left(\frac{\mu_N \mu_S}{r^3}\right)^2 f(\omega, \tau_c)$$

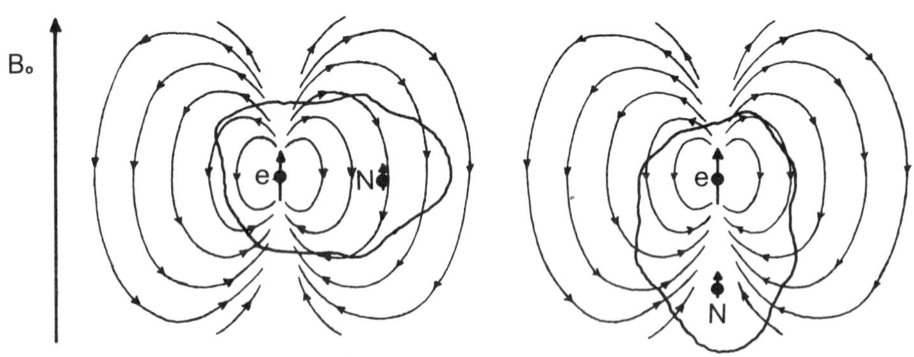

Figure 3.1 The magnetic field induced at the nucleus by an unpaired electron in the same molecule changes with the orientation of the molecule with respect to the external magnetic field.

* Throughout this chapter the transition frequencies of electrons and nuclei will be indicated as ω (rad s^{-1}) rather than ν (Hz). The relation is: $\omega = 2\pi\nu$.

This is the basic form of the equations for the dipolar interaction that will be encountered throughout this book. The magnetic field produced at the nucleus by the electron varies with time as discussed in Section 3.1.1. τ_c can be viewed as the time interval in which it is reasonable to expect the magnetic field at the nucleus to be constant. If an ensemble of electron-nucleus pairs is considered, the percentage of pairs in which the magnetic field at the nucleus has not changed after a time, t, will exponentially decrease with increasing t. This is of course true if the process is purely statistical. The correlation time, τ_c, is defined as the time constant of this exponential process.

In real systems, the unpaired electron(s) is delocalized through chemical bonds in the neighborhood of the metal, to an extent depending on the chemical properties of the system. Therefore, the dipolar interaction with the resonating nucleus should be evaluated by integrating over all the points where there is a finite unpaired spin density—approximate procedures will be given in Section 3.1.2. An account of the nuclear relaxation times based on this general approach requires details of the electronic structure of the system, including the unpaired spin densities on the various nuclei obtained through an analysis of the isotropic shifts. While keeping the general picture in mind, it is appropriate to consider first the simpler case in which the electron is localized on the metal atom.

3.1.1 Dipolar Coupling in a Simple Limit Case

Consider a single electron and a single nucleus separated by a distance r, with associated motions to be described. The energy levels of the coupled spin system in a magnetic field, $\mathbf{B}_0$, are given by the usual Hamiltonian:

$$\mathscr{H} = g_e\mu_B\mathbf{B}_0 \cdot \hat{\mathbf{S}} - \hbar\gamma_N\mathbf{B}_0 \cdot \hat{\mathbf{I}} + \hat{\mathbf{I}} \cdot \mathbf{A}_{\text{dip}} \cdot \hat{\mathbf{S}}$$

where the first two terms refer to the Zeeman interaction for the electron and nucleus,† respectively (see Section 1.4), and the $\hat{\mathbf{I}} \cdot \mathbf{A}_{\text{dip}} \cdot \hat{\mathbf{S}}$ term is the dipolar part of the coupling between the two—Equation (2.17) with $r_N = r$ and the $\hat{I}_N$-containing term equal to zero. The energy levels, the transition frequencies, ω, and the relative transition probabilities per unit time, W, are shown in Figure 3.2. The enhancement of the nuclear longitudinal relaxation rate can be expressed in terms of the probabilities of the transitions involving nuclear spin flipping:

$$T_{1M}^{-1} = W_{(\omega_S - \omega_I)} + 2W_{\omega_I} + W_{(\omega_S + \omega_I)} \tag{3.1}$$

Such transition probabilities can be computed [2], and so Equation (3.1) becomes

$$T_{1M}^{-1} = \frac{1}{10}\left(\frac{\mu_0}{4\pi}\right)^2 \frac{\hbar^2\gamma_N^2\gamma_S^2}{r^6}\left[\frac{\tau_c}{1 + (\omega_I - \omega_s)^2\tau_c^2} + \frac{3\tau_c}{1 + \omega_I^2\tau_c^2} + \frac{6\tau_c}{1 + (\omega_I + \omega_s)^2\tau_c^2}\right] \tag{3.2}$$

† The electronic and nuclear Zeeman terms have opposite signs due to the choice of positive g_e.

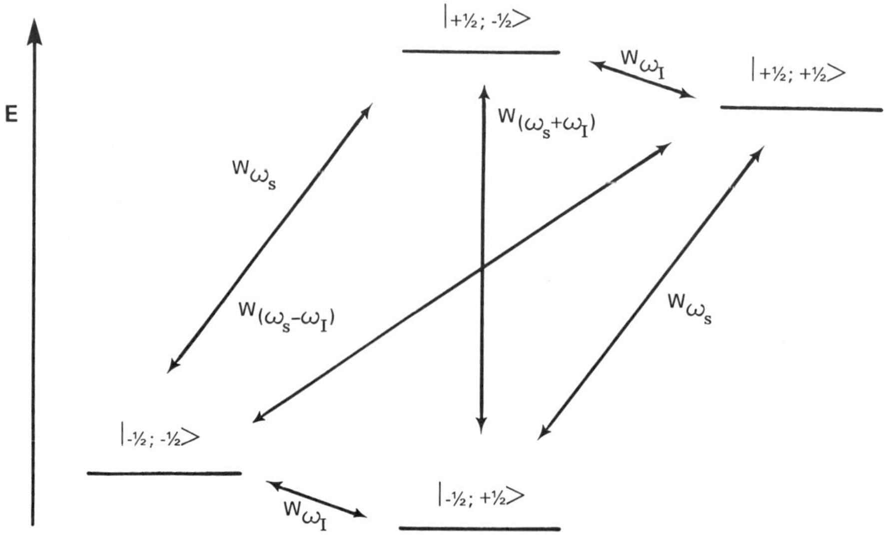

Figure 3.2 Energy levels, transition probabilities, W, and transition frequencies, ω, for an $S = \frac{1}{2}$, $I = \frac{1}{2}$ dipolarly coupled system.

which is the Solomon equation [3, 4]. The three terms in parentheses are proportional to the transition probabilities. This equation can be generalized [5] to systems with $S > \frac{1}{2}$ and $I > \frac{1}{2}$:

$$T_{1M}^{-1} = \frac{2}{15}\left(\frac{\mu_0}{4\pi}\right)^2 \frac{\gamma_N^2 g_e^2 \mu_B^2 S(S+1)}{r^6}\left[\frac{\tau_c}{1+(\omega_I-\omega_s)^2\tau_c^2} + \frac{3\tau_c}{1+\omega_I^2\tau_c^2} + \frac{6\tau_c}{1+(\omega_I+\omega_s)^2\tau_c^2}\right] \tag{3.3}$$

Note that Equation (3.3) contains S but not I; T_{1M}^{-1} is independent of I. Analogously [3–5], an equation can be derived for T_{2M}^{-1}:

$$T_{2M}^{-1} = \frac{1}{15}\left(\frac{\mu_0}{4\pi}\right)^2 \frac{\gamma_N^2 g_e^2 \mu_B^2 S(S+1)}{r^6}\left[4\tau_c + \frac{\tau_c}{1+(\omega_I-\omega_s)^2\tau_c^2} + \frac{3\tau_c}{1+\omega_I^2\tau_c^2} + \frac{6\tau_c}{1+(\omega_I+\omega_s)^2\tau_c^2} + \frac{6\tau_c}{1+\omega_s^2\tau_c^2}\right] \tag{3.4}$$

Since $\omega_s \pm \omega_I$ is about equal to ω_s, when $\tau_c \ll \omega_s^{-1}, \omega_I^{-1}$ (fast-motion limit, typical of many liquid solutions), the denominators of the fractions between brackets become unity, and Equations (3.3) and (3.4) reduce to

$$T_{1M}^{-1} = T_{2M}^{-1} = \frac{4}{3}\left(\frac{\mu_0}{4\pi}\right)^2 \frac{\gamma_N^2 g_e^2 \mu_B^2 S(S+1)}{r^6}\tau_c \tag{3.5}$$

At a magnetic field of 2.35 T the proton transition frequency (Larmor frequency) is: $\nu_0 = 100$ MHz; $\omega_I = 6.28 \times 10^8$ rad s^{-1}; $\omega_s = 4.13 \times 10^{11}$ rad s^{-1}. The ratio ω_s/ω_I is 658 if the nucleus is a proton, and higher for other nuclei (except tritium).

It should be recalled that the longitudinal relaxation process requires energy exchange with the lattice. In this case, the lattice is represented by the electron spin system, which in turn exchanges energy with the nuclear spin system. As was pointed out in Section 2.1, every process that causes T_1 relaxation also contributes to T_2 relaxation. In the fast-motion limit, the coupling with the lattice has the same effect on the magnetization along the z axis and on that in the xy plane, and $T_1 = T_2$. However, outside this limit the two functions $2f_1(\omega, \tau_c)$ for T_{1M}^{-1} and $f_2(\omega, \tau_c)$ for T_{2M}^{-1}, corresponding to the expressions in parentheses in Equations (3.3) and (3.4), diverge, leading to $T_{1M}^{-1} < T_{2M}^{-1}$. $2f_1(\omega, \tau_c)$ has its maximum value when the nuclear Larmor precession frequency, ω_I, and the reorientation frequency, τ_c^{-1}, are equal (Figure 3.3). This makes the coupling of the electron and nuclear spins most efficient. The function $f_2(\omega, \tau_c)$ increases monotonically with τ_c and differs from $2f_1(\omega, \tau_c)$ for $\tau_c > \omega_I^{-1}$ (Figure 3.3) owing to the presence of the frequency-independent term $4\tau_c$.

T_{1M} and T_{2M} values also depend on the applied magnetic field, as ω_I and ω_s are determined by it ($\omega_I = \gamma_N B_0$; $\omega_s = \gamma_S B_0 = 658\omega_I$ for protons). The magnetic fields currently used range from that of the earth to $14T(\omega_I/2\pi = 600$ MHz). The

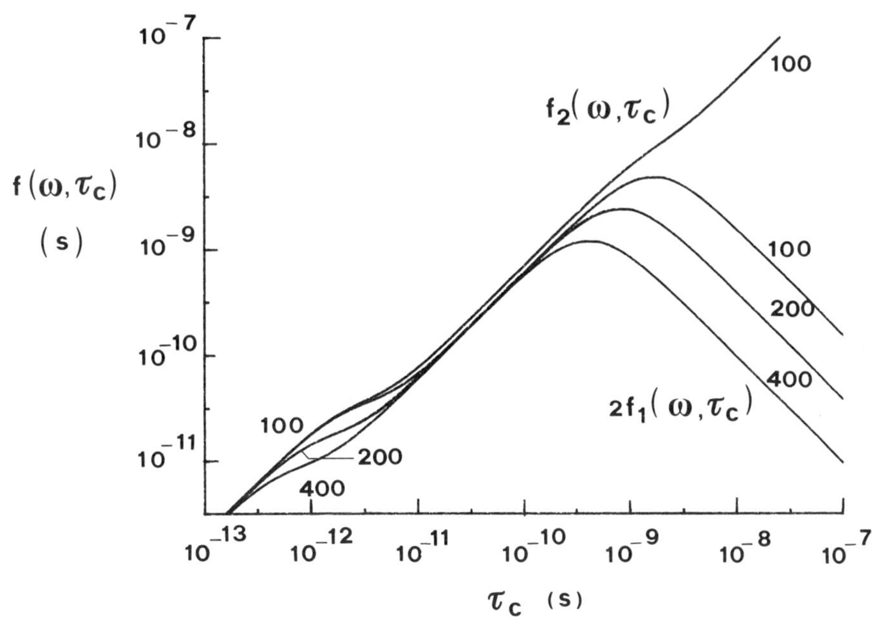

Figure 3.3 Plot of $f(\omega, \tau_c)$ (Equations 3.3 and 3.4) against τ_c for various proton Larmor frequencies: $2f_1(\omega, \tau_c)$ at 100, 200, and 400 MHz; $f_2(\omega, \tau_c)$ at 100 MHz.

field dependence of $2f_1(\omega, \tau_c)$ (Figure 3.4) shows two plateaus whose heights are in a 10/3 ratio and depend on the value of τ_c. The latter parameter is considered constant within the magnetic field range used, but this may not always be so (Section 3.7). The field regions in which $2f_1(\omega, \tau_c)$ is field dependent are called dispersions. The inflection point of the first dispersion is halfway between the two plateaus and corresponds to a magnetic field value at which $\omega_S = \tau_c^{-1}$; the inflection point of the second dispersion is halfway between the second plateau and zero, and corresponds to $\omega_I = \tau_c^{-1}$.

The field dependence of $f_2(\omega, \tau_c)$ is similar to that of $2f_1(\omega, \tau_c)$ (Figure 3.4), except that the relative height of the two plateaus is 20/7 and at high field the function levels off at a finite value that is one-fifth of the low-field limit, owing to the presence of the field-independent term $4\tau_c$.

It is appropriate now to further discuss the physical meaning of the dipolar correlation time, τ_c. The rate constant, τ_c^{-1}, is the sum of a number of different rate constants, each representing a different time-dependent process. Referring back to the three mechanisms discussed in Section 3.1, the following correlation rates should be considered:

1. The electronic relaxation rate, indicated as τ_S^{-1} when the longitudinal and transverse electronic relaxation rates are equal;

2. The appropriate rotational correlation rate, τ_r^{-1}, of the molecule bearing the pair of nuclear and electron spins;

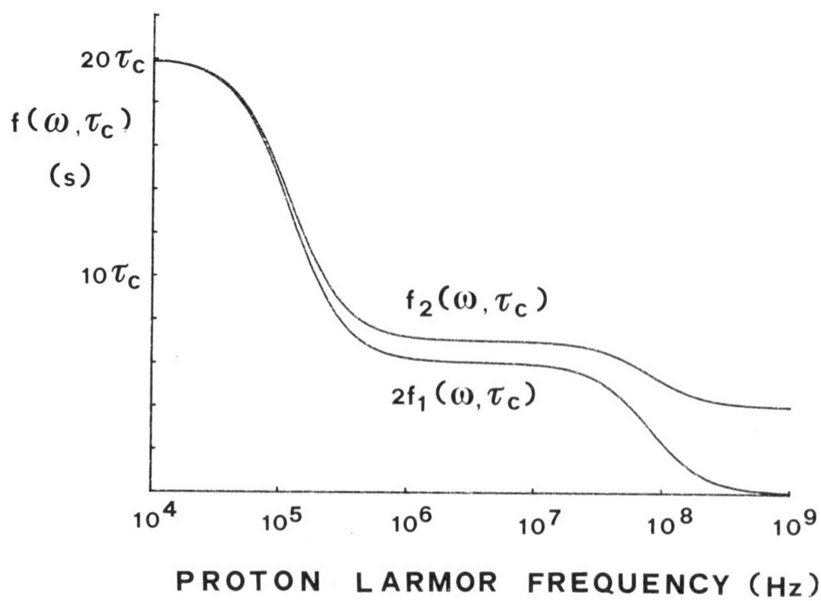

Figure 3.4 Plot of $f(\omega, \tau_c)$ (Equations 3.3 and 3.4) as a function of the proton Larmor frequency; $\tau_c = 2 \times 10^{-9}$ s.

3. The chemical exchange rate, τ_M^{-1}, which is present if the nucleus of interest belongs to a ligand capable of exchanging with the bulk solution.

The rate constant is therefore expressed as

$$\tau_c^{-1} = \tau_s^{-1} + \tau_r^{-1} + \tau_M^{-1} \tag{3.6}$$

Chemical exchange has further impact on the nuclear relaxation processes, and will be treated in detail in Chapter 4. The chemical exchange rate rarely exceeds 10^8 s^{-1}, and is usually much smaller (Figure 3.5).

The rotational correlation time, τ_r, may be qualitatively estimated from the Stokes-Einstein equation, which is valid for spherical Brownian particles isotropically rotating in the medium:

$$\tau_r = \frac{4\pi\eta a^3}{3kT} \tag{3.7}$$

where η is the viscosity of the solvent* and a is the radius of the molecule. As τ_r is dependent on the term a^3, its value is roughly proportional to the molecular weight of the complex. Values for τ_r in water solutions at room temperature range from 10^{-10} s for hexaaqua complexes, to 10^{-8} s for small proteins (MW $\sim 10^4$), to 10^{-6} s for macromolecules (MW $\sim 10^6$) (Figure 3.5). The value of τ_r may also be estimated from T_1 measured at several magnetic fields on a diamagnetic analog of the metal complexes of interest [6, 7]. In fact, in diamagnetic systems the

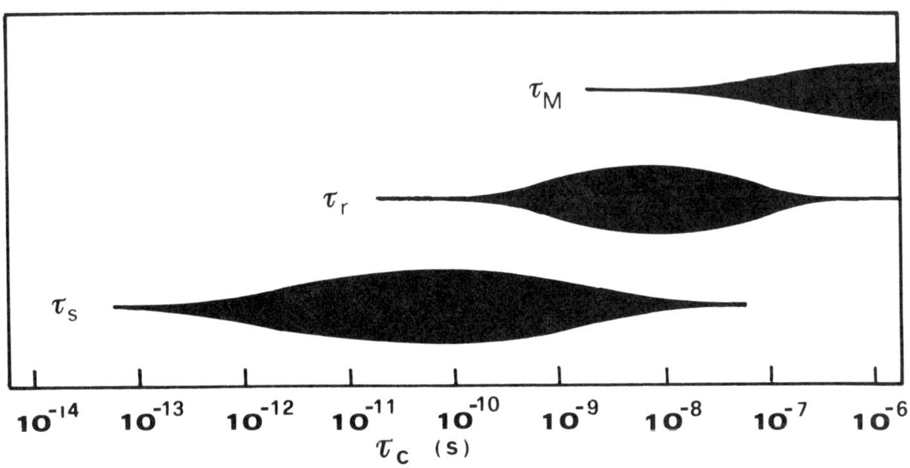

Figure 3.5 Common ranges of values for τ_s, τ_r, and τ_M, as defined in Equation (3.6). τ_r values are those for biological molecules.

* The SI units for η are kg s^{-1} m^{-1}. For water at 25°C, $\eta = 0.8904 \times 10^{-3}$ kg s^{-1} m^{-1}.

relaxation times are usually determined by dipolar coupling with other nuclei in the same molecule, the correlation time for the interaction being the rotational correlation time of the molecule itself, if chemical exchange is not the dominant term.

In the foregoing discussion, the electron-nucleus spin system was assumed to be rigidly held within a molecule isotropically rotating in solution. If the molecule cannot be treated as a sphere, its motion is in general anisotropic and three different rotational correlation times should be considered. Furthermore, if the molecule is not rigid, the correlation times for rotations about single bonds should also be taken into account [8]. For the case of a metal ion rigidly held to a macromolecule, but with a coordinated water molecule freely rotating about the M—O bond, the effect of neglecting the latter rotation on the calculation of T_{1M}^{-1} for the protons is estimated to be about 20% [8]. The effect may be greater if such rotation alters the electronic levels and the anisotropy of the metal ion (Section 3.5). It should be recalled, however, that in metalloproteins the coordinated water molecules are often hydrogen bonded to nearby protein residues and not free to rotate fast about the M—O axis.

The electronic relaxation time, τ_s, is also extremely variable, ranging from 10^{-8} to 10^{-13} s depending on the nature of the metal ion, the type of coordination polyhedron, the type of donor atoms in the coordination sphere, and $\mathbf{B}_0$ (see Section 3.7.1). There is no simple way to predict τ_s, or to obtain a satisfactory measure of it through other spectroscopic techniques, especially when it is shorter than 10^{-10} s. On the other hand, in most of the biological systems of interest, τ_s is the rate constant that dominates τ_c; therefore, its value is needed if nuclear relaxation times are to be used for obtaining accurate structural information. When magnetic field dependence is used to obtain τ_s, a further complication arises: τ_s may be by itself field dependent, since electron relaxation can be often described by equations similar to those for nuclear relaxation (see Section 3.7.1). Owing to this dependence, longitudinal (T_{1e}) and transverse (T_{2e}) electronic relaxation times may not be equal and, strictly speaking, when electron relaxation is the dominant correlation time, Equations (3.3) and (3.4) should be written as

$$T_{1M}^{-1} = \frac{2}{15}\left(\frac{\mu_0}{4\pi}\right)^2 \frac{\gamma_N^2 g_e^2 \mu_B^2 S(S+1)}{r^6}\left(\frac{7T_{2e}}{1+\omega_s^2 T_{2e}^2} + \frac{3T_{1e}}{1+\omega_I^2 T_{1e}^2}\right) \tag{3.8}$$

$$T_{2M}^{-1} = \frac{1}{15}\left(\frac{\mu_0}{4\pi}\right)^2 \frac{\gamma_N^2 g_e^2 \mu_B^2 S(S+1)}{r^6}\left(\frac{13T_{2e}}{1+\omega_s^2 T_{2e}^2} + \frac{3T_{1e}}{1+\omega_I^2 T_{1e}^2} + 4T_{1e}\right) \tag{3.9}$$

where the terms containing ω_s, $\omega_s + \omega_I$, and $\omega_s - \omega_I$ have been collected for simplicity, since $\omega_I \ll \omega_s$. In the low-field (or fast-motion) limit, and for $T_{2e} = T_{1e}$, Equations (3.8) and (3.9) reduce to

$$T_{1M}^{-1} = T_{2M}^{-1} = \frac{4}{3}\left(\frac{\mu_0}{4\pi}\right)^2 \frac{\gamma_N^2 g_e^2 \mu_B^2 S(S+1)}{r^6} T_{1e}$$

justifying the qualitative statement that the longer the electronic relaxation times the broader the NMR lines.

3.1.2 Generalized Dipolar Coupling

As outlined in Section 3.1, the unpaired electrons should not be considered as point dipoles centered on the metal ion because they are extended over a region around it. Even if the unpaired electrons were not delocalized over the donor atoms, they would not be exactly on the metal nucleus but delocalized in atomic orbitals. In the latter case, the effect of the deviation from the point dipole approximation on nuclei a few chemical bonds away is estimated to be negligible. Electron delocalization, on the other hand, may heavily affect the overall relaxation phenomena.

The case of a metal-coordinated water molecule is of special interest in this book. When looking at the proton relaxation times, the simplest way to account for spin delocalization is to consider separately the effect of unpaired electrons centered at the metal and at the oxygen nucleus. In such cases, the value of $1/r^6$ in the Solomon equation can be substituted by an effective distance:

$$\left\langle \frac{1}{r^3} \right\rangle_{\text{eff}}^2 = \frac{1}{1+\lambda^2}\left(\frac{1}{r^3_{\text{M-H}}} + \frac{\lambda^2}{6}\frac{1}{r^3_{\text{O-H}}}\right)^2 \tag{3.10}$$

where λ is the coefficient of the ligand orbitals in the molecular orbital. When $\lambda = 0$ the equation reduces to the point-dipole value. As the distance $r_{\text{O-H}}$ is much shorter than $r_{\text{M-H}}$ (96 versus 280 pm), even small values of λ significantly increase $\langle 1/r^3 \rangle^2_{\text{eff}}$ (Figure 3.6A). It has been clearly pointed out [9], however, that this approach heavily overestimates the deviations from the point-dipole model, mainly because

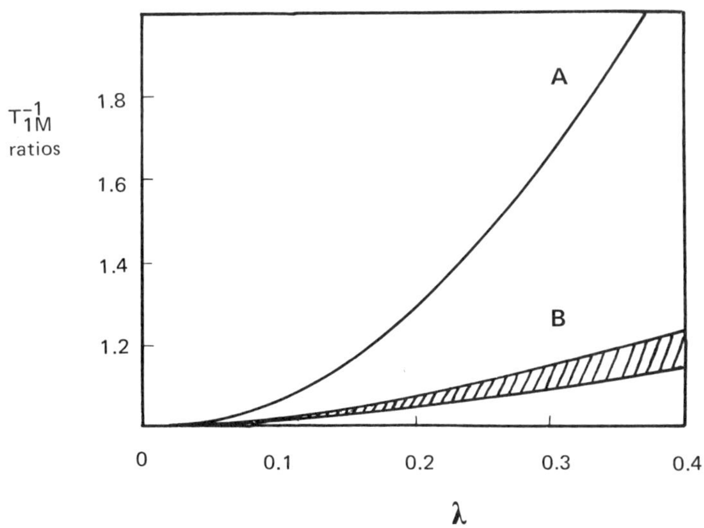

Figure 3.6 Expected dipolar ^{1}H T^{-1}_{1M} ratios determined by considering unpaired spin delocalization and pure metal-centered dipolar coupling for $Mn(H_2O)_6^{2+}$, as a function of the covalency parameter λ, according to: **A** Equation (3.10); **B** Semiempirical molecular orbital calculations [9].

it neglects cross-terms containing both metal and ligand orbitals, and the actual distribution of the unpaired spin density among the various orbitals of the oxygen atom. If such contributions are taken into account [9] the effect of increasing λ is much smaller (Figure 3.6B). More sophisticated calculations of this type have been performed for both ^{1}H and ^{17}O nuclei in a series of hexaaqua complexes [10] (Table 3.1). Whereas the deviations of $\langle 1/r^3 \rangle^2_{\text{eff}}$ from the crystallographic values are small for protons, they are sizeable for ^{17}O, and dramatically increase from $3d^5$ to $3d^9$ metal ions.

T_{1M}^{-1} calculations based on molecular orbital calculations are also available for metal complexes with ligands having dominant π spin density delocalization mechanisms [12]. With certain approximations, the relaxation rates of protons and carbon atoms in sp^2 CH moieties can be expressed [12, 13] as the sum of a metal-centered term (given by the Solomon equation) and a ligand-centered term. For the latter contribution, the effective distance from the metal is given by

$$\left\langle \frac{1}{r^3} \right\rangle^2_{\text{eff}} = \rho_{\text{C}}^{\pi 2}\, \Delta \tag{3.11}$$

where Δ is calculated to be 2.8×10^{-13} pm^{-6} for protons and 1.0×10^{-11} pm^{-6} for carbon atoms. In general, ρ_{C}^{π} can be evaluated from the contact part of the proton isotropic shift. This can be an easy-to-follow procedure for estimating ligand-centered contributions to the relaxation rates for π delocalized spin density.

Table 3.1 Crystallographic Values of Metal-Hydrogen and Metal-Oxygen Distances in Hexaaqua Complexes of Divalent $3d$ Metal Ions^a Compared with Calculated Effective Distances.^{b}

	Metal Ions				
Measurements	Mn	Fe	Co	Ni	Cu
^{1}H Data					
r_{cryst} (pm)	290.6	290.6	278.1	275.5	270.4
r_{eff} (pm)	290.0	289.5	276.3	273.1	268.4
$r^6_{\text{cryst}} \cdot \langle 1/r^3 \rangle^2_{\text{eff}^b}$	1.01	1.02	1.04	1.05	1.05
^{17}O Data					
r_{cryst} (pm)	222.0	222.0	209.0	206.1	201.0
r_{eff} (pm)	219.8	211.7	190.5	174.9	172.3
$r^6_{\text{cryst}} \cdot \langle 1/r^3 \rangle^2_{\text{eff}^b}$	1.06	1.33	1.74	2.67	2.52

^{a} From Reference [10]. M—H distances were calculated using $r_{\text{O—H}} = 95.7$ pm and $\hat{\text{HOH}} = 104.52°$. M—O distances were taken from Reference [11].
^{b} The effective distances account for the experimental T_{1M}^{-1} by considering the electronic distribution predicted from *ab initio* molecular orbital calculations. $r^6_{\text{cryst}} \cdot \langle 1/r^3 \rangle^2_{\text{eff}}$ is the predicted enhancement coefficient of T_{1M}^{-1} with respect to a pure metal-centered dipolar relaxation mechanism.

It should be noted [10], however, that this procedure may lead to an overestimate of the effect.

3.2 CONTACT RELAXATION

As with the paramagnetic shift, nuclear relaxation may also arise from contact interactions of the nuclear magnetic moments with unpaired spin density delocalized at the nuclei themselves [14].

The transition probabilities as defined in Figure 3.2, where the splitting of the levels is now due to the contact contribution, are found to be [14]

$$W_{\omega_I} = W_{\omega_S} = W_{(\omega_S + \omega_I)} = 0 \tag{3.12}$$

$$W_{(\omega_S - \omega_I)} = \frac{1}{2}\left(\frac{A_c}{\hbar}\right)^2 \frac{\tau_e}{1 + (\omega_s - \omega_I)^2 \tau_e^2} \tag{3.13}$$

Here the correlation time is labelled τ_e. This differs from τ_c of Equation (3.6) and is given by

$$\tau_e^{-1} = \tau_s^{-1} + \tau_M^{-1} \tag{3.14}$$

The correlation rate does not contain the rotational time, τ_r, since the contact interaction, by its nature, is not altered by reorientation of the molecules.

In the absence of chemical exchange phenomena ($\tau_M^{-1} \ll \tau_s^{-1}$), the correlation time equals the electronic relaxation time, τ_s.

The enhancement of the longitudinal contact relaxation rate, after extension to the general case $S > \frac{1}{2}$, is

$$T_{1M}^{-1} = \frac{2}{3} S(S+1)\left(\frac{A_c}{\hbar}\right)^2 \frac{T_{2e}}{1 + \omega_s^2 T_{2e}^2} \tag{3.15}$$

The equation for the transverse contact relaxation rate is

$$T_{2M}^{-1} = \frac{1}{3} S(S+1)\left(\frac{A_c}{\hbar}\right)^2 \left(T_{1e} + \frac{T_{2e}}{1 + \omega_s^2 T_{2e}^2}\right) \tag{3.16}$$

In the fast-motion limit, and for $T_{1e} = T_{2e} = \tau_s$, Equations (3.15) and (3.16) reduce to

$$T_{1M}^{-1} = T_{2M}^{-1} = \frac{2}{3} S(S+1)\left(\frac{A_c}{\hbar}\right)^2 \tau_s \tag{3.17}$$

As in the case of dipolar relaxation, the presence of a field-independent term in the equation for T_{2M}^{-1} causes the two nuclear relaxation values to diverge for $\omega_s^2 T_{2e}^2 > 1$ (Figures 3.7, 3.8).

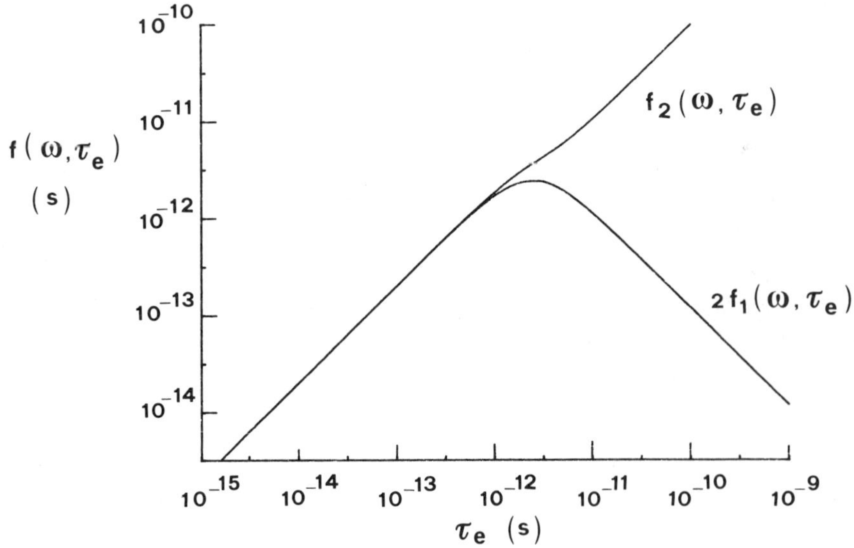

Figure 3.7 Plot of $2f_1(\omega, \tau_e)$ and $f_2(\omega, \tau_e)$ (Equations 3.15 and 3.16) against τ_e for a nuclear Larmor frequency of 100 MHz. It is assumed that $T_{1e} = T_{2e} = \tau_s = \tau_e$.

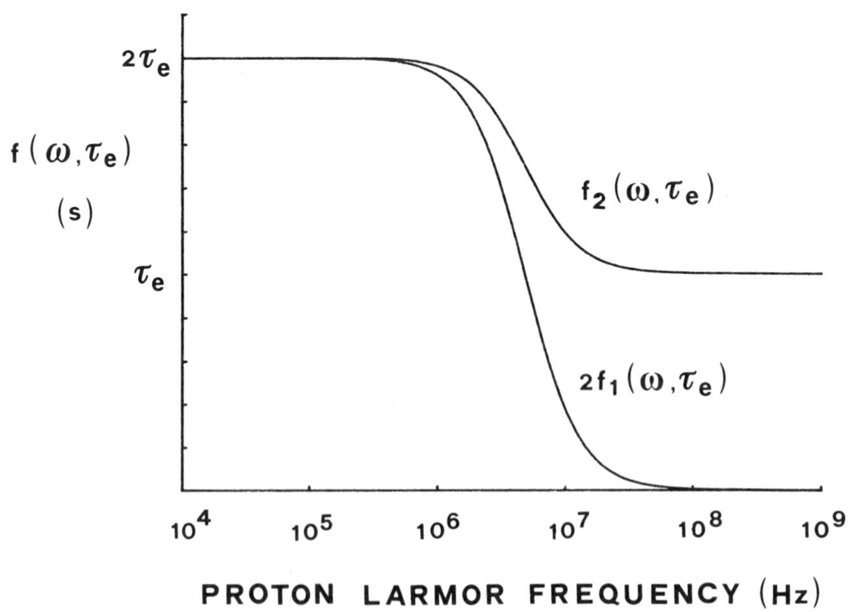

Figure 3.8 Plot of $2f_1(\omega, \tau_e)$ and $f_2(\omega, \tau_e)$ (Equations 3.15 and 3.16) as a function of the proton Larmor frequency. It is assumed that $T_{1e} = T_{2e} = \tau_s = \tau_e = 5 \times 10^{-11}$ s.

3.3 A PICTORIAL ANALYSIS OF THE RELAXATION EQUATIONS

The $f(\omega, \tau)$ functions in the equations analyzed so far contain several terms. More can be learned about their physical meaning by considering the case in which τ is determined by the appropriate electronic relaxation time, T_{1e} or T_{2e}. In the Solomon [3] and Bloembergen [14] approaches, these terms arise from the probabilities of the various nuclear transitions. These transitions are induced by *random* fluctuations of the magnetic field at the nucleus generated by the paramagnetic center.

From a different point of view, the nuclear T_1^{-1} can be calculated in terms of the power dissipated by the nucleus to the electron spin system constituting the lattice. Consider a continuous wave NMR experiment performed on a set of nuclei at a radiofrequency power weak enough to produce a stationary population of nuclear energy levels. Under such conditions, the power dissipated to the environment is proportional to the nuclear T_1^{-1}. If the lattice is constituted of unpaired electrons, the energy per unit time—that is, the power—absorbed by the lattice can be calculated [15, 16]. In the presence of a magnetic field, $\mathbf{B}_0$, the nuclear magnetization vector precesses at a frequency, ω_I, about the z axis, and therefore its projection in the xy plane also rotates at frequency ω_I. It is the *periodic* fluctuation of the magnetic field at an electron site generated in such a way by the nucleus that allows the energy transfer, through mechanisms that can be related to the various $f(\omega, \tau)$ terms.

If the metal ion is located on the nuclear z axis, either above or below the nucleus itself (Figure 3.9A), the component of the nuclear magnetization in the xy plane rotates clockwise round the nucleus at frequency ω_I, whereas, due to the same field $\mathbf{B}_0$, the electron magnetization rotates counterclockwise round the metal at frequency ω_s. (The opposite rotations of the electron and proton magnetizations are due to the signs of the gyromagnetic ratios.) The additional magnetic field produced by the nucleus at the metal site also rotates clockwise in the xy plane at frequency ω_I (Figure 3.9A). The electron spin in this particular spatial orientation can therefore be regarded as responding to a radiofrequency, ω_I, applied in the xy plane. The power absorbed, P', is proportional to what would be the fractional intensity of the EPR signal when the electron spin is irradiated at frequency ω_I; that is, far-off resonance. If the EPR line is Lorentzian, its lineshape function, v, is

$$v = \frac{T_{2e}}{1 + (\omega_s - \omega)^2 T_{2e}^2}$$

where ω_s is the electron resonance frequency and ω is the irradiating frequency. In this case, $\omega = \omega_I$, and therefore

$$P' \propto \frac{T_{2e}}{1 + (\omega_s - \omega_I)^2 T_{2e}^2}$$

If the metal ion is now located in the nuclear xy plane (Figure 3.9B), the field induced at the metal site by the xy component of the nuclear magnetization

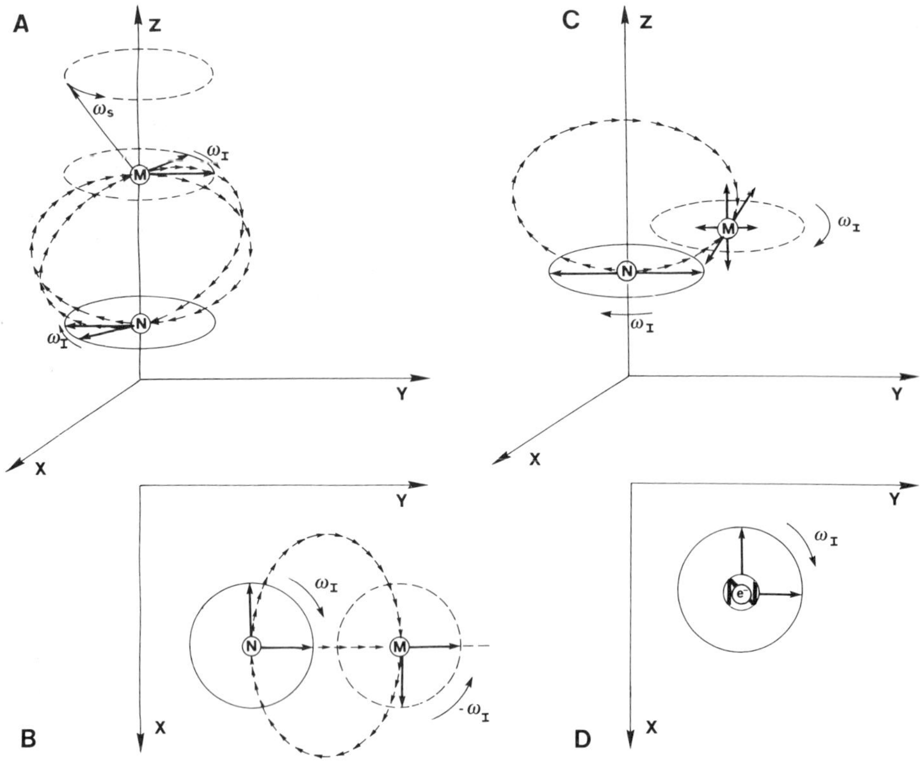

Figure 3.9 Periodic magnetic field induced at the metal site by the *xy* component of the nuclear magnetic moment rotating at frequency ω_I. **A** When the metal is on the *z* axis above the nucleus. **B** When the metal is in the nuclear *xy* plane. **C** When the metal is in a general position with respect to the nuclear coordinates. **D** When a fraction of the electron spin density (e^-) is *on* the nucleus (contact interaction).

is again in the metal xy plane but now rotates counterclockwise. Therefore, $\omega = -\omega_I$ and the power absorbed from this geometrical arrangement is

$$P'' \propto \frac{T_{2e}}{1 + (\omega_S + \omega_I)^2 T_{2e}^2}$$

For a general orientation of the metal ion with respect to the nuclear z axis, contributions of the type P' or P'' are present in addition to a further contribution, which is discussed below. Integration over all the orientations shows that

$$P' = K \frac{T_{2e}}{1 + (\omega_s - \omega_I)^2 T_{2e}^2} \qquad P'' = 6K \frac{T_{2e}}{1 + (\omega_s + \omega_I)^2 T_{2e}^2}$$

These expressions correspond to the first and third terms in Equation (3.3). Note that these terms are constant as long as $\omega_s \pm \omega_I$ ($\simeq \omega_s$) is relatively small so that $(\omega_s \pm \omega_I)^2 T_{2e}^2 \ll 1$ or $\omega_s \pm \omega_I \ll T_{2e}^{-1}$, and decrease when $\omega_s \pm \omega_I \gg T_{2e}^{-1}$ (Figure 3.4). Since T_{2e}^{-1} is a measure of the EPR linewidth ($T_{2e}^{-1} = \pi \Delta\nu$), these terms are significant when the EPR linewidth is relatively large.

If the metal ion is outside both the nuclear xy plane and the z axis, the magnetic field induced at the metal also has a component along the z axis (Figure 3.9C). This component oscillates at frequency ω_I, algebraically summing to the static magnetic field $\mathbf{B}_0$. The power absorbed by the electron spin system from this component, after integration over all the orientations, is

$$P_z = 3K \frac{T_{1e}}{1 + \omega_I^2 T_{1e}^2}$$

which corresponds to the second term in Equation (3.3). Unlike the other two terms, this term contains T_{1e}; in fact, T_{1e} is the time constant for the electron magnetization along z to follow variations of the magnetic field along z. This term is large when the oscillation is slow compared to T_{1e}^{-1}($\omega_I \ll T_{1e}^{-1}$); when the oscillation is too fast ($\omega_I \gg T_{1e}^{-1}$), the response of the electron spin system is less and less effective, and P_z is quenched.

It is now straightforward to predict which terms will appear in the expression for T_{1M}^{-1} for contact interaction [16]. Since in this case the nucleus and the electron (or a fraction of the electron density) reside in the same place, the electron is only affected by a magnetic field rotating at frequency ω_I in the xy plane (Figure 3.9D). Therefore the only power absorbed by the electron spin system is given by a term of the P' type—see Equation (3.13).

This analysis can be further extended to the terms in the equations for the dipolar and contact contributions to describe T_{2M}^{-1} [16].

3.4 CURIE SPIN RELAXATION

In deriving the dipolar and contact relaxation contributions due to the presence of unpaired electrons, as shown in Equations (3.3), (3.4), (3.15), and (3.16), the small difference in the population of the electron spin levels according to the Boltzmann distribution has been neglected. Such a difference gives rise to a finite static magnetic moment, which is related to the $\langle S_z \rangle$ value given in Equation (2.3) for a simplified case:

$$\langle S_z \rangle = -S(S+1)\frac{g_e \mu_B B_0}{3kT} \tag{3.18}$$

The interaction of the nuclear spins with the static magnetic moment related to $\langle S_z \rangle$ provides a further relaxation contribution [17, 18]. Such interaction, of course, cannot be modulated by electron relaxation since $\langle S_z \rangle$ is already an average over the electron spin states. The correlation time for the coupling is only determined by τ_r (or possibly by τ_M). This relaxation mechanism is usually called magnetic susceptibility relaxation or Curie spin relaxation—to reflect its relationship with the magnetic susceptibility of a sample (Curie law).

The dipolar contributions to T_{1M}^{-1} and T_{2M}^{-1} provided by this mechanism in the limit $\omega_s\tau_r \gg 1$ are

$$
\begin{aligned}
T_{1M}^{-1} &= \frac{2}{5}\left(\frac{\mu_0}{4\pi}\right)^2 \frac{\gamma_N^2 g_e^2 \mu_B^2}{r^6} \langle S_z \rangle^2 \frac{3\tau_r}{1+\omega_I^2\tau_r^2} \\
&= \frac{2}{5}\left(\frac{\mu_0}{4\pi}\right)^2 \frac{\omega_I^2 g_e^4 \mu_B^4 S^2(S+1)^2}{(3kT)^2 r^6} \frac{3\tau_r}{1+\omega_I^2\tau_r^2}
\end{aligned}
\tag{3.19}
$$

$$
\begin{aligned}
T_{2M}^{-1} &= \frac{1}{5}\left(\frac{\mu_0}{4\pi}\right)^2 \frac{\gamma_N^2 g_e^2 \mu_B^2}{r^6} \langle S_z \rangle^2 \left(4\tau_r + \frac{3\tau_r}{1+\omega_I^2\tau_r^2}\right) \\
&= \frac{1}{5}\left(\frac{\mu_0}{4\pi}\right)^2 \frac{\omega_I^2 g_e^4 \mu_B^4 S^2(S+1)^2}{(3kT)^2 r^6} \left(4\tau_r + \frac{3\tau_r}{1+\omega_I^2\tau_r^2}\right)
\end{aligned}
\tag{3.20}
$$

Although the value of $\langle S_z \rangle^2$ is much smaller than $S(S+1)/3$, which appears in the Solomon equations (Equations 3.3 and 3.4), Curie spin relaxation may be significant when the dipolar coupling described by the Solomon equations is governed by the electronic relaxation times; that is, when the latter are much smaller than the rotational correlation times. Furthermore, since $\langle S_z \rangle^2$ depends on the square of the external magnetic field, Curie spin relaxation increases with the square of the nuclear Larmor frequency ω_I—see Equations (3.19) and (3.20), and Figure 3.10. For example, when $\tau_r = 10^{-7}$ s, $\tau_c = T_{1e} = 10^{-12}$ s, and $B_0 = 6T$, the Curie spin relaxation for T_{2M}^{-1} is one order of magnitude larger than the Solomon contribution. Such conditions can be easily met for biological systems of interest, and the effects of Curie spin relaxation are very often seen as an additional line broadening of isotropically shifted resonances. On the other hand, Curie spin relaxation does not significantly contribute to T_{1M}^{-1}; at high magnetic field ($\omega_I^2\tau_r^2 \gg 1$) the dependence on ω_I^2 in the numerator is cancelled out. T_{1M}^{-1} levels off at a value that is always small compared to the dipolar contribution.

The occurrence of this particular relaxation mechanism can be recognized through the field and temperature dependence of the linewidths. In the absence of other effects, the line broadening is expected to increase with the square of the magnetic field (Figure 3.10) and to decrease with η/T^3, the latter effect arising from the $1/T^2$ dependence of $\langle S_z \rangle^2$ and the η/T dependence of τ_r, as shown in Equation (3.7).

In principle there may also be a Curie relaxation contribution of the contact type whenever there is chemical exchange or intramolecular rotation to modulate the coupling [17]. The contribution to T_{2M}^{-1} would then be [17–19]

$$
T_{2M}^{-1} = \frac{4}{3}\left(\frac{A_c}{\hbar}\right)^2 \langle S_z \rangle^2 \tau_M = \frac{4}{3}(\Delta\omega)^2 \tau_M
$$

where $\Delta\omega$ is the observed contact shift ($\Delta\omega = 2\pi\,\Delta\nu$).

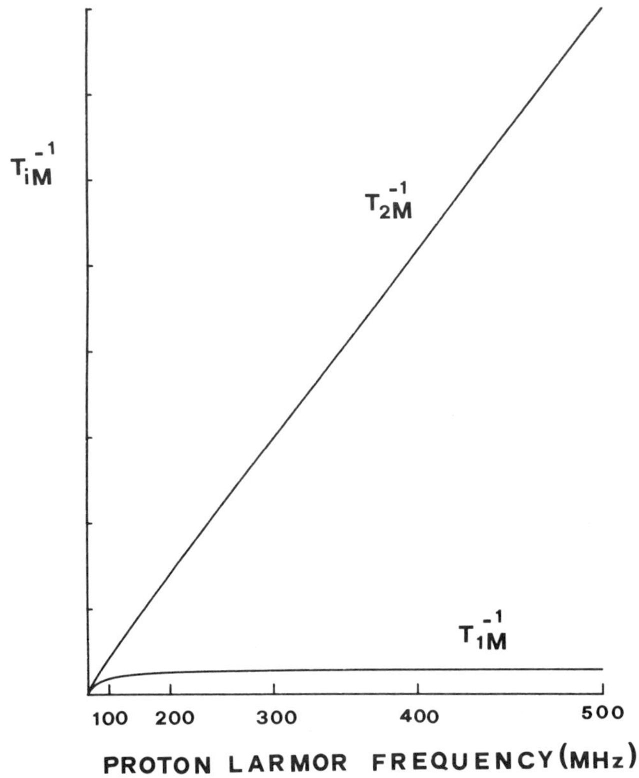

Figure 3.10 Field dependence of the Curie spin relaxation contributions to T_{1M}^{-1} and T_{2M}^{-1} (arbitrary scale); $\tau_r = 2 \times 10^{-9}$ s.

3.5 THE EFFECT OF g ANISOTROPY AND ZERO FIELD SPLITTING

In the development of the basic equations for nuclear relaxation in paramagnetic solutions it has been assumed that (1) the system has an isotropic g value; and (2) for $S > \frac{1}{2}$ the spin transitions are degenerate. Figure 3.11 provides an example of the latter statement; the lack of zero field splitting justifies the degeneracy of the spin transitions.

In real systems these requirements are generally not fulfilled, and some caution should be used in handling the equations. Attempts have been made to derive equations for more complicated cases, but the available theories still do not cover many of the experimental situations. Furthermore, the theoretical results are rather complex and of limited applicability. Nevertheless, it may be useful to give an account of the type of complications that may arise, so that the degree of deviation from the simple behavior can be judged in each case.

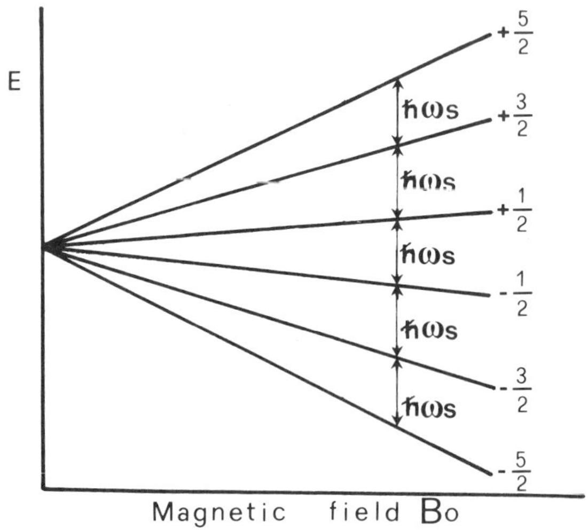

Figure 3.11 Effect of a magnetic field on the electron spin levels of an $S = \frac{5}{2}$ system in the absence of zero field splitting and g anisotropy.

When g is anisotropic, for each orientation of the molecule in the magnetic field, the system experiences a particular g value and then a particular ω_s value, since $g\mu_B B_0 = \hbar\omega_s$. From the point of view of the nucleus, this will result in fluctuating g and ω_s values, the fluctuation depending on the rotational time of the molecule. For axial systems, the g_e^2 factor in Equations (3.8) and (3.9) should be substituted for by a function of $g_\parallel$, $g_\perp$, and of the angle θ between the $g_\parallel$ direction and the metal-nucleus vector. As far as ω_s is concerned, two separate cases should be considered, depending on whether the rotation of the molecule does or does not average the **g** tensor anisotropy at a given external magnetic field. In the former case we can define $\omega_z = g_\parallel \mu_B B_0/\hbar$.

The following equations have been derived [20] for an axial system in the fast-rotation limit:

$$\begin{aligned} T_{1M}^{-1} = {} & \frac{2}{5}\left(\frac{\mu_0}{4\pi}\right)^2 \frac{\gamma_N^2 \mu_B^2 S(S+1)}{r^6} \left\{\left[\frac{1}{9}(16g_\parallel^2 + 9g_\perp^2 - 4g_\parallel g_\perp) + (g_\parallel - g_\perp)^2 \sin^4 \vartheta \right.\right. \\ & \left. + \frac{7}{3} g_\parallel (g_\perp - g_\parallel) \sin^2 \vartheta \right] \frac{\tau_c}{1 + \omega_z^2 \tau_c^2} + \left[\frac{1}{9}(4g_\parallel^2 + g_\perp^2 + 4g_\parallel g_\perp) \right. \\ & \left.\left. - (g_\parallel - g_\perp)^2 \sin^4 \vartheta - \frac{1}{3}(2g_\parallel - 5g_\perp)(g_\perp - g_\parallel) \sin^2 \vartheta \right] \frac{\tau_c}{1 + \omega_I^2 \tau_c^2} \right\} \\ & + \frac{4}{9}\left(\frac{\mu_0}{4\pi}\right) \frac{\gamma_N \mu_B S(S+1)}{r^3} \frac{A_c}{\hbar} (g_\parallel - g_\perp)(3\cos^2 \vartheta - 1)\delta(\tau_c, \tau_e) \frac{\tau_c}{1 + \omega_z^2 \tau_c^2} \\ & + \frac{2}{3}\left(\frac{A_c}{\hbar}\right)^2 S(S+1) \frac{\tau_e}{1 + \omega_z^2 \tau_e^2} \end{aligned} \tag{3.21}$$

$$T_{2M}^{-1} = \frac{4}{15}\left(\frac{\mu_0}{4\pi}\right)^2 \frac{\gamma_N^2\mu_B^2S(S+1)}{r^6}\left\{\left[\frac{1}{3}(12g_{\parallel}^2 + 8g_{\perp}^2 - 8g_{\parallel}g_{\perp}) + 6(g_{\parallel} - g_{\perp})^2 \sin^4 \vartheta\right.\right.$$

$$\left. + (9g_{\parallel} - 5g_{\perp})(g_{\perp} - g_{\parallel}) \sin^2 \vartheta\right]\tau_c + \left[\frac{1}{3}(24g_{\parallel}^2 + 11g_{\perp}^2 + 4g_{\parallel}g_{\perp})\right.$$

$$\left. - 3(g_{\parallel} - g_{\perp})^2 \sin^4 \vartheta + (3g_{\parallel} + 10g_{\perp})(g_{\perp} - g_{\parallel}) \sin^2 \vartheta\right]\frac{\tau_c}{1 + \omega_z^2\tau_c^2}$$

$$+ \left[\frac{1}{3}(4g_{\parallel}^2 + g_{\perp}^2 + 4g_{\parallel}g_{\perp}) - 3(g_{\parallel} - g_{\perp})^2 \sin^4 \vartheta\right.$$

$$\left.\left. - (2g_{\parallel} - 5g_{\perp})(g_{\perp} - g_{\parallel}) \sin^2 \vartheta\right]\frac{\tau_c}{1 + \omega_I^2\tau_c^2}\right\}$$

$$+ \frac{2}{9}\frac{\mu_0}{4\pi}\frac{\gamma_N\mu_B S(S+1)}{r^3}\frac{A_c}{\hbar}(g_{\parallel} - g_{\perp})(3\cos^2\vartheta - 1)\delta(\tau_c, \tau_e)\left(\tau_c + \frac{\tau_c}{1 + \omega_z^2\tau_c^2}\right)$$

$$+ \frac{1}{3}\left(\frac{A_c}{\hbar}\right)^2 S(S+1)\left(\tau_e + \frac{\tau_e}{1 + \omega_z^2\tau_e^2}\right) \tag{3.22}$$

where the first terms are the dipolar contributions, the third are the contact contributions, and the second are the cross-terms. The dimensionless quantities in square brackets replace g^2 in the Solomon equation.

In the slow-rotation limit, which is of interest in paramagnetic macromolecules, the ω_z terms are substituted by a spatial average over a range of ω_s values [21]. For relatively small g anisotropies, such as those found in copper(II) complexes and $3d$ metal ions, Equations (3.21) and (3.22) are still essentially valid.

In the limit $\omega_s\tau_c \ll 1$, Equations (3.21) and (3.22) reduce to Equation (3.23) for both fast- and slow-rotation limit [20]:

$$T_{1M}^{-1} = T_{2M}^{-1} = \left(\frac{\mu_0}{4\pi}\right)^2 \frac{2\gamma_N^2\mu_B^2S(S+1)}{3r^6}\left[\frac{1}{3}(g_{\parallel}^2 + 2g_{\perp}^2) + g_{\parallel}^2\cos^2\vartheta + g_{\perp}^2\sin^2\vartheta\right]T_{1e}$$

$$+ \frac{\mu_0}{4\pi}\frac{4\gamma_N\mu_B S(S+1)A_c(g_{\parallel} - g_{\perp})}{9\hbar r^3}(3\cos^2\vartheta - 1)T_{1e}$$

$$+ \frac{2S(S+1)A_c^2}{3\hbar^2}T_{1e} \tag{3.23}$$

Equations (3.21), (3.22), and (3.23) may be of limited applicability since the directions of the magnetic axes are not generally known. Nevertheless, an estimate of the effect of magnetic anisotropy in a real system is possible, and the degree of deviation from isotropic behavior may be evaluated. In most cases of interest the dipolar term is dominant, and therefore both the contact and cross-terms can be neglected. Under such conditions sample calculations show that, even for relatively large anisotropy, the overall value of g^2 obtained from the term in square brackets in the $\omega_s\tau_c \ll 1$ limit does not deviate more than 20 to 30% from the isotropic value of four. A treatment is also available for introducing the magnetic anisotropy in the equations for the Curie spin relaxation mechanism [18]. Analogous to the effect of g anisotropy on the dipolar and contact terms, magnetic anisotropy introduces in Curie relaxation a dependence on the nuclear coordinates in the molecular framework [18].

When $S > \frac{1}{2}$, the effect of spin-orbit coupling results in zero field splitting which alters the spacing of the spin levels. If the zero field splitting is smaller than the electron Zeeman energy, it will only introduce a small perturbation; even if the zero field splitting is larger than the Zeeman energy, its effect is negligible if the spacing is smaller than the EPR linewidth, $\hbar T_{2e}^{-1}$. When the zero field splitting is large compared to both the Zeeman energy and the EPR linewidth, its presence may drastically alter the relaxation equations. Under these conditions, it is necessary to distinguish between integer and half-integer S systems. When S is a half-integer there is always at least one transition within the S multiplet that does not contain the zero field splitting energy. Such a transition will always be effective for nuclear relaxation, whereas other transitions will be ineffective, since the denominator in the $f(\omega, \tau)$ terms contains $\omega^2\tau^2$ and therefore the zero field splitting energy ZFS, as it is included in ω. As a consequence, the contribution of the ω_s term in Equation (3.8) will be smaller [21, 22]. Furthermore, values of T_{1M}^{-1} will depend on the angle θ. Integration over all the orientations required by the slow-rotation condition results in a smoother dispersion curve [23]. As may be predicted by analysis of Equation (3.8), zero field splitting does not substantially affect the ω_I contribution to T_{1M}^{-1}.

When S is an integer and D is large, all the ω_s transitions contain D and are ineffective for nuclear relaxation. For example, in high spin nickel(II), which is an $S = 1$ system, the two transitions will occur at $D + \hbar\omega_s$ and $D - \hbar\omega_s$ (Figure 3.12), respectively, and the dispersive terms containing the electron transition frequency

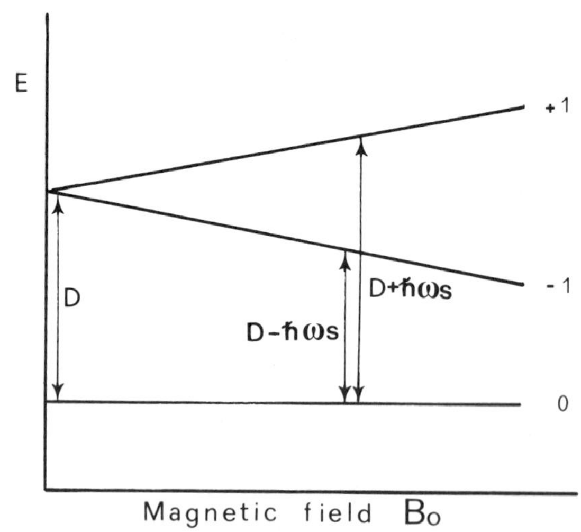

Figure 3.12 Effect of a magnetic field on the electron spin levels of an $S = 1$ system in the presence of zero field splitting (D). The magnetic field is applied along the z axis of the **D** tensor.

will be of the type $\tau_s/[1 + (D/\hbar \pm \omega_s)^2\tau_s^2]$. Since $(D/\hbar \pm \omega_s)\tau_s$ is much larger than unity, the dispersive terms will be quenched independently of the applied magnetic field. Lindner [24] has derived equations for an $S = 1$ system in axial symmetry in the presence of zero field splitting under the slow-rotation condition, which is appropriate for macromolecules. If the rotation is slow the effect on $T_{1M,2M}^{-1}$ for one particular orientation can be computed and then integrated over all the orientations. The expression for the dipolar contribution to T_{1M}^{-1} in the presence of axial zero field splitting as derived from Lindner's treatment with inclusion of the angle ϑ, as for the case of g anisotropy, is [21]

$$T_{1M}^{-1} = \frac{K}{12}\tau_s\left\{\ln\left(\frac{1 + \tau_s^2\omega^2 + 2\omega\omega_s\tau_s^2}{1 + \tau_s^2\omega^2 - 2\omega\omega_s\tau_s^2}\right)\left[\frac{\mathrm{C}}{\tau_s^2\omega\omega_s} + \mathrm{E}\frac{(1 + \tau_s^2\omega^2)^2}{4\omega^3\omega_s^3\tau_s^6}\right] - \mathrm{E}\frac{1 + \tau_s^2\omega^2}{\omega^2\omega_s^2\tau_s^4} + 4\left(\mathrm{B} - \frac{\mathrm{A}}{3}\right)\frac{1}{1 + \omega_I^2\tau_s^2}\right\} \tag{3.24}$$

where $\omega = ZFS/\hbar$; $\mathrm{A} = 27y^4 - 21y^2 + 2$; $\mathrm{B} = 9y^4 - 3y^2 + 2$; $\mathrm{C} = -9y^4 + 6y^2 + 5$; $\mathrm{E} = 27y^4 - 30y^2 + 5$; $y = \cos\vartheta$; and $K = (\mu_0/4\pi)^2\gamma_N^2\mu_B^2 r^{-6}$. Equation (3.24) is valid as long as $\mathrm{ZFS} \gg g_e\mu_B B_0,\ \hbar T_{2e}^{-1}$.

An analogous equation is available for the $S = \frac{3}{2}$ case [23]:

$$\begin{aligned} T_{1\mathrm{M}}^{-1} = \frac{K}{32}\tau_s\Bigg\{ & \frac{8}{\omega_S\tau_s}tg^{-1}(\omega_S\tau_s)\left(\mathrm{C} - \mathrm{D}\frac{\omega_I}{\omega_S} - \mathrm{E}\frac{1 - \omega_I^2\tau_s^2}{\omega_S^2\tau_s^2}\right) \\ & + 8\left(\mathrm{D}\frac{\omega_I}{\omega_S}\frac{1}{1 + \omega_S^2\tau_s^2} + \mathrm{E}\frac{1}{\omega_S^2\tau_s^2}\right) - 3\mathrm{E}\frac{1 + \omega^2\tau_s^2}{\omega^2\omega_S^2\tau_s^4} \\ & + 3\ln\frac{1 + \omega^2\tau_s^2 + 2\omega\omega_S\tau_s^2}{1 + \omega^2\tau_s^2 - 2\omega\omega_S\tau_s^2} \\ & \times\left(\mathrm{C}\frac{1}{\omega\omega_S\tau_s^2} + \mathrm{E}\frac{(1 + \omega^2\tau_s^2)^2}{4\omega^3\omega_S^3\tau_s^6}\right) + 20\left(\mathrm{B} - \frac{\mathrm{A}}{3}\right)\frac{1}{1 + \omega_I^2\tau_s^2}\Bigg\} \end{aligned} \tag{3.25}$$

where $\mathrm{D} = 12y^2 - 8$ and all the other symbols are as defined in Equation (3.24).

When $\mathrm{S} = \frac{1}{2}$, an effect similar to the zero-field splitting is provided by the $\hat{\mathbf{I}}\cdot\mathbf{A}\cdot\hat{\mathbf{S}}$ term, where this time **I** refers to the metal nucleus. Examples of such effects will be given in Sections 6.5.1 and 9.4.

It should be stressed that all the equations presented in this section contain the implicit assumption that electron relaxation is an exponential process described by a time constant, τ_s. This point will be further discussed in Section 3.7.

3.6 A COMPARISON OF DIPOLAR, CONTACT, AND CURIE SPIN RELAXATION

Once T_{1M} and T_{2M} have been measured, it is useful to try to understand the relative weight of dipolar, contact, and Curie spin contributions to the overall relaxation effect. Indeed, each of the three contributions is independently capable of providing valuable information.

The dipolar and Curie relaxation mechanisms—Equations (3.8), (3.9), and (3.20)—provide experimental values that are related to the distance of the nucleus from the paramagnetic metal and τ_c or τ_r, assuming that ligand-centered effects are neglected. Measurements at various magnetic fields lead, in principle, to an estimation of these parameters. The field dependence of the dipolar contribution can be observed in any region of the accessible magnetic field range—that is, from 0 to 600 MHz; whereas the Curie relaxation can usually only be observed in the region above 100 MHz. Even at a single magnetic field value, measurements of dipolar T_1 contributions for various nuclei of the same moiety give information on their relative distance from the metal, as long as τ_c is the same. This is also true for T_2 measurements when they are determined by dipolar and/or Curie contributions.

The presence of contact relaxation indicates that a given moiety is covalently bound to a paramagnetic metal ion and provides an estimate of the absolute value of A_c—Equations (3.15) and (3.16). Sometimes the contact coupling constant can be evaluated by shift measurements, and it is therefore possible to predict whether the contact relaxation contributions to T_{1M}, T_{2M}, or both, are negligible or sizeable.

Unlike isotropic shifts, which often contain dipolar and contact contributions of the same order of magnitude, relaxation times can often be recognized to be dominated by only one of the possible contributions.

Some qualitative guidelines can be given to make an *a priori* estimate of the relative weight of dipolar, contact, and Curie relaxation contributions. Consider first the fast-motion limit where $T_{1M} = T_{2M}$ and none of the frequency-dependent terms is dispersed. The equations then take the simple form already noted:

$$T_{1M}^{-1} = T_{2M}^{-1} = \frac{4}{3}\left(\frac{\mu_0}{4\pi}\right)^2 \frac{\gamma_N^2 g_e^2 \mu_B^2 S(S+1)}{r^6}\tau_c + \frac{2}{3}\left(\frac{A_c}{\hbar}\right)^2 S(S+1)\tau_s$$
$$+ \left(\frac{\mu_0}{4\pi}\right)^2 \frac{4\omega_I^2 g_e^4 \mu_B^4 S^2(S+1)^2}{(3kT)^2 r^6}\tau_r$$

Under these conditions the third term, which arises from Curie spin relaxation, is always negligible (see Section 3.4). If $\tau_c = \tau_s$ it is only necessary to compare the following two expressions:

$$\frac{4}{3}\frac{\gamma_N^2}{r^6} 8.6 \times 10^{-61} \quad \text{and} \quad \frac{2}{3}\left(\frac{A_c}{\hbar}\right)^2$$

where 8.6×10^{-61} is $(\mu_0/4\pi)^2 g_e^2 \mu_B^2$ in units of T^2 m^6. For hydrogen nuclei, γ_N^2 is 7.16×10^{16} rad^2 s^{-2} T^{-2}; whereas $(A_c/\hbar)^2$ does not usually exceed 10^{12} rad^2 s^{-2}, and is often found to be much lower. Thus, for metal-nucleus distances of 500 pm or smaller, the dipolar term is largely dominant. The value of the dipolar term drops dramatically with increasing r owing to its r^{-6} dependence, and for larger distances it could become smaller than the contact term. However, the $(A_c/\hbar)^2$ value is also qualitatively expected to decrease with increasing distance. Therefore, unless some efficient unpaired spin delocalization pathway is operative, the dipolar term in proton relaxation is usually dominant for most systems of chemical interest.

The situation can be different when the rotational correlation time is shorter than τ_s and therefore determines the overall correlation time, τ_c, in the dipolar term. In this case, the relative importance of the dipolar term is decreased by a factor τ_s/τ_c, which can be as large as 10^3 for small complexes with rotational correlation times of 10^{-11} to 10^{-12} s and electronic relaxation times of 10^{-8} to 10^{-9} s—for example, Cu^{2+}, Mn^{2+}, and VO^{2+}. In macromolecular complexes, rotational correlation times are much larger, and situations of this type do not occur.

For nuclei other than protons the dipolar term is smaller, owing to the smaller values of γ_N (see Table 1.1). On the other hand, $(A_c/\hbar)^2$ may be much larger, being as high as 10^{17} rad^2 s^{-2} for directly coordinated nuclei like ^{17}O, ^{19}F, ^{14}N, and ^{15}N. In such cases, contact interaction may easily be the dominant mechanism for nuclear relaxation, especially for T_{2M}^{-1}.

Outside the fast-motion limit the relative weight of contact and dipolar interactions on T_{1M}^{-1} and T_{2M}^{-1} may also be different. The following considerations are particularly relevant to proton relaxation. The presence of a Curie spin relaxation contribution to T_{2M}^{-1} should be checked and factorized out. The contribution of contact and dipolar interactions can be estimated by examining the correlation time dependence of the paramagnetic relaxation depicted in Figures 3.3 and 3.7. It appears that the maximum for T_{1M}^{-1} occurs at $\omega_I\tau_c \sim 1$ in the dipolar term and at $\omega_S\tau_S \sim 1$ in the contact term. Taking for simplicity $\tau_c = \tau_S$, this means that in the intermediate situation where $\omega_S\tau_S > 1 > \omega_I\tau_S$ the relative importance of the contact term is even smaller than that estimated in the fast-motion limit. The equation for T_{2M}^{-1} has nondispersive terms in both the dipolar and contact contributions (accounting for one-fifth and one-half of the total effect measured in the fast-motion limit, respectively), and therefore the conclusions drawn in the fast-motion limit are still qualitatively correct.

A comparison of T_{1M}^{-1} and T_{2M}^{-1} values may thus be useful to evaluate the occurrence of contact relaxation. Taking again $\tau_c = \tau_S$, and keeping in mind the τ_S or field dependence of T_{1M}^{-1} and T_{2M}^{-1}, the T_{1M}/T_{2M} ratios are expected to be as follows:

	$\omega_S\tau_S < 1$	$\omega_S\tau_S > 1 > \omega_I\tau_S$	$\omega_I\tau_S > 1$
T_{1M}/T_{2M} (dipolar)	1	7/6	$\gg 1$
T_{1M}/T_{2M} (contact)	1	$\gg 1$	$\gg 1$

When the estimate of τ_S is such that the intermediate situation ($\omega_S\tau_S > 1 > \omega_I\tau_S$) occurs, dominant dipolar relaxation will still give T_{1M}/T_{2M} ratios close to unity, whereas dominant contact relaxation will give $T_{2M}^{-1} \gg T_{1M}^{-1}$. In the latter case, no information is obtained concerning the mechanism controlling T_{1M}^{-1}; however, an idea of what happens to T_{1M}^{-1} can be perceived by the following procedure:

1. Use T_{2M}^{-1} and a reasonable estimate for τ_s to calculate $(A_c/\hbar)^2$;
2. Use the values of τ_s and $(A_c/\hbar)^2$ in the contact equation for T_{1M}^{-1} to estimate the contact contribution to the longitudinal relaxation;
3. Compare this value with the experimental value of T_{1M}^{-1} to check whether the contact contribution is negligible.

3.6.1 Secondary Relaxation Phenomena

The Solomon dipolar coupling scheme discussed in Section 3.1.1 is not limited to the case of electron-nucleus coupling. It also holds for nucleus-nucleus coupling and, indeed, important contributions to the overall diamagnetic relaxation rates of nuclei arise from such coupling. In the case of coupling between nuclei of the same type the $W(\omega_s - \omega_I)$ transition (see Figure 3.2) will not change the energy of the system and the corresponding term in Equation (3.1) will therefore be of the nondispersive type. At high fields, a term of this type is responsible for the so-called cross-relaxation, which can be described as a mutual spin flip of the two coupled nuclei [3]. This becomes important when the two nuclei interact to different extents with a nearby paramagnetic center undergoing relaxation rate enhancements. The relaxation behavior of the nuclei can be influenced by cross-relaxation in addition to the direct interaction with the unpaired electrons [25–27]. A practical consequence of cross-relaxation is that the nucleus that is less affected by the paramagnetic center tends to relax quicker than expected; in principle the reverse holds for the other nucleus.

In a classical nonselective T_1 experiment (Section 5.3.1), the magnetizations of both spins are perturbed and subsequently sampled along their recovery towards equilibrium. In the absence of cross-relaxation the two decays are exponential (Section 3.1) and can be described by time constants T_1^A and T_1^B. When cross-relaxation is present, the two decays are only exponential—with time constants T_1^A and T_1^B—for a time shorter than the cross-relaxation time. The slower decay is then accelerated while the faster one is slowed down. The two spins relax together towards equilibrium with a single time constant intermediate between T_1^A and T_1^B. Therefore, each decay can in principle be described by the sum of two exponentials [25, 26], and its fitting to a single exponential may give a T_1 value somewhat different from the one characterizing the electron-nucleus interaction. Consequently, it is important to estimate the relative importance of cross-relaxation when analyzing nuclear relaxation data in paramagnetic systems.

In practice, estimates of the paramagnetic contributions to the nuclear relaxation rates—Equations (3.8) and (3.15)—should be compared with those of the cross-relaxation rate, σ, which, for τ_r values typical of proteins, is given by [26]

$$\sigma = \frac{2}{15}\left(\frac{\mu_0}{4\pi}\right)^2 \gamma_N^4 \hbar^2 I(I+1) r^{-6} \tau_r \tag{3.26}$$

The σ value is therefore expected to increase with the size of the protein. Furthermore, since it depends on γ_N^4, it will be essentially negligible for nuclei other than protons. For a τ_r of 10^{-8} s, σ ranges from 1 to $50\,s^{-1}$ for pairs of protons with r values between 300 and 150 pm. Paramagnetic enhancements are generally higher than these values; however, in particular situations cross-relaxation effects may be, and have been [27], detected. In such cases, information on the paramagnetic contributions can still be extracted with the aid of suitable experimental techniques, such as selective T_1 (Section 5.3.1) or NOE (Section 5.3.2) [28], and proper analysis of the data [26].

3.7 ELECTRON RELAXATION

Another assumption implicit in the derivation of Equations (3.3), (3.4), (3.15), and (3.16), as well as of their modifications discussed in Section 3.5, is that the correlation time for the interaction, τ_c or τ_e, is shorter than the nuclear relaxation times, T_{1M} and T_{2M}. This limit is generally called the Redfield [29] limit. For virtually all the cases of chemical interest this limit holds and therefore the equations are of general applicability—at least from this point of view.

When considering electron relaxation it should be borne in mind that the electron-lattice coupling is also related to its own correlation time, τ_c' or τ_v. Electronic relaxation times are typically in the range 10^{-8} to 10^{-12} s, so are the time constants for the fluctuations of the random magnetic fields generated by the lattice [30, 31], depending on the specific relaxation mechanism. Therefore, only metal ions with long electronic relaxation times may fulfill the condition that the correlation time τ_v is shorter than τ_s; that is, the general requirement of the Redfield limit is also satisfied for electrons.* (These cases will be discussed in some detail in Section 3.7.1.) On the other hand, metal ions with shorter electronic relaxation times than τ_v will be in the so-called strong-field limit. A completely general treatment of this case is not available, although various attempts can be found in the literature [33–41]. When the electronic relaxation time is faster than the associated correlation time the decay mode may no longer be exponential; indeed, a multiexponential decay would probably be closer to the real situation [37, 38]. Since Equations (3.3) and (3.4) were derived with the assumption that the decay described by the time constant τ_c is exponential, when τ_c is dominated by the electronic relaxation time, and the latter cannot be associated to a single exponential decay, Equations (3.3) and (3.4) would no longer be valid. In this case, it is preferable to develop new expressions for T_{1M} and T_{2M} in terms of a more complex theory which does not require that lattice and electron spin motions be separable [39]. In particular, the field dependence expected as a result of the presence of the various dispersive terms would probably be smoother [37, 38, 41]. Fitting of the field-dependent data to the equations discussed up to now is proba-

* The conditions of validity of the Redfield limit may be different for T_{1e} and T_{2e} [32].

bly still reasonable, except that the value of τ_c obtained would no longer strictly represent the electronic relaxation time [33, 37, 38].†

3.7.1 The Main Mechanisms for Electron Relaxation

As pointed out in the preceding section, no comprehensive theory is available for electron relaxation in solution. The main reason for this is probably due to the fact that room temperature EPR spectra in solution can only be obtained for a few metal ions. Nevertheless, several possible mechanisms for electron relaxation have been successfully identified, and equations have been derived for many of them in the Redfield limit approximation. These mechanisms are discussed here, bearing in mind that they might still be qualitatively valid outside the limits of their derivation.

Solid-state theories ascribe electron relaxation to the coupling of electrons with lattice vibrations, or more generally with phonons. (A phonon is a quasi-particle with the energy of a crystal vibration.) It is customary to separate the coupling into direct processes [42], Raman processes [42], and Orbach processes [43], all of which are shown in Figure 3.13. In the direct process a phonon with the same energy as the electron spin transition is required, whereas in the Raman process two phonons of higher energy may simultaneously interact with the spin system so that their energy difference equals the electron Zeeman energy. This phenomenon can also be regarded as the scattering of a single phonon at a different energy upon collision with the electron spin. The Orbach process requires an excited level with different m_s accessible through coupling with a single phonon. Both the Raman and, more predominantly, the Orbach processes are enhanced by the presence of a relatively close excited electronic level. The efficiency of the mechanisms depends on the magnitude of the spin-orbit coupling constant. The Orbach process requires the presence of a relatively large proportion of phonons of appropriate energy, so that the energy separation between the ground and first excited state should not be much larger than kT ($kT \simeq 200\ \text{cm}^{-1}$ at room temperature).

Following the solid-state approach, equations have been derived [44] for the electron spin relaxation of $S = \frac{1}{2}$ ions in solution determined by the aforementioned processes. The contribution from the direct process has been shown to be negligible around room temperature; that from the Orbach and Raman processes is directly proportional to the collision frequency, which is of the order of 10^{11} to $10^{12}\ \text{s}^{-1}$, of solvent molecules with small metal complexes—this may also be the mechanism holding for metalloproteins. A further relaxation pathway of solid-state

† It has been recently shown that every system can be treated as if it were in the Redfield limit for electron relaxation as far as nuclear relaxation is concerned. Calculated field dependences of T_{1M}^{-1} in high spin nickel(II) and cobalt(II) systems are remarkably similar independently of whether the Redfield limit holds or not [67]. In the latter case the obtained τ_c may be considered as a fictitious τ_s for the system. A comparison of the experimental T_{1M}^{-1} data for a given system under different rotational correlation times may in principle permit discrimination between Redfield and strong field limit situations [68].

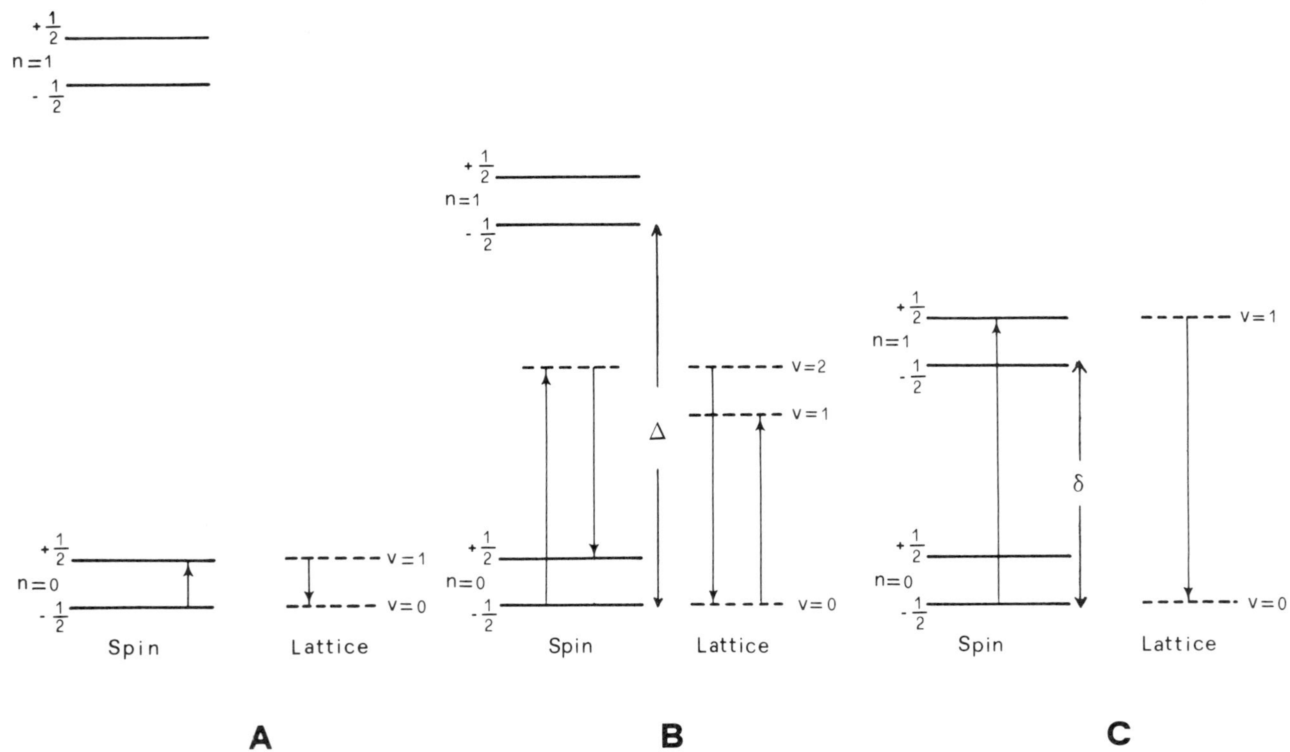

Figure 3.13 Phonon-electron spin coupling is capable of exciting a spin transition (for example, $-\frac{1}{2} \rightarrow \frac{1}{2}$) through: **A** a direct process, **B** a Raman two-phonon process, or **C** an Orbach process. The proximity of the excited electron level $n = 1$ favors both the Raman and the Orbach processes. Δ and δ indicate the energy separations with excited electronic states coupled to the ground state via spin-orbit coupling.

type is related to the presence of hyperfine coupling with the metal nucleus [45]. The contact coupling constant, A_c, is different for the ground and excited states; when the system is brought to an excited state and back to the ground state during an Orbach process the modulation of A_c provides a further relaxation mechanism. According to Freed *et al*, the presence of nearby degenerate orbital levels may be a further cause of electron relaxation through dynamic Jahn-Teller distortions [46, 47].

Two other mechanisms proposed for $S = \frac{1}{2}$ ions are typical of solutions [48] and are determined by the rotation of the molecule. The first is due to the anisotropy of g, and possibly of A, and is proportional to τ_r [48–52]. If g or A are anisotropic, their values are orientation dependent, and are therefore modulated by rotation. The second is the so-called spin-rotation mechanism, which is proportional to the departure of $\bar{g}$ from g_e [53, 54]. This mechanism arises from the fact that, upon rotation, the electron cloud does not rigidly follow the molecular skeleton, and there is a coupling between the spin angular momentum and the rotational angular momentum of the molecule, which causes relaxation. The faster the rotation, the larger the effect; thus, the contribution to the relaxation rate of this mechanism is inversely proportional to τ_r. Consequently, this mechanism is probably negligible for macromolecules.

Additional relaxation mechanisms, for ions with $S > \frac{1}{2}$, arise from modulation of the quadratic zero field splitting in solution; these mechanisms have been discussed in detail by Bloembergen and Morgan [55] and Rubinstein *et al* [56]. The following equations have been derived for T_{1e} and T_{2e}:

$$T_{1e}^{-1} = C\left(\frac{\tau_v}{1 + \omega_s^2\tau_v^2} + \frac{4\tau_v}{1 + 4\omega_s^2\tau_v^2}\right) \tag{3.27}$$

$$T_{2e}^{-1} = \frac{1}{2}C\left(3\tau_v + \frac{5\tau_v}{1 + \omega_s^2\tau_v^2} + \frac{2\tau_v}{1 + 4\omega_s^2\tau_v^2}\right) \tag{3.28}$$

where C is proportional to the mean squared fluctuation of the zero field splitting and τ_v is the correlation time for the instantaneous distortions of the chromophore, which may be caused by collisions with solvent molecules.

As was pointed out in Section 3.7, the functional form of Equations (3.27) and (3.28), as well as of the equations derived for the other electron relaxation mechanisms discussed before, is surely correct only in the Redfield limit—that is, when the correlation time is shorter than τ_s. But even in the strong-field limit some kind of analogous mechanisms must be operative.

From the foregoing, the importance of the contribution of the various electronic relaxation mechanisms can be considered qualitatively for various cases. The least efficient mechanisms are the rotational mechanisms, which account for T_{1e}^{-1} values up to 10^9 s^{-1}, and are only detected in the absence of other mechanisms. Contributions from the modulation of the quadratic zero field splitting to the relaxation rates can range from 10^8 to 10^{11} s^{-1}, depending on the extent of the geometrical distortions; the magnitude of the static zero field splitting can give a rough idea of the efficiency of such a mechanism. The contributions of the Raman and Orbach processes are extremely variable, depending on the availability of low-lying excited states. In particular, the Orbach process may be the only process that can account for relaxation rates higher than 10^{12} s^{-1}. It should be borne in mind that the availability of low-lying excited states increases the zero field splitting of an orbitally nondegenerate S level. Thus, irrespective of the particular mechanism, the availability of low-lying excited states contributes to the shortening of electronic relaxation times. Finally, it should also be remembered that the extent of the splitting of an S manifold depends on the spin-orbit coupling constant, which increases from left to right and from top to bottom in the periodic table.

$S = \frac{1}{2}$ ions like Cu^{2+} have no zero field splitting, small magnetic anisotropy and, in general, far-off excited states. Electronic relaxation times are therefore long. The main relaxation mechanism operative for $Cu(OH_2)_6^{2+}$ is the spin-rotation mechanism, although there are probably some contributions from the Raman process and modulation of g by tumbling [31, 57]. For copper(II) in a macromolecule, the spin-rotation mechanism is probably quenched and, since in low-symmetry environments the first excited state is higher in energy than in the aquaion, the Raman process is also reduced; thus, the tumbling modulation of g is probably of comparable efficiency [58]. Contributions from the Raman process may, however, be again sizeable for a four-coordinate copper(II) chromophore distorted towards a pseudotetrahedral geometry, since the electron levels become closer [59, 60] (Figure 3.14). $S = \frac{1}{2}$ ions like low spin iron(III) with ground levels deriving from an orbitally degenerate ground level in cubic symmetry (Figure 1.1) may have low-lying excited states; therefore, Orbach processes are likely to be very efficient and short relaxation times are expected [31]. The same holds for pseudooctahedral cobalt(II) and pseudotetrahedral nickel(II) chromophores. $S = \frac{5}{2}$ ions, like high spin manganese(II) and high spin iron(III), have small g anisotropy and first excited levels much higher than the ground level, since they arise from a different free ion term. In these cases, modulation of the quadratic zero field splitting is probably the most efficient relaxation mechanism [55, 56]; such a mechanism is much more efficient for iron(III) than for manganese(II) since in the former the static zero field splitting is larger [61]. Pseudooctahedral nickel(II) and pseudotetrahedral cobalt(II) have excited states higher in energy than those in the pseudotetrahedral and pseudooctahedral analogs already discussed, so the Orbach and Raman relaxation mechanisms are expected to be relatively less efficient and probably comparable with the modulation of the quadratic zero field splitting. All the lanthanide ions, with the exception of gadolinium(III), are likely to be relaxed by Orbach type processes [43]. For gadolinium(III), which has a ^8S free ion ground term, the modulation of the quadratic zero field splitting is again the dominant mechanism [31].

Considerations of this type account for the electronic relaxation times reported in Table 3.2. Such values can be used to give a first estimate of the nuclear relaxation properties produced by various metal ions.

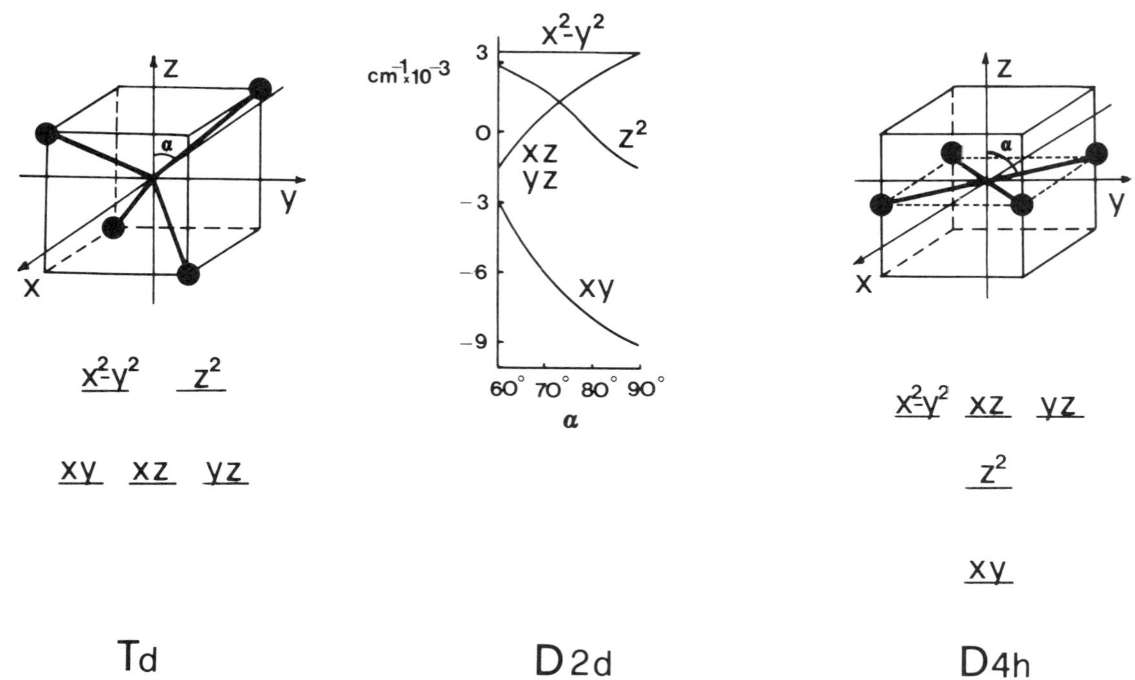

Figure 3.14 Energy-level diagrams (hole formalism) for a D_{2d} tetracoordinated copper(II) complex as a function of the angle α at constant ligand field strength [60].

Table 3.2 Estimated Electronic Relaxation Times for Some Paramagnetic Metal Ions Commonly Used in NMR

Metal Ion	τ_s (s)
Ti^{3+}	10^{-9}–10^{-10}
V^{2+}	5×10^{-10}
V^{3+}	5×10^{-12}
VO^{2+}	10^{-8}–10^{-9}
Cr^{3+}	5×10^{-10}
Cr^{2+}	10^{-11}
Mn^{3+}	10^{-10}–10^{-11}
Mn^{2+}	10^{-8}–10^{-9}
Fe^{3+} (H.S.)	10^{-10}–10^{-11}
Fe^{3+} (L.S.)	10^{-11}–10^{-12}
Fe^{2+} (H.S.)	10^{-12}
Co^{2+} (H.S.)	10^{-11}–10^{-12}
Co^{2+} (L.S.)	10^{-9}–10^{-10}
Ni^{2+}	10^{-10}–10^{-12}
Cu^{2+}	1–3×10^{-9}
Ru^{3+}	10^{-11}–10^{-12}
Re^{3+}	10^{-11}
Gd^{3+}	10^{-8}–10^{-9}
Dy^{3+}	8×10^{-13}
Ho^{3+}	7.5×10^{-13}
Tb^{3+}	8×10^{-13}
Tm^{3+}	8×10^{-13}
Yb^{3+}	1×10^{-12}

3.8 NUCLEAR RELAXATION IN THE PRESENCE OF MAGNETICALLY COUPLED METAL IONS

So far, only electron-nucleus interactions in systems where the metal ion is well isolated from other metal ions have been considered. Such an arrangement always holds for dilute solutions of molecules containing a single metal ion, especially for macromolecules whose dimensions prevent one metal ion approaching another. However, the situation is different if more than one paramagnetic center, of the same or different type, is present in the same molecule. Here the interaction with a nucleus will generally give rise to a multiple-center interaction which is extremely difficult to analyze; indeed, it is only recently that attempts have been made to theoretically treat the problem [21, 62]. Furthermore, the experimental data available are still few.

When metal ions A and B are not interacting with each other, but are both interacting with nucleus N, the nuclear relaxation enhancement, T_{iM}^{-1} $(i = 1, 2)$, may be expressed by [63]

$$T_{iM}^{-1} = T_{iM(A)}^{-1} + T_{iM(B)}^{-1}$$

where $T_{iM(A)}^{-1}$ and $T_{iM(B)}^{-1}$ are the effects that each metal ion would produce if the other metal ion were diamagnetic.

The nuclear relaxation enhancement is expressed as the sum of the contributions from each metal ion even in magnetically coupled systems; however, the equations for T_{iM}^{-1} are expected to differ from Equations (3.3), (3.4), (3.15), and (3.16) because the effect of the coupling on the spin levels of the system has to be considered [21, 62]. Changes in electronic relaxation times may also occur.

In this context, three limiting cases can be analyzed [21]:

1. The magnetic coupling is so small that $|J|$ in the Hamiltonian Equation (2.25) is smaller than $\hbar\tau_s^{-1}$ for both metal ions:

$$|J| \ll \hbar\tau_{s(A)}^{-1}, \hbar\tau_{s(B)}^{-1}$$

The electronic relaxation times are thus not expected to be affected by the magnetic coupling, and the relaxation equation valid for isolated metal ions still holds for both $T_{iM(A)}^{-1}$ and $T_{iM(B)}^{-1}$.

2. $|J|$ is larger than $\hbar\tau_s^{-1}$ for both metal ions:

$$|J| \gg \hbar\tau_{s(A)}^{-1}, \hbar\tau_{s(B)}^{-1}$$

Presumably, the metal ions can be treated as having a single electronic relaxation time, which is likely to be

$$\tau_s^{-1} = \tau_{s(A)}^{-1} + \tau_{s(B)}^{-1}$$

This is only a lower limit for τ_s^{-1} if the interaction between the metal ions gives rise to additional relaxation mechanisms, such as those due to new excited energy levels arising from magnetic coupling. The resulting τ_s should be used for both contributions to the nuclear relaxation rate. The magnetic coupling produces electron levels with total spin quantum numbers both larger and smaller than the starting quantum numbers. For example, for two $S = \frac{1}{2}$ metal ions magnetic coupling produces a singlet and a triplet state. As long as the two levels are equally populated (otherwise the population distribution should be considered), the effect of strong magnetic coupling is to reduce the nuclear relaxation caused by each metal ion by a half. The sign of J—ferro- or antiferromagnetic coupling—is also presumably unimportant as long as $J < kT$.

3. $|J|$ is intermediate between $\hbar\tau_{s(A)}^{-1}$ and $\hbar\tau_{s(B)}^{-1}$:

$$\hbar\tau_{s(A)}^{-1} > |J| > \hbar\tau_{s(B)}^{-1}$$

In this case, the electronic relaxation time of the slow-relaxing ion is expected to be shortened, up to the limit of about $\hbar|J|^{-1}$. A possible equation would be

$$\tau_{s(B)}^{-1}(J) = \tau_{s(B)}^{-1}(0) + \frac{J^2}{\hbar^2}\tau_{s(B)}(0) \tag{3.29}$$

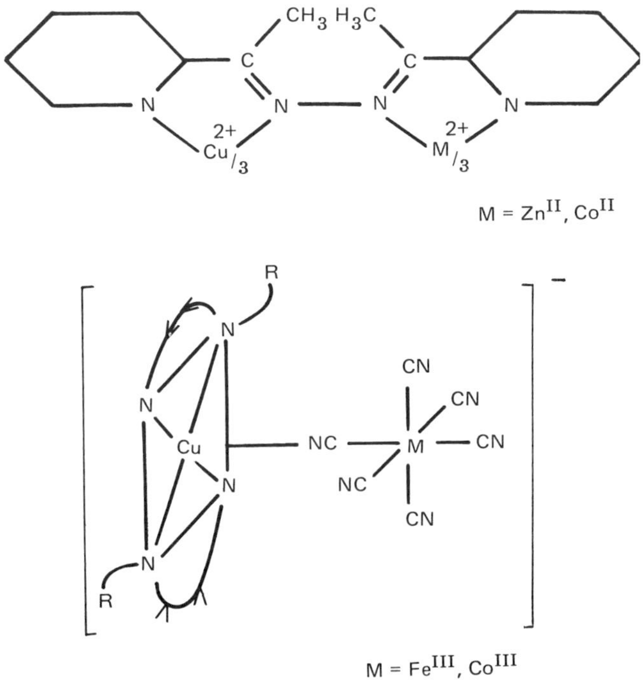

Figure 3.15 Bimetallic species for which shortening of the copper(II) electronic relaxation times when *M* is paramagnetic has been observed [62–64].

Again, the nuclear relaxation contribution, $T_{iM(B)}^{-1}$, would be smaller than that in the absence of magnetic coupling.

Data available in the literature on coupled systems in solution considers the narrowing of the proton linewidth in copper(II) complexes as a consequence of the interaction with fast-relaxing systems, such as high spin cobalt(II) or low spin iron(III) (Figure 3.15) [62–64]. More extensive data are being obtained [65, 66].

3.9 THE EFFECT OF TEMPERATURE ON THE ELECTRON-NUCLEUS SPIN INTERACTION

No explicit temperature dependence is included in the equations for T_{1M}^{-1} and T_{2M}^{-1}, except for cases where Curie spin relaxation is the dominant term (Sections 3.1, 3.2, and 3.4). However, all the correlation times modulating the electron-nucleus coupling are generally temperature dependent, although in different ways, and their variation will therefore be reflected in the values of T_{1M}^{-1} and T_{2M}^{-1}.

In the limits of validity for the Solomon and Bloembergen equations, the correlation time for dipolar interaction and contact relaxation were given by Equations (3.6) and (3.14), respectively. The exchange time, τ_M, is rarely short enough to dominate in Equations (3.6) and (3.14); however, when this is the case,

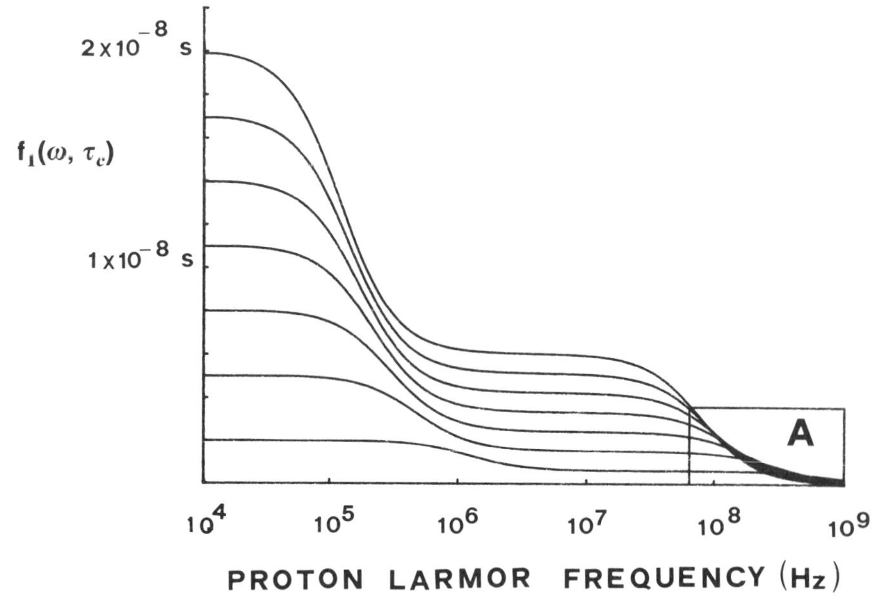

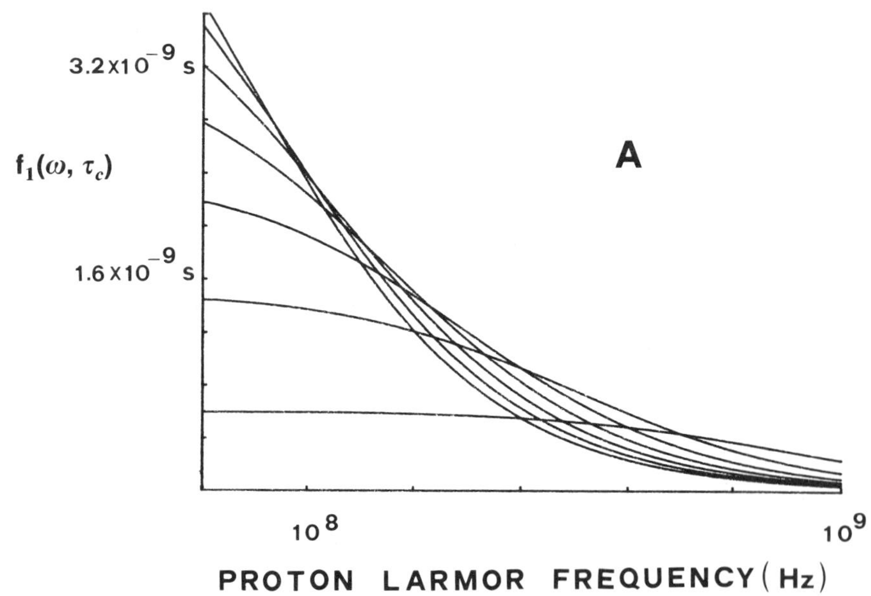

Figure 3.16 Plot of $f_1(\omega, \tau_c)$, according to Equation (3.3), as a function of the proton Larmor frequency. The curves are calculated for τ_c values decreasing from 2×10^{-9} s (upper curve on the left) in steps of 3×10^{-10} s. Inset **A** shows the increase of $f_1(\omega, \tau_c)$ at high fields upon decreasing τ_c.

a strong temperature dependence of the nuclear relaxation times is expected, because of the exponential variation of the exchange rate with temperature. When the rotational correlation time is the dominant term, the effect of temperature would be anticipated to be less, since, as already seen, τ_r may be approximated by the Stokes Equation (3.7). The effect is enhanced however by the change in viscosity of the solvent, η; in the case of water, η decreases as much as 2.7 times from 0 to 40°C. Qualitatively, both τ_r and τ_M decrease with increasing temperature; this causes a decrease in the nuclear relaxation rates in the nondispersive regions and an increase in the second half of the ω_I dispersive region (Figure 3.16).

The temperature dependence of τ_s, including its sign, may be a matter of extensive speculation, mainly because the origin of the electron-spin relaxation mechanism in each particular case may be different, and in general is not known. When Equations (3.27) and (3.28) hold—by assuming that the time constant, τ_v, associated with the relaxation process has an exponential dependence on temperature

$$\tau_v^{-1} = \tau_{v0}^{-1} \exp\left(\frac{-E_v}{RT}\right) \tag{3.30}$$

—τ_s would be expected to increase with increasing temperature in the nondispersive region ($\omega_s \tau_v \ll 1$) and to decrease in the dispersive region (high magnetic field). However, this picture may be further complicated by a possible temperature dependence of the C term in Equations (3.27) and (3.28), making it difficult to predict the overall behavior of τ_s. The electronic relaxation mechanism may then vary at different temperatures, although the possibility of switching from one mechanism to another in a few tens of degrees around room temperature is rather unlikely.

In the types of coupled metal systems discussed in Section 3.8, the effect of temperature will obviously depend on the sign and magnitude of the exchange coupling constant, J.

The effects of chemical exchange on the NMR parameters and hence the influence of temperature through modulation of the exchange process will be considered in Chapter 4.

REFERENCES

1. Koenig, S. H., Prodell, A. G., and Kusch, P. *Phys. Rev.* **88** (1952) 191.
2. Bloembergen, N., Purcell, E. M., and Pound, R. V. *Phys. Rev.* **73** (1948) 679.
3. Solomon, I. *Phys. Rev.* **99** (1955) 559.
4. Kubo, R., and Tomita, K. *J. Phys. Soc. Japan* **9** (1954) 888.
5. Abragam, A. In *The Principles of Nuclear Magnetism*, Oxford University Press, Oxford, 1961, Chapter VIII.
6. Doddrell, D. M., Pegg, D. T., Bendall, M. R., and Gregson, A. K. *Chem. Phys. Lett.* **40** (1976) 142.
7. Hallenga, K., and Koenig, S. H. *Biochemistry* **15** (1976) 4255.
8. Woessner, D. E. *J. Chem. Phys.* **36** (1962) 1.

9. Waysbort, D., and Navon, G. *J. Chem. Phys.* **68** (1978) 3704.
10. Nordenskiöld, L., Laaksonen, A., and Kowalewski, J. *J. Am. Chem. Soc.* **104** (1982) 379.
11. Drakin, S. I. *Zh. Strukt. Khim.* **4** (1963) 514.
12. Gottlieb, H. P. W., Barfield, M., and Doddrell, D. M. *J. Chem. Phys.* **67** (1977) 3785.
13. Doddrell, D. M., Healy, P. C., and Bendall, M. R. *J. Magn. Reson.* **29** (1978) 163.
14. Bloembergen, N. *J. Chem. Phys.* **27** (1957) 575.
15. Koenig, S. H. *J. Magn. Reson.* **47** (1982) 441.
16. Koenig, S. H. *J. Magn. Reson.* **31** (1978) 1.
17. Gueron, M. *J. Magn. Reson.* **19** (1975) 58.
18. Vega, A. J., and Fiat, D. *Mol. Phys.* **31** (1976) 347.
19. La Mar, G. N., and Van Hecke, G. R. *J. Chem. Phys.* **52** (1970) 5676.
20. Sternlicht. H. *J. Chem. Phys.* **42** (1965) 2250.
21. Bertini, I., Luchinat, C., Mancini, M., and Spina, G. In *Magneto-Structural Correlations in Exchange Coupled Systems*, D. Gatteschi, O. Kahn, and R. D. Willett (eds.), Reidel, Dordrecht, 1985, p. 421
22. Koenig, S. H., Brown, R. D., and Studebaker, J. *Cold Spring Harbor Symp. Quant. Biol.* **36** (1971) 551.
23. Bertini, I., Luchinat, C., Mancini, M., and Spina, G. *J. Magn. Reson.* **59** (1984) 213.
24. Lindner, U. *Ann. Phys.* **16** (1965) 319.
25. Andree, P. J. *J. Magn. Reson.* **29** (1978) 419.
26. Granot, J. *J. Magn. Reson.* **49** (1982) 257.
27. Sletten, E., Jackson, J. T., Burns, P. D., and La Mar, G. N. *J. Magn. Reson.* **52** (1983) 492.
28. Morris, G. A., and Freeman, R. *J. Magn. Reson.* **29** (1978) 433.
29. Redfield, A. G. *IBM Res. Dev.* **1** (1957) 19.
30. Al'Tshuler, A., and Valiev, K. A. *Sov. Phys. Jetp* **35** (1959) 661.
31. Lewis, W. B., and Morgan, L. O. *Trans. Met. Chem.* **4** (1968) 33.
32. Freed, J. H., Bruno, G. V., and Polnaszek, C. F. *J. Phys. Chem.* **75** (1971) 3385.
33. Pegg, D. T., and Doddrell, D. M. *Chem. Phys. Lett.* **42** (1976) 607.
34. Hwang, L. P., Anderson, C. F., and Friedman, H. L. *J. Chem. Phys.* **62** (1975) 2098.
35. Anderson, C. F., Hwang, L. P., and Friedman, H. L. *J. Chem. Phys.* **64** (1976) 2806.
36. Friedman, H. L., Holz, M., and Hertz, H. G. *J. Chem. Phys.* **70** (1979) 3369.
37. Pegg, D. T., and Doddrell, D. M. *Aust. J. Chem.* **29** (1976) 1869.
38. Pegg, D. T., and Doddrell, D. M. *Aust. J. Chem.* **29** (1976) 1885.
39. Benetis, N., Kowalewski, J., Nordenskiöld, L., Wennerström, H., and Westlund, P.-O. *Mol. Phys.* **48** (1983) 329.
40. Benetis, N., Kowalewski, J., Nordenskiöld, L., Wennerström, H., and Westlund, P.-O. *Mol. Phys.* **50** (1983) 515.
41. Westlund, P.-O., Wennerström, H., Nordenskiöld, L., Kowalewski, J., and Benetis, N. *J. Magn. Reson.* **59** (1984) 91.
42. Van Vleck, J. H. *Phys. Rev.* **57** (1940) 426.
43. Orbach, R. *Proc. Phys. Soc. (London)* **A77** (1961) 821.
44. Kivelson, D. *J. Chem. Phys.* **45** (1966) 1324.
45. Atkins, P. W., and Kivelson, D. *Mol. Phys.* **12** (1967) 201.
46. Freed, J. H., and Kooser, R. G. *J. Chem. Phys.* **49** (1968) 4715.
47. Kooser, R. G., Volland, W. V., and Freed, J. H. *J. Chem. Phys.* **50** (1969) 5243.
48. Muus, L. T., and Atkins, P. W. *Electron Spin Relaxation in Liquids*, (eds.), Plenum Press, New York and London, 1972.

49. McConnell, H. M. *J. Chem. Phys.* **25** (1956) 709.
50. Kivelson, D. *J. Chem. Phys.* **33** (1960) 1094.; Kivelson, D. *J. Chem. Phys.* **41** (1964) 1904.
51. Wilson, R., and Kivelson, D. *J. Chem. Phys.* **44** (1966) 154.
52. Wilson, R., and Kivelson, D. *J. Chem. Phys.* **44** (1966) 4440.
53. Hubbard, P. S. *Phys. Rev.* **131** (1963) 1155.
54. Hoel, D., and Kivelson, D. *J. Chem. Phys.* **62** (1975) 4535.
55. Bloembergen, N., and Morgan, L. O. *J. Chem. Phys.* **34** (1961) 842.
56. Rubinstein, M., Baram, A., and Luz, Z. *Mol. Phys.* **20** (1971) 67.
57. Wilson, R. "Electron Paramagnetic Resonance Studies of the Relaxation of Copper(II), Titanium(III), and Vanadyl Ions in Solution," Ph.D. Thesis, Xum, 76–15, 448, Ann Arbor, 1975.
58. Koenig, S. H., and Brown, R. D. *Ann. N.Y. Acad. Sci.* **222** (1973) 752.
59. Bertini, I., Dei, A., and Scozzafava, A. *Inorg. Chem.* **14** (1975) 1526.
60. Bertini, I., Canti, G., Grassi, R., and Scozzafava, A. *Inorg. Chem.* **19** (1980) 2198.
61. La Mar, G. N., and Walker, F. A. *J. Am. Chem. Soc.* **95** (1973) 6950.
62. Bertini, I., Lanini, G., Luchinat, C., Mancini, M., and Spina, G., *J. Magn. Reson.* **63** (1985) 56.
63. Benelli, C., Dei, A., and Gatteschi, D. *Inorg. Chem.* **21** (1982) 1284.
64. Bertini, I., Luchinat, C., Mani, F., and Scozzafava, A. *Inorg. Chem.* **19** (1980) 1333.
65. Owens, C., Drago, R. S., Bertini, I., Luchinat, C., and Banci, L., submitted.
66. Bertini, I., Luchinat, C., Owens, C., and Drago, R. S., submitted.
67. Bertini, I., Luchinat, C., and Kowalewski, J., *J. Magn. Reson.* **62** (1985) 235.
68. Banci, L., Bertini, I., and Luchinat, C., *Inorg. Chim. Acta* **100** (1985) 173.

CHAPTER 4

Chemical Exchange and Chemical Equilibria

4.1 DEFINITION OF CHEMICAL SITES

A nucleus under investigation may change its position relative to the paramagnetic metal ion during the course of an experiment. When this change occurs at an appropriate rate the observed effects of the electron-nucleus coupling are altered in several ways.

In the most general case, the nuclei of interest will spend a fraction of time residing in each of n different chemical environments. The probability of finding a chosen nucleus in a particular site i, P_i, will be equal to the molar fraction, f_i, of nuclei in that environment:

$$P_i = f_i = \frac{N_i}{\sum_{i=1}^{n} N_i}$$

where N_i is the number of nuclei in the ith site, $\sum_{i=1}^{n} N_i$ is the total number of nuclei in the sample, and $\sum_{i=1}^{n} P_i = 1$ by definition.

Here, the only approximation is to neglect the time spent by a nucleus *in moving* from one environment to another. For example, consider a 0.1 mM water solution of a paramagnetic metalloenzyme, EM, to which a glycinate ion is bound—through its carboxylic end, $EM—O—CO—CH_2—NH_2$, for instance. The molarity of free glycinate ions is 0.1 M. If there is chemical exchange between bound and free glycinate, each of the two CH_2 protons will spend

$$\frac{10^{-4}}{10^{-1} + 10^{-4}} = \frac{1}{1001}$$

of the time in the bound position, and

$$\frac{10^{-1}}{10^{-1} + 10^{-4}} = \frac{1000}{1001}$$

of the time in the bulk solution.

The NH_2 protons can also exchange with water molecules. The probability of finding such a proton in each of the three environments will be

$$P_{\text{bound}} = \frac{10^{-4}}{10^{-4} + 10^{-1} + 55} \simeq 1.8 \times 10^{-6}$$

$$P_{\text{free}} = \frac{10^{-1}}{10^{-4} + 10^{-1} + 55} \simeq 1.8 \times 10^{-3}$$

$$P_{H_2O} = \frac{55}{10^{-4} + 10^{-1} + 55} \simeq 0.998$$

where 55 is the approximate molarity of water. Therefore, the exchangeable protons will only spend 1/550000 of their time in a bound position.

The probabilities, or molar fractions, are equal to the fraction of time a nucleus will spend in a particular environment when observed for a suitably long period of time. They don't reveal *how long* the nucleus will stay, on average, in each particular position; in other words, the average lifetime of the nucleus in each environment cannot be calculated. As will be seen, the chemical lifetime or its reciprocal, the chemical dissociation rate, are crucial parameters governing the influence of chemical exchange on NMR parameters.

4.2 TWO-SITE EXCHANGE

4.2.1 Chemical Shifts

Consider, for simplicity, the exchange of nucleus N between two sites, A and B [1, 2]:

$$N_A \underset{\tau_B^{-1}}{\overset{\tau_A^{-1}}{\rightleftharpoons}} N_B$$

Although the dissociation rate constants, τ_A^{-1} and τ_B^{-1}, are not known, it is known that at equilibrium $\tau_A/\tau_B = P_A/P_B$.

The nuclei in the A site will resonate at a Larmor precession frequency ω_A, and those in the B site will resonate at a frequency ω_B.* Under very slow chemical exchange conditions, two signals will be observed, centered at frequency ω_A and ω_B, with intensities proportional to P_A and P_B, respectively. When $P_A = P_B$ the two signals will be of equal intensity and $\tau_A^{-1} = \tau_B^{-1} = \tau^{-1}$ (Figure 4.1A). When the exchange rate is about equal to the difference, $\Delta\omega$, between the two resonance frequencies ω_A and ω_B ($\tau^{-1} \sim \Delta\omega$), the shape of the spectrum will alter as a single nucleus resonating, say, at frequency ω_A, after jumping into the B site will tend

* Throughout this chapter chemical shifts will be expressed as absolute resonance frequencies in rad s^{-1} (ω) instead of Hz (ν). The relation is: $\omega = 2\pi\nu$. Frequencies expressed in rad s^{-1} are directly comparable with exchange rates τ^{-1}.

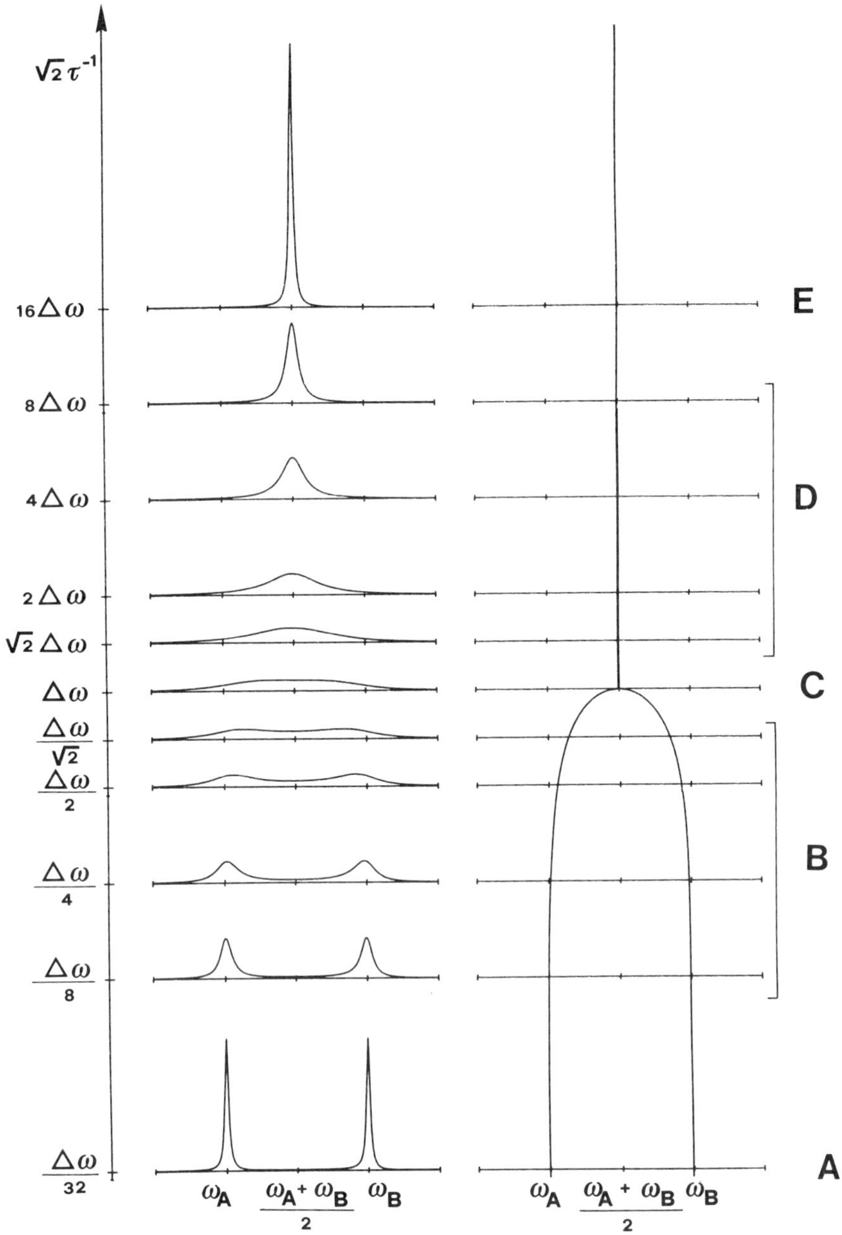

Figure 4.1 Calculated lineshape for the NMR signals of nuclei equally distributed between two sites, **A** and **B**, as a function of the exchange rate τ^{-1} (left-hand pictures); and calculated dependence of the chemical shift on the exchange rate τ^{-1} (right-hand picture). The various exchange regions are indicated as: **A**: slow exchange, $\tau^{-1} \ll \Delta\omega$; **B**: quasi-slow exchange, $\tau^{-1} < \Delta\omega/\sqrt{2}$; **C**: coalescence, $\tau^{-1} = \Delta\omega/\sqrt{2}$; **D**: quasi-fast exchange, $\tau^{-1} > \Delta\omega/\sqrt{2}$; **E**: fast exchange, $\tau^{-1} \gg \Delta\omega$.

to modify its resonance frequency to the new value ω_B. The time for this change is presumably of the order of $1/\Delta\omega$, and in this case is about equal to the residence time, τ, in the site. The nucleus will then probably jump again into the A site and try to adjust its resonance frequency to the new situation. In other words, the nucleus will spend most of its time varying its resonance frequency within the values ω_A and ω_B, without being in either of the two extremes for a significant amount of time. As all the nuclei in the sample will behave in the same way, a very broad signal extending symmetrically from the ω_A to the ω_B positions will result (Figure 4.1C).

If the exchange rate is further increased, the nuclei will never be able to reach the values ω_A and ω_B, and will have a narrower range of oscillation around $(\omega_A + \omega_B)/2$ (Figure 4.1D). Finally, for $\tau^{-1} \gg \Delta\omega$, all the nuclei, from the point of view of an NMR experiment, will be equivalent and will be located in an average environment characterized by the chemical shift $(\omega_A + \omega_B)/2$. The result will be a sharp signal at this frequency (Figure 4.1E).

In Figure 4.1B the exchange rate, τ^{-1}, is still small, although not negligible with respect to $\Delta\omega$. The peaks are therefore slightly moved towards the center, since the population of nuclei in the A site includes some that have just come from the B site bearing a partial memory of their previous resonance frequency ω_B. The same reasoning applies to the nuclei at the B site. Furthermore, the lines are broader than when there is an absence of exchange (see later) because of the reduced lifetime in each state. In this context, it is important to recall that the transverse relaxation time, T_2, which is related to the linewidth, $\Delta\nu$, through the relation $T_2 = 1/\pi\,\Delta\nu$, is just a measure of the lifetime in a given nuclear spin state.

It has been shown that by increasing τ^{-1} the resonance frequency of each signal passes from ω_A or ω_B to the average value $(\omega_A + \omega_B)/2$. The right-hand picture of Figure 4.1 shows that this transition is not smooth, and that the two signals collapse for τ^{-1} values approaching $\Delta\omega/\sqrt{2}$. The situation $\sqrt{2}\tau^{-1} = \Delta\omega$ represents the borderline between the so-called slow- and fast-exchange regions. Thus, slow- or fast-exchange rates can only be defined with respect to the difference in Larmor frequency, $\Delta\omega$.

In the example given at the beginning of this section, the difference in population between bound and free molecules was very large—this is a case often encountered in biological studies and therefore deserves particular attention. The effects of increasing the exchange rate between the two sites will be qualitatively similar to those just described. When the exchange rate is very slow there will be two signals centered at ω_M and ω_d, where ω_M is the chemical shift at the paramagnetic metal site and ω_d is the chemical shift of the free ligand in solution, which is assumed to experience negligible paramagnetic effects. The difference in chemical shift, $\Delta\omega_M = \omega_M - \omega_d$, will usually be very large (see Chapter 2); but the intensity of the signal in the paramagnetic site will be much smaller ($\sim 1/1000$ of the other site in the example), and the signal itself will be intrinsically broadened because of paramagnetic effects on the transverse relaxation rate, T_{2M}^{-1}, discussed in Chapter 3. Such effects may be so large that the paramagnetic signal is not observable in many cases.

As for the diamagnetic signal, this will be broadened and shifted as the exchange time of the molecule from the paramagnetic site, τ_M, is decreased. At the very fast exchange limit the signal will be centered at a different position, ω_p, which in this case is the weighted average position between ω_d and ω_M:

$$\omega_p = f_d\omega_d + f_M\omega_M = \omega_d + f_M \Delta\omega_M \tag{4.1}$$

where f_M and f_d are the molar fractions of nuclei in the paramagnetic and diamagnetic sites. In the case where there is a large excess of free ligand, the general formula for the observed shift of the signal from the diamagnetic position as a function of τ_M^{-1}, is given by [3]

$$\Delta\omega_p = \omega_p - \omega_d = \frac{f_M}{\tau_M^2} \frac{\Delta\omega_M}{(T_{2M}^{-1} + \tau_M^{-1})^2 + (\Delta\omega_M)^2} \tag{4.2}$$

For $\tau_M^{-1} \gg T_{2M}^{-1}, \Delta\omega_M$, Equation (4.2) reduces to Equation (4.1).

Figure 4.2 shows the variation of $\Delta\omega_p$ as a function of τ_M^{-1} for the two limiting cases: $T_{2M}^{-1} \gg \Delta\omega_M$ and $T_{2M}^{-1} \ll \Delta\omega_M$.

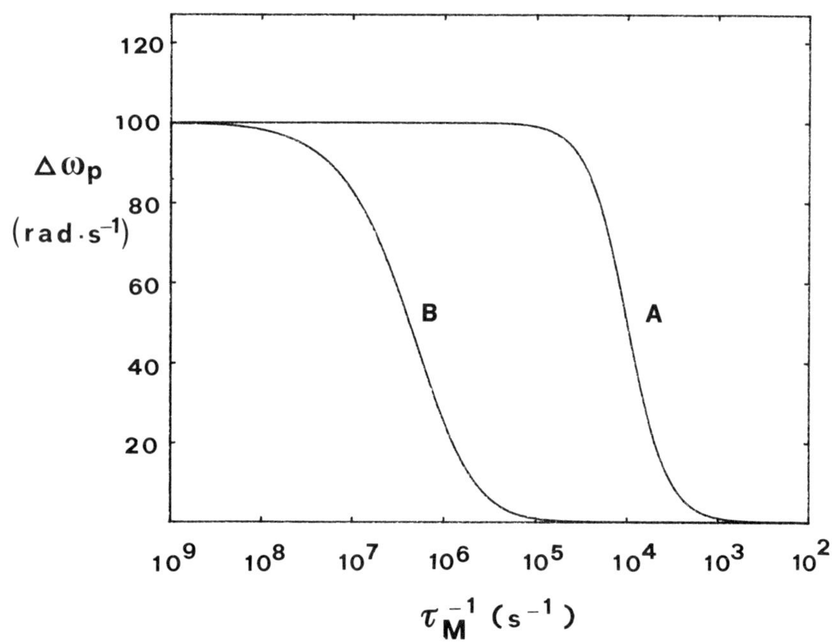

Figure 4.2 Variation of the paramagnetic contribution to the chemical shift, $\Delta\omega_p$, as a function of the exchange rate of the nucleus, τ_M^{-1}, from the paramagnetic site. The curves are calculated using the following values for the parameters in Equation (4.2): $\Delta\omega_M = 10^4$ rad s^{-1}, $f_M = 10^{-2}$, $T_{2M}^{-1} = 10^2$ (**A**) or 10^6 s^{-1} (**B**).

4.2.2 Relaxation Rates

The analysis of the relaxation rates will be limited to the cases of nuclei belonging to a ligand or inhibitor, for example, that is in large excess with respect to the metalloprotein. In particular, the way in which the chemical exchange can propagate the paramagnetic enhancement of the nuclear relaxation rates, T_{1M}^{-1} and T_{2M}^{-1}, of the bound nuclei to the nuclei in the bulk solution will be investigated. In analogy with the chemical shift, the relaxation rates of the bulk nuclei will be termed T_{1d}^{-1} and T_{2d}^{-1}. In the presence of chemical exchange with nuclei in the paramagnetic site, such rates will be enhanced by an amount T_{1p}^{-1} or T_{2p}^{-1}, and the measured values will be

$$T_{1\text{meas}}^{-1} = T_{1d}^{-1} + T_{1p}^{-1}$$

$$T_{2\text{meas}}^{-1} = T_{2d}^{-1} + T_{2p}^{-1}$$

The longitudinal relaxation rate enhancement, T_{1p}^{-1}, will go asymptotically from zero to $f_M T_{1M}^{-1}$, as a function of τ_M^{-1} (Figure 4.3), according to the following equation:

$$T_{1p}^{-1} = f_M(T_{1M} + \tau_M)^{-1} \tag{4.3}$$

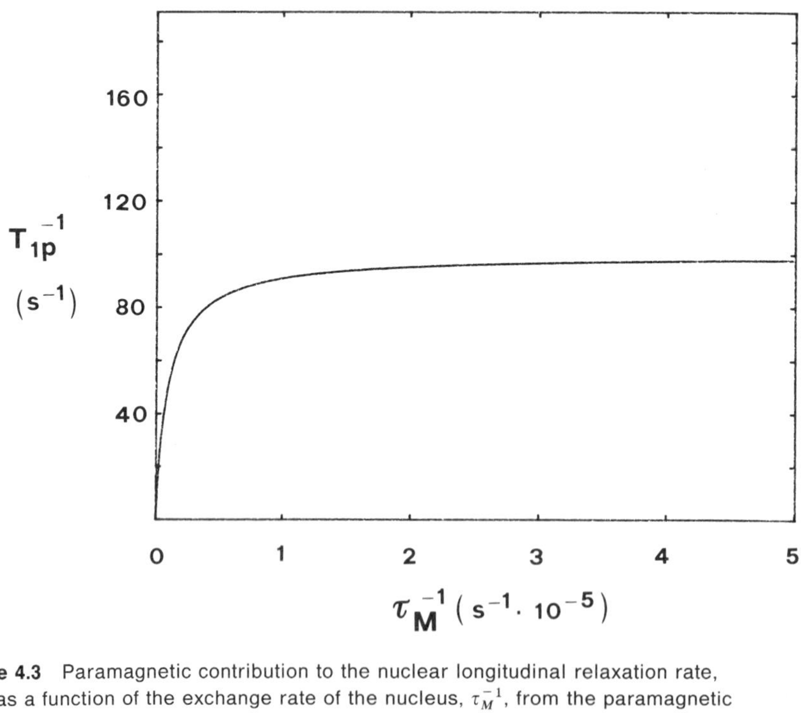

Figure 4.3 Paramagnetic contribution to the nuclear longitudinal relaxation rate, T_{1p}^{-1}, as a function of the exchange rate of the nucleus, τ_M^{-1}, from the paramagnetic site. The curve is calculated according to Equation (4.3) with $T_{1M}^{-1} = 10^4\ s^{-1}$ and $f_M = 10^{-2}$.

In other words, in the absence of exchange there will be no paramagnetic contribution to the relaxation rate of the bulk nuclei, whereas in the fast-exchange limit the measured relaxation rate will be the weighted average between the diamagnetic, T_{1d}^{-1}, and paramagnetic, T_{1M}^{-1}, rates:

$$T_{1\,\mathrm{meas}}^{-1} = T_{1d}^{-1} + f_M T_{1M}^{-1}$$

where the molar fraction of bound nuclei, f_M, is the weighting factor. It should be remembered that, as in the case of the paramagnetic shift, T_{1M}^{-1} can be very large as compared to T_{1d}^{-1}, so that even for small f_M values a noticeable relaxation rate enhancement can be measured.

The transverse relaxation rate enhancement, T_{2p}^{-1}, could be treated in the same way if it were not for the difference in chemical shift, $\Delta\omega_M$, between the paramagnetic and diamagnetic species. As has already been shown (Section 4.2.1), the difference in chemical shift causes a line broadening—an increase in T_{2M}^{-1}—even when the exchange is between two diamagnetic sites. Furthermore, the effect is expected to be much larger in this case since much larger chemical shift differences are involved. The general formula relating T_{2p}^{-1} to $\Delta\omega_M$ and τ_M^{-1} of the bulk nuclei is

$$T_{2p}^{-1} = \frac{f_M}{\tau_M} \frac{T_{2M}^{-2} + T_{2M}^{-1}\tau_M^{-1} + (\Delta\omega_M)^2}{(T_{2M}^{-1} + \tau_M^{-1})^2 + (\Delta\omega_M)^2} \tag{4.4}$$

Figure 4.4 shows the variation of T_{2p}^{-1} as a function of τ_M^{-1} for different values of $\Delta\omega_M$.

Note that in the absence of exchange ($\tau_M^{-1} = 0$) $T_{2p}^{-1} = 0$, and for very fast exchange ($\tau_M^{-1} \gg T_{2M}^{-1}, T_{2M}(\Delta\omega_M)^2$), $T_{2p}^{-1} = f_M T_{2M}^{-1}$, as was found for the longitudinal relaxation rate.

In the intermediate region, where τ_M^{-1} is comparable to $\Delta\omega_M$, T_{2p}^{-1} may be larger than T_{1p}^{-1} even if $T_{2M}^{-1} = T_{1M}^{-1}$ because of exchange broadening effects. If little or no isotropic shift is present ($\Delta\omega_M \ll T_{2M}^{-1}$), Equation (4.4) reduces to

$$T_{2p}^{-1} = f_M(T_{2M} + \tau_M)^{-1}$$

which has the same form as that for T_{1p}^{-1} in Equation (4.3).

4.2.3 The Effect of Temperature on the Exchange-Controlled Parameters

In the very fast exchange regions, the observed paramagnetic effects on the chemical shift, $\Delta\omega_p$, and the relaxation rates, T_{1p}^{-1}, T_{2p}^{-1}, are simply proportional to the full paramagnetic effects in the metal binding site:

$$\begin{aligned} \Delta\omega_p &= f_M \Delta\omega_M \\ T_{1p}^{-1} &= f_M T_{1M}^{-1} \\ T_{2p}^{-1} &= f_M T_{2M}^{-1} \end{aligned} \tag{4.5}$$

Therefore, if the molar fraction of bound ligand, f_M, is known, the full paramagnetic

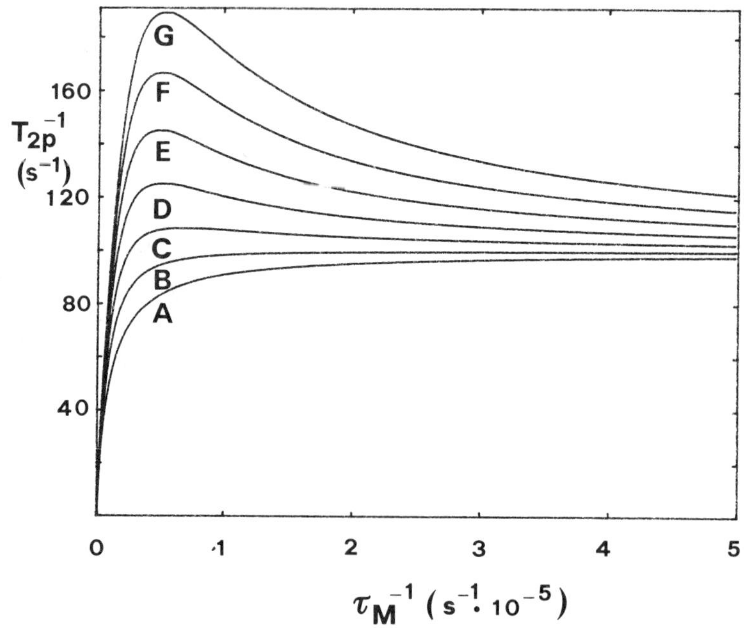

Figure 4.4 Paramagnetic contribution to the nuclear transverse relaxation rate, T_{2p}^{-1}, as a function of the exchange rate of the nucleus, τ_M^{-1}, from the paramagnetic site. The curves are calculated according to Equation (4.4) with $T_{2M}^{-1} = 10^4\ s^{-1}$, $f_M = 10^{-2}$, and $\Delta\omega_M = 0$ (**A**), 1×10^{-4} (**B**), 1.5×10^{-4} (**C**), 2×10^{-4} (**D**), 2.5×10^{-4} (**E**), 3×10^{-4} (**F**), and 3.5×10^{-4} (**G**) rad s^{-1}. Note that curve **A** ($\Delta\omega_M = 0$) is the same as that for T_{1p}^{-1} in Figure 4.3.

effects can be calculated. Of course, if no exchange takes place the ligand in the bulk solution is completely blind to the presence of the metal ion.

In the intermediate situation the measured parameters are complicated functions of the full effects experienced in the paramagnetic site and the exchange rate τ_M^{-1}. The τ_M^{-1} dependence of such functions has been summarized in Figures 4.2, 4.3, and 4.4. There are, however, particular regions in which the measured parameters are simpler functions of τ_M^{-1}. These regions correspond to cases in which Equations (4.2), (4.3), and (4.4) can be approximated by the following:

$$\text{I.} \quad \Delta\omega_p = \frac{f_M}{\tau_M^2 \Delta\omega_M} \qquad (\Delta\omega_M \gg T_{2M}^{-1}, \tau_M^{-1})$$

$$\text{II.} \quad \Delta\omega_p = \frac{f_M \Delta\omega_M}{\tau_M^2 T_{2M}^{-2}} \qquad (T_{2M}^{-1} \gg \tau_M^{-1}, \Delta\omega_M)$$

$$\text{III.} \quad T_{1p}^{-1} = f_M \tau_M^{-1} \qquad (T_{1M}^{-1} \gg \tau_M^{-1})$$

$$\text{IV.} \quad T_{2p}^{-1} = f_M \tau_M^{-1} \qquad (T_{2M}^{-1} \gg \tau_M^{-1}, \Delta\omega_M)$$

$$\text{V.} \quad T_{2p}^{-1} = f_M \Delta\omega_M^2 \tau_M \qquad (\tau_M^{-1} \gg (\Delta\omega_M)^2 \tau_M \gg T_{2M}^{-1})$$

In cases III and IV, T_{1p}^{-1} and T_{2p}^{-1} are a direct measure of the exchange rate, τ_M^{-1}. The structural information on the paramagnetic site is lost, but τ_M may provide useful kinetic information on the system. Kinetic parameters for the dissociation process may be obtained by performing relaxation measurements as a function of temperature. If it is assumed that the dissociation of the ligand from the paramagnetic site is a first-order kinetic process, the dissociation rate constant, τ_M^{-1}, is given by the Eyring relationship:

$$\tau_M^{-1} = \frac{kT}{h} \exp\left(\frac{-\Delta G^{\ddagger}}{RT}\right) \tag{4.6}$$

where $\Delta G^{\ddagger}$ is the free energy of activation for the dissociation process.

In the normal range of temperatures used to study biological samples, the major source of variation of τ_M^{-1} with temperature is contained in the exponential part. In other words, a plot of log τ_M against $1/T$ (Arrhenius plot) will give a fairly straight line. Given the linear relationship between τ_M^{-1} and the relaxation rates, straight lines will also be obtained by plotting log T_{1p}^{-1} or log T_{2p}^{-1} against $1/T$ ($\propto$ log τ_M) (Figure 4.5). Note that a straight line of opposite slope is obtained in case V. The appropriate equations can then be found by simply taking

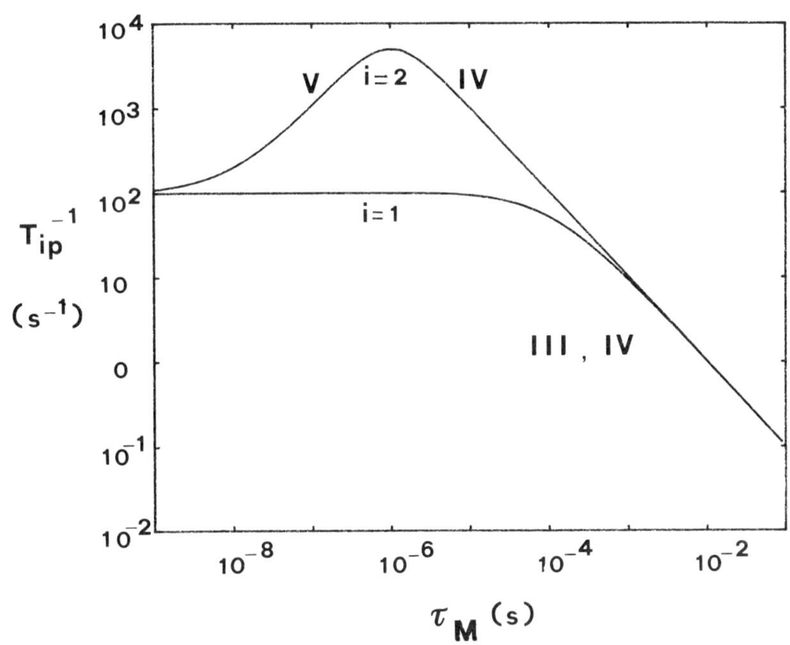

Figure 4.5 Linear dependence of log T_{1p}^{-1} and log T_{2p}^{-1} on log τ_M in the exchange regions corresponding to cases III to V. The curves are calculated according to Equations (4.3) (1) and (4.4) (2) with $T_{iM}^{-1} = 10^4\ \text{s}^{-1}$, $f_M = 10^{-2}$, and $\Delta\omega_M = 10^6$ rad s^{-1}.

the log of expressions V and III,IV with τ_M given by Equation (4.6):

$$\begin{aligned}\log T_{2p}^{-1} &= \log\left(f_M \Delta\omega_M^2 \frac{h}{kT}\right) + \frac{\Delta H^\ddagger}{RT} - \frac{\Delta S^\ddagger}{R} \cong \text{cost} + \frac{\Delta H^\ddagger}{RT} \\ \log T_{1p}^{-1} = \log T_{2p}^{-1} &= \log\left(f_M \frac{kT}{h}\right) - \frac{\Delta H^\ddagger}{RT} + \frac{\Delta S^\ddagger}{R} \cong \text{cost}' - \frac{\Delta H^\ddagger}{RT}\end{aligned} \tag{4.7}$$

The slope of the line in regions V and III,IV is $\pm\Delta H^\ddagger/R$, where $\Delta H^\ddagger$ is the enthalpy of activation for the dissociation process. The entropy of activation, $\Delta S^\ddagger$, may then be calculated by substituting the $\Delta H^\ddagger$ value obtained in Equation (4.7). In this kind of analysis care should be taken to evaluate the extent and sign of the effects of temperature on the full paramagnetic effects, $\Delta\omega_M$, T_{1M} and T_{2M}, as discussed in Sections 2.8 and 3.9.

4.3 EQUILIBRIUM CONSTANTS FOR TWO-SITE EXCHANGE

4.3.1 *EM* + **L* ⇌ *EM***L* (NMR of *L* Nuclei)

Consider a metalloenzyme, *EM*, whose stability is high enough so that the amount of free metal ions, *M*, in solution is negligible at the working concentrations of *EM*. The paramagnetic effects observed on a nucleus of a ligand *L* can then be used to obtain information on its dissociation constant:

$$K = \frac{[EM][L]}{[EML]} = \frac{(C_E - [EML])(C_L - [EML])}{[EML]} \tag{4.8}$$

where C_E and C_L are the total concentrations of all the enzyme-containing and ligand-containing species, respectively, and $[EML]$ is the equilibrium concentration of the adduct.

The molar fraction of bound ligand, f_M, can be expressed in terms of Equations (4.5) and (4.8) as

$$\begin{aligned} f_M &= \frac{[EML]}{C_L} = \frac{\Delta\omega_p}{\Delta\omega_M} = \frac{T_{1p}^{-1}}{T_{1M}^{-1}} = \frac{T_{2p}^{-1}}{T_{2M}^{-1}} \\ &= \frac{K + C_E + C_L - \sqrt{(K + C_E + C_L)^2 - 4C_E C_L}}{2C_L} \end{aligned} \tag{4.9}$$

Therefore, measurements of either $\Delta\omega_p$, T_{1p}^{-1}, or T_{2p}^{-1} at various concentrations of C_L and/or C_E allow a two-parameter fitting of the data through Equation (4.9) in terms of K and $\Delta\omega_M$, T_{1M}^{-1}, or T_{2M}^{-1}.

If the experimental conditions are such that $[EML]$ is always much smaller than C_L, then Equation (4.8) becomes

$$K = \frac{(C_E - [EML])C_L}{[EML]} \tag{4.10}$$

and

$$f_M = \frac{[EML]}{C_L} = \frac{C_E}{K + C_L} \tag{4.11}$$

Expressing f_M in the form of Equations (4.5) gives

$$C_L = C_E \frac{\Delta\omega_M}{\Delta\omega_p} - K = C_E \frac{T_{1p}}{T_{1M}} - K = C_E \frac{T_{2p}}{T_{2M}} - K \tag{4.12}$$

For constant C_E, a plot of C_L against $1/\Delta\omega_p$, T_{1p}, or T_{2p}, gives a straight line; $\Delta\omega_M$, T_{1M}, or T_{2M} can then be obtained from the slope, and K can be found from the intercept on the y axis.

Equation (4.12), like Equation (4.9), is valid when the ligand is in fast exchange; it is also true under exchange-limited conditions, which are described in Equations (4.2), (4.3), and (4.4), provided that τ_M is not dependent on C_L. In this case, K can still be obtained, in addition to the limit values of T_{1p}, T_{2p}, or $\Delta\omega_p$ at $f_M = 1$.

4.3.2 $*E + M \rightleftharpoons *EM$ (NMR of E Nuclei)

This condition is relevant when considering metal-activated enzymes and, more generally, in cases where the dissociation constant of the metal from the enzyme is such that, at the usual working conditions, there are appreciable amounts of both E and EM in solution. The dissociation constant is

$$K = \frac{[E][M]}{[EM]} = \frac{(C_E - [EM])(C_M - [EM])}{[EM]} \tag{4.13}$$

where C_M, in analogy to C_L and C_E, is the total concentration of all the metal-containing species. If the paramagnetic effect on an enzyme resonance is observed, the NMR technique can be used to determine K. The molar fraction of metalloenzyme, f_M, is

$$f_M = \frac{[EM]}{C_E}$$

and deriving expressions analogous to Equations (4.9) and (4.12), for which solutions are obtained with the same procedures, gives

$$f_M = \frac{[EM]}{C_E} = \frac{K + C_M + C_E - \sqrt{(K + C_M + C_E)^2 - 4C_M C_E}}{2C_E} \tag{4.14}$$

or, if $[EM] \ll C_E$

$$C_E = C_M \frac{\Delta\omega_M}{\Delta\omega_p} - K = C_M \frac{T_{1p}}{T_{1M}} - K = C_E \frac{T_{2p}}{T_{2M}} - K \tag{4.15}$$

4.4 EQUILIBRIUM CONSTANTS FOR $EM \rightleftharpoons E + M$ MEASURED THROUGH WATER PROTON RELAXATION: THE ENHANCEMENT FACTOR

When one or more water molecules interact with a metal ion in a metal–macromolecule complex, EM, in the presence of free metal ions, M, in solution, then the exchange of water among three sites should be considered. The molar fractions of water nuclei are given by

$$f_M = \frac{p[M]}{C_{H_2O}} \qquad f_{EM} = \frac{q[EM]}{C_{H_2O}} \qquad f_{\text{bulk}} = \frac{[H_2O]}{C_{H_2O}} \simeq 1 \tag{4.16}$$

where p and q are the numbers of water molecules interacting with free and bound metal ions, respectively. The water proton longitudinal relaxation rate enhancement, T_{1p}^{-1}, is given by the contribution of the metal, $(T_{1p}^{-1})_M$, and the complex, $(T_{1p}^{-1})_{EM}$, both of which can be expressed through Equation (4.3):

$$\begin{aligned} T_{1p}^{-1} &= (T_{1p}^{-1})_M + (T_{1p}^{-1})_{EM} = f_M \frac{1}{T_{1M} + \tau_M} + f_{EM} \frac{1}{T_{1EM} + \tau_{EM}} \\ &= \frac{p[M]}{C_{H_2O}} \frac{1}{T_{1M} + \tau_M} + \frac{q[EM]}{C_{H_2O}} \frac{1}{T_{1EM} + \tau_{EM}} \end{aligned} \tag{4.17}$$

The two limiting cases—where the metal is completely free ($K = \infty$) or completely bound ($K = 0$) to the macromolecule—would then give

$$T_{1p(\infty)}^{-1} = \frac{pC_M}{C_{H_2O}} \frac{1}{T_{1M} + \tau_M} \qquad T_{1p(0)}^{-1} = \frac{qC_M}{C_{H_2O}} \frac{1}{T_{1EM} + \tau_{EM}} \tag{4.18}$$

where C_M is the total metal concentration, and ∞ and 0 refer to the limiting values of the dissociation constant, K. The first case ($K = \infty$) can be simulated by simply not adding any macromolecule to the solution. It is then customary to define an enhancement factor, ε, as [4]

$$\varepsilon = \frac{T_{1p}^{-1}}{T_{1p(\infty)}^{-1}} \tag{4.19}$$

where T_{1p}^{-1} is measured in a solution containing both the metal ion and the macromolecule, and $T_{1p(\infty)}^{-1}$ is obtained from a solution containing the same concentration of metal ion but in the absence of the macromolecule. From Equations (4.17) and (4.18):

$$\begin{aligned} \varepsilon &= \frac{[M]}{C_M} + \frac{q[EM]}{pC_M} \frac{(T_{1EM} + \tau_{EM})^{-1}}{(T_{1M} + \tau_M)^{-1}} = \frac{[M]}{C_M} + \frac{[EM]}{C_M} \varepsilon_0 \\ &= f_f + f_b \varepsilon_0 = 1 + f_b(\varepsilon_0 - 1) \end{aligned} \tag{4.20}$$

where f_f and f_b are the molar fractions of free and bound metal ion and

$$\varepsilon_0 = \frac{q}{p} \frac{(T_{1EM} + \tau_{EM})^{-1}}{(T_{1M} + \tau_M)^{-1}} = \frac{T_{1p(0)}^{-1}}{T_{1p(\infty)}^{-1}}$$

When all the metal is bound to the macromolecule, $[M] = 0$ and $[EM] = C_M$; therefore, from Equation (4.20), $\varepsilon = \varepsilon_0$. ε_0 is thus defined as the enhancement factor measured in a solution where all the metal is bound to the macromolecule. It should be noted that, since in general $q < p$, ε_0 should be smaller than unity if the intrinsic nuclear relaxation times are the same in the metal complex with the macromolecule and the aquaion. However, as often $T_{1EM} \ll T_{1M}$ owing to a longer correlation time τ_c in the macromolecule (see Chapter 3), ε_0 is usually larger than unity.

In Equation (4.20), ε is defined only in terms of the molar fraction of bound metal ion; that is, independently of the actual concentrations. By making use of an equation for f_b, which is analogous to Equation (4.14), the enhancement factor can be expressed as

$$\varepsilon = 1 + \frac{K + C_M + C_E - \sqrt{(K + C_M + C_E)^2 - 4C_M C_E}}{2C_M}(\varepsilon_0 - 1) \tag{4.21}$$

Both K and ε_0 can be obtained through a two-parameter fitting of the ε data obtained at various C_E and/or C_M concentrations.

In the more general case of a macromolecule with n equivalent metal binding sites, the equilibrium is

$$E + nM \rightleftharpoons EM_n \quad \text{or} \quad \frac{1}{n}E + M \rightleftharpoons E_{1/n}M$$

and the dissociation constant can be written as

$$K = \frac{(nC_E - [E_{1/n}M])(C_M - [E_{1/n}M])}{[E_{1/n}M]}$$

where nC_E is the total concentration of metal binding sites. Therefore $f_b = [E_{1/n}M]/C_M$, and Equation (4.21) becomes

$$\varepsilon = 1 + \frac{K + C_M + nC_E - \sqrt{(K + C_M + nC_E)^2 - 4C_M nC_E}}{2C_M}(\varepsilon_0 - 1) \tag{4.22}$$

In this case, a three-parameter fitting of the ε data obtained over a suitable range of C_E and C_M concentrations gives K, ε_0, and n.

It is worthy to note that Equations (4.21) and (4.22) are valid irrespective of the rate at which water exchanges from the two paramagnetic sites. In fact, the enhancement factor is defined in terms of the quantities $T_{1M} + \tau_M$ and $T_{1EM} + \tau_{EM}$, which are likely to be constant and, in particular, independent of the concentration of the various species in solution. Furthermore, the exchanging ligand here is water, and its bulk concentration is fairly constant under all circumstances.

More complex equilibria—inequivalent sites and ternary complexes, for example—can be, and have been, studied using measurements of the enhancement factor. The treatment for these equilibria is of the same type as that outlined and does not differ in principle from the analysis of several simultaneous equilibria detected by spectroscopic techniques. The same general rule also holds here: the larger the number of equilibria, the more accurate and extensive the experimental data need to be to allow a meaningful analysis.

4.5 OUTER-SPHERE RELAXATION

In the discussion of chemical exchange so far, it has been assumed that the time taken for the nucleus to pass from the paramagnetic site to the diamagnetic site is negligible compared to the time spent in each site. Although the paramagnetic site is most often the first coordination sphere of the metal, it can in principle be any other binding site with a definite lifetime and definite distance, r, from the metal ion. When no such site exists, the nuclear relaxation rates may still be enhanced by dipolar interaction with the paramagnetic center simply because the nucleus and the metal ion may randomly come close to each other owing to free diffusion in solution.

The correlation time for such an interaction may be either the electronic relaxation time, τ_s, or the diffusional correlation time, τ_D. The latter is defined as $2d^2/(D_N + D_M)$, where d is the distance of closest approach, and D_N and D_M are the diffusion coefficients of the molecules containing the nucleus under investigation and the metal ion, respectively. Since the metal ion and the interacting nucleus are not held together in the same molecular framework (see Section 3.1 and Figure 3.1), the rotational time, τ_r, is ineffective in modulating the electron-nucleus interaction and need not be considered further. τ_D is typically in the range 10^{-9} to 10^{-10} s; therefore, τ_s is still the dominant correlation time when the metal ion undergoes fast electron relaxation. In such a case, the Solomon Equations (3.8) and (3.9) need only be modified [5] by substituting $N_M \int_d^\infty 4\pi r^2/r^6 \, dr$ for $1/r^6$; that is, $1/r^6$ is averaged over the volume between a sphere of radius d and infinity. Here, N_M represents the number of metal ions per cubic meter of solution. The result is then

$$\left(\frac{1}{r^6}\right)_{\text{av}} = N_M \frac{4\pi}{3d^3} \tag{4.23}$$

When the access to the metal is limited, as in a metalloprotein that has the metal ion lying at the bottom of a deep cavity, it is appropriate to consider a cone, rather than a sphere, and introduce a proper reduction factor in Equation (4.23) [6]. Whether the access to the metal is limited or not, analysis of the nuclear relaxation data involves replacing the metal-nucleus distance, r, as a parameter, by the distance of closest approach, d.

For metal ions with long electronic relaxation times, τ_s is no longer the correlation time. Furthermore, it has been shown [7–11] that the shape of the $f(\omega, \tau)$ function in the Solomon equation varies for this case because of the different nature of the modulation of the electron-nucleus interaction brought about by diffusion. The following equations have been derived:

$$T_{1p}^{-1} = N_M \left(\frac{\mu_0}{4\pi}\right)^2 \frac{16\pi}{225} \frac{\gamma_N^2 g_e^2 \mu_B^2 S(S+1)}{d(D_N + D_M)} [7f(\omega_S\tau_D) + 3f(\omega_I\tau_D)] \tag{4.24}$$

$$T_{2p}^{-1} = N_M \left(\frac{\mu_0}{4\pi}\right)^2 \frac{8\pi}{225} \frac{\gamma_N^2 g_e^2 \mu_B^2 S(S+1)}{d(D_N + D_M)} [13f(\omega_S \tau_D) + 3f(\omega_I \tau_D) + 4] \quad (4.25)$$

where $f(\omega, \tau_D) = \frac{15}{2} I(u)$; $u = |\omega\tau_D|^{1/2}$; $I(u) = u^{-5}\{u^2 - 2 + e^{-u}[(u^2 - 2)\sin u + (u^2 + 4u + 2)\cos u]\}$. Values for D_N and D_M can be estimated by assuming that the motions can be described by the diffusion of rigid spheres in a medium of viscosity η:

$$D = \frac{kT}{6\pi a \eta}$$

where a is the radius of the molecule.

A comment is needed here concerning the presence of N_M in Equations (4.23), (4.24), and (4.25). The Solomon Equations (3.8) and (3.9) give T_{1M}^{-1} for a single nucleus interacting with a single metal ion. When there is fast exchange of excess nuclei the measured effect is weighted for the molar fraction of bound nuclei, $f_M = C_M/C_N$—see Equations (4.5). When no specific site is available each nucleus may interact with more than one metal at a time, independently of the concentration of nuclei; T_{1p}^{-1} is therefore proportional to the number of metal ions per unit volume, which is expressed by N_M.

The above theory is of general applicability to every kind of nuclei, although in practice the effect is measurable only on protons of water or of excess ligands.

An estimate of the relative importance of outer-sphere relaxation on T_{1p}^{-1} of water protons in slow-relaxing paramagnetic hexaaquaion solutions may be made by comparing Equations (3.8) and (4.24) through (4.5). At 60 MHz, for $r = 280$ pm, $d = 450$ pm, $\tau_D = 10^{-10}$ s, $\tau_r = 3 \times 10^{-11}$ s, $a_M = 300$ pm, and $a_N = 150$ pm, the outer-sphere contribution to the relaxation is about 6%. Larger contributions ($\sim$ 10%) are predicted for fast-relaxing metal ions by comparing the Solomon Equation (3.8) with that modified according to Equation (4.23). Other equations are available which give similar estimates of outer sphere effects. These equations are based on different models for the motions of the particles [12–14].

When the first coordination sphere is accessible to a fast-exchanging ligand, the outer-sphere relaxation effect on ligand nuclei is not usually the dominant contribution to T_{1p}^{-1}. However, when the exchange from the first coordination sphere is slow, outer-sphere effects may be the only measurable source of relaxation enhancement. Therefore, attention should always be paid, in analyzing relaxation data, to the possibility of mistaking outer-sphere relaxation for a weak inner-sphere interaction.

4.6 MAPPING

Electron-nucleus interactions detected through dipolar shift and relaxation measurements depend on geometrical parameters: dipolar relaxation rates are proportional to $1/r^6$—Equations (3.3), (3.4), and (3.20), whereas dipolar shifts are more or less complicated functions of two (r, θ) or all three (r, θ, Ω) of the polar coordinates of the nucleus with respect to the principal directions of the magnetic susceptibility tensor of the molecule—Equations (2.20) and (2.18).

When a molecule interacts with a paramagnetic metal, the measurement of NMR parameters of several assigned signals from nuclei of the molecule may in principle allow the determination of their coordinates with respect to the metal ion center. In the simple case of a rigid molecule, T_1 or T_2 measurements of four or more signals are often enough to exactly locate the molecule with respect to the metal ion, independently of the correlation time. In the case of two or three nuclei being detected, with fixed internuclear distances, nuclear relaxation measurements give the metal-nuclei distance ratios, and thus provide preliminary structural information. If an estimate of the correlation time (τ_c for classical dipolar relaxation rates and τ_r for Curie relaxation) is available, and Equations (3.3), (3.4), and (3.20) are assumed to hold, then the absolute distances are obtained. Alternatively, if one of the metal-nucleus distances is assumed, and this is reasonable when one of the resonating nuclei is the donor atom itself or is rigidly attached to it, then all the atoms are fixed with respect to the metal atom. For four or more signals detected in a rigid molecule, no further assumption is needed to locate the molecule with respect to the metal atom, since three independent distance ratios identify a unique set of metal coordinates.*

If the molecule is not rigid, but is assumed to have a single conformation, the minimum number of signals required to obtain the solution dramatically increases with the total number of rotational degrees of freedom of the fragments of the molecule, and the number of signals available is usually less than required. In this case, chemical sense may help to solve the problem. For example, if a molecule is directly bound to the metal ion, all the possible donor groups can be tested by placing the metal ion at a reasonable bond distance from each of them; one position should reproduce the experimental distance ratios far better than the others, unless metal binding occurs at more than one site.

Dipolar shift measurements for n nuclei provide a set of n experimental data; however, these require $3n + 2$ parameters, $3n$ representing r, θ, and Ω for each nucleus with respect to the principal axes of the magnetic anisotropy tensor of the metal—see Equation (2.18), and 2 being the parameters needed to define the anisotropy tensor itself. Clearly, the problem cannot be solved unless structural restrictions are assumed to hold. The general approach makes use of computer-generated conformations which are used to search for the orientations of the molecule with respect to the metal ion that best fit both dipolar shift and relaxation data [15–25]. Of course, further data—NOE, cross-relaxation, and nuclear spin-spin coupling, for example—from NMR of analogous diamagnetic systems should be exploited. The analysis may also be complicated by the existence of several, almost equi-energetic, conformations, and/or of several metal binding sites. The NMR approach in these cases only gives a result averaged over all the possibilities, and should therefore be interpreted with great care [26].

* If the nuclei under observation all lie in a plane, the ambiguity in placing the metal above or below the plane cannot be resolved.

REFERENCES

1. Gutowski, H. S., McCall, D. M., and Slichter, C. P. *J. Chem. Phys.* **21** (1953) 279.
2. Gutowski, H. S., and Holm, C. H. *J. Chem. Phys.* **25** (1956) 1228.
3. Swift, T. J., and Connick, R. E. *J. Chem. Phys.* **37** (1962) 307. Swift, T. J. In *NMR of Paramagnetic Molecules*, G. N. La Mar, W. DeW. Horrocks, Jr., and R. H. Holm (eds.), Academic Press, New York, 1973, p. 53.
4. Dwek, R. A. In *Nuclear Magnetic Resonance in Biochemistry*, Oxford University Press, Oxford, 1973, Chapter 11.
5. Luz, Z., and Meiboom, S. *J. Chem. Phys.* **40** (1964) 2686.
6. Koenig, S. H., and Brown, R. D. *Ann. N.Y. Acad. Sci.* **222** (1973) 752.
7. Pfeifer, H. *Ann. Phys. Leipzig* **8** (1961) 1.
8. Pfeifer, H. *Z. Naturforsch.* **17a** (1962) 279.
9. Pfeifer, H. *Biochim. Biophys. Acta* **66** (1963) 434.
10. Hausser, R., and Noack, F. *Z. Naturforsch.* **20a** (1965) 1668.
11. Hubbard, P. S. *Proc. Roy. Soc. London* **A291** (1966) 537.
12. Hwang, L. P., and Freed, J. H. *J. Chem. Phys.* **63** (1975) 4017.
13. Ayant, Y., Belorizky, E., Alizon, J., and Gallice, J. *J. Phys. (Paris)* **36** (1975) 991
14. Freed, J. H. *J. Chem. Phys.* **68** (1978) 4034
15. Mildvan, A. S., Granot, J., Smith, G. S., and Liebman, M. N. In *Advances in Inorganic Biochemistry*, D. W. Darnall and R. G. Wilkins (eds.), Elsevier North Holland, New York, 1980, Volume 2, Chapter 8.
16. Mildvan, A. S. *Acc. Chem. Res.* **10** (1977) 246.
17. Mildvan, A. S., and Gupta, R. K. *Methods Enzymol.* **F49** (1978) 322.
18. Williams, R. J. P. *Struct. Bonding* **50** (1982) 79.
19. Barry, C. D., Glasel, J., Williams, R. J. P., and Xavier, A. V. *J. Mol. Biol.* **84** (1974) 371.
20. Barry, C. D., Martin, D. R., Williams, R. J. P., and Xavier, A. V. *J. Mol. Biol.* **84** (1974) 491.
21. Furie, B., Griffen, J. H., Feldmann, R. J., Sokoloski, A., and Schechter, A. N. *Proc. Natl. Acad. Sci. U.S.A.* **71** (1974) 2344.
22. Sloan, D. L., Loeb, L. A., Mildvan, A. S., and Feldmann, R. J. *J. Biol. Chem.* **250** (1975) 8913.
23. Fung, C. H., Feldmann, R. J., and Mildvan, A. S. *Biochemistry* **15** (1976) 75.
24. Lee, L., and Sykes, B. D. In *Advances in Inorganic Biochemistry*, D. W. Darnall and R. G. Wilkins (eds.), Elsevier North Holland, New York, 1980, Volume 2, Chapter 7.
25. Ascenso, J. R., and Xavier, A. V. In Systematic Properties of Lanthanides, S. Sinha (ed.), Reidel, Dordrecht, 1983, p. 501.
26. Jardetzky, O. *Biochim. Biophys. Acta* **621** (1980) 227.

CHAPTER 5

Experimental Techniques

5.1 DETECTION OF NMR SIGNALS: CONTINUOUS WAVE AND FOURIER TRANSFORM TECHNIQUES

NMR spectroscopy is concerned with transitions among nuclear spin levels whose degeneracy is removed by a static magnetic field. Such transitions are detected by applying an electromagnetic radiation of suitable energy. As mentioned in Chapter 2, detection of an NMR signal is made possible by the different populations of the various spin states, which arise from thermal distribution of the individual spins among the energy levels.

In a continuous wave (CW) instrument, an NMR spectrum can be obtained by slowly sweeping a radiofrequency signal over the frequency range of interest and detecting the energy absorption by the sample as a function of frequency. Alternatively, the external magnetic field can be swept, at fixed radiofrequency, over a range that allows the resonance conditions of each different set of nuclei to be fulfilled. In both methods, the sets of nuclei resonate one at a time.

Of the NMR parameters, chemical shifts and linewidths can be easily obtained on CW instruments. However, care must be taken to avoid saturation of the signals as this would result in distorted lineshapes, and would make linewidth and intensity measurements unreliable. The homogeneity is usually adjusted by maximizing the internal lock signal, and an internal standard is used as a reference for linewidths.

Longitudinal relaxation times can also be measured through rather indirect and inaccurate techniques developed by the pioneers of nuclear spin relaxation in the preFT days. Qualitatively, these techniques use variable radiofrequency power to increase and then saturate the signal under the usual working conditions.

The main limitations of CW techniques, besides their intrinsic low sensitivity, arise because only a relatively narrow range of T_1 values can be explored.

Pulsed NMR instruments are based on the Fourier Transform (FT) technique. The FT is a mathematical operation that transforms a function, $f(t)$, in the time domain to a corresponding function, $F(\omega)$, in the frequency domain and vice versa according to the following equations:

$$F(\omega) = \int_{-\infty}^{+\infty} f(t)e^{-i\omega t}\,dt \qquad f(t) = \frac{1}{2\pi}\int_{-\infty}^{+\infty} F(\omega)e^{i\omega t}\,d\omega$$

In a pulsed instrument, a single frequency is generated at a fixed magnetic field by a master oscillator in the radiofrequency region where resonances are expected to occur. This radiofrequency (carrier frequency) is then applied to the sample at very high power (10^2 to 10^3 W compared to 10^{-1} to 1 W in a CW instrument) for a short time (typically a few microseconds). The result is a square pulse containing a few hundreds of cycles (10^3 for a 100-MHz pulse 10 μs long) (Figure 5.1). A square pulse can be mathematically analyzed in terms of, and physically built up by, an infinite number of simple sine waves of different frequencies. Those sine waves that have frequencies near the pulsed frequency (carrier) will have higher amplitude; whereas those with frequencies far away from the carrier will give vanishingly small contributions (Figure 5.1). A single pulse is thus equivalent to simultaneously irradiating the sample with all the frequencies needed to excite nuclear transitions in all the spin sets.

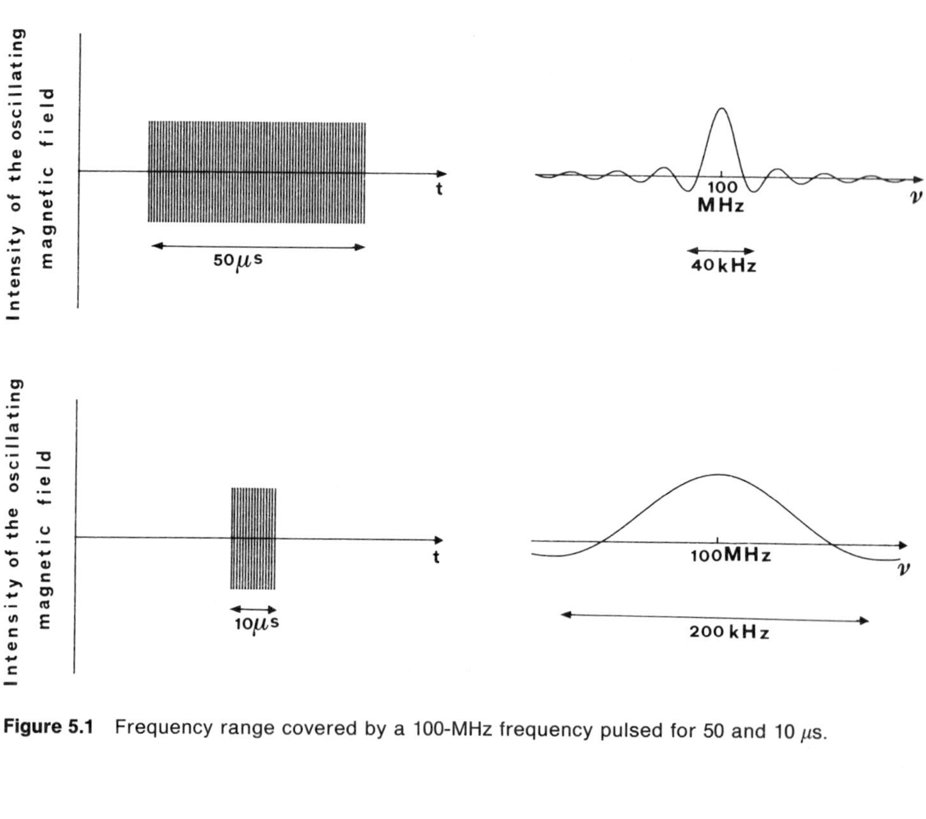

Figure 5.1 Frequency range covered by a 100-MHz frequency pulsed for 50 and 10 μs.

The thermal equilibrium of the sample is greatly perturbed by the pulse and the spins will tend to relax to their initial state. If the sample contains a single set of nuclei, there will be a macroscopic magnetic moment of finite intensity along the z axis before irradiation. This moment is a consequence of the difference between the vectorial sum of all the individual magnetic moments precessing with characteristic frequency, ν_i, about the z axis on the positive side and those precessing on the negative side. Since the frequency, ν_0, of the exciting pulse has been chosen relatively close to ν_i the pulse will also have a strong component at ν_i. This will cause a flip of the overall magnetization vector, whose angle depends on the intensity and duration of the pulse. If the latter parameters are adjusted in such a way to obtain a 90° flip angle, for example, the z component of the spin magnetization will vanish and reappear on the xy plane rotating at frequency ν_i. Therefore, an oscillating magnetic field is generated that will induce a radiofrequency, ν_i, in a properly tuned coil, which may also be used for irradiating the sample. In the receiver, such a frequency is compared to the carrier frequency, ν_0, and the difference, $\nu_0 - \nu_i$, is displayed. After the pulse, the magnetization in the xy plane will decrease due to (1) progressive reestablishment of magnetization along the z axis and (2) the possible additional dephasing of the precessing spins. The sum of these two phenomena can often be described as a first-order kinetic process with rate constant T_2^{-1}. The detected signal will appear as a damped oscillation called free induction decay (FID) (Figure 5.2A).

The signal in the time domain (Figure 5.2A) is recorded by the computer through an analog-to-digital converter (ADC) and the data are Fourier transformed, yielding a signal in the frequency domain (Figure 5.2B). If more than one set of nuclei is present, the FID is the sum of the individual FIDs and the spectrum contains several signals (Figure 5.3).

The main advantage of FT over CW techniques is that the time required to obtain a spectrum is of the order of T_2; that is, of the order of seconds or less, compared to minutes. A second FID can then be digitized and summed to the first, and the process repeated many times. The final FID is then Fourier transformed

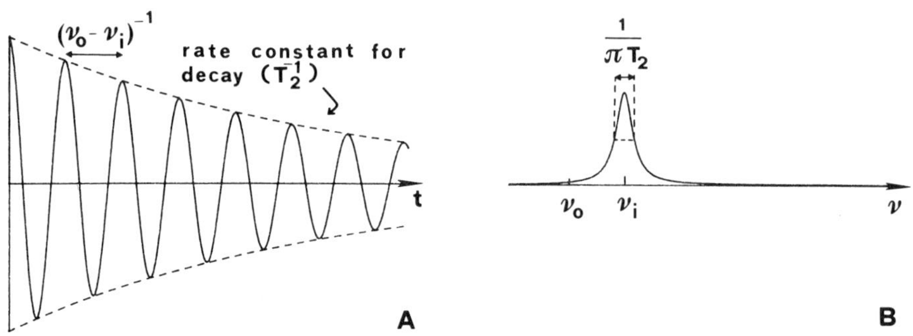

Figure 5.2 **A** Free induction decay in the time domain and **B** corresponding signal in the frequency domain for a single set of equivalent nuclei.

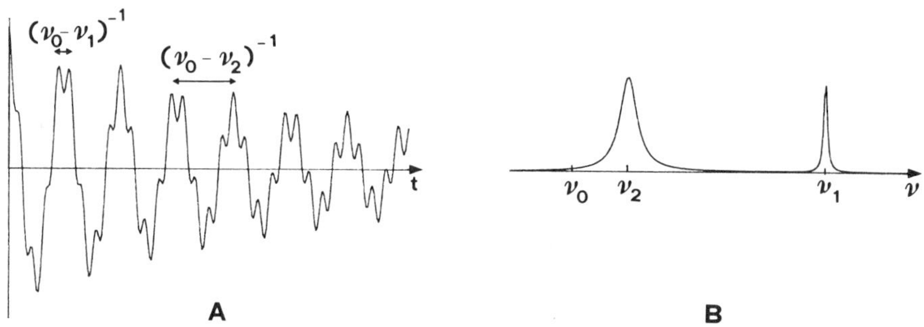

Figure 5.3 **A** Free induction decay in the time domain and **B** corresponding signals in the frequency domain for two sets of nuclei.

in the usual way to obtain the spectrum, with a signal-to-noise ratio after N scans of $\sqrt{N}$ times better than after a single scan. The repetition time has to be of the order of T_1 to allow a reasonable recovery of magnetization along the z axis.

5.2 FEATURES OF FT

5.2.1 Pulse Power and Pulse Length

As has already been discussed, the bandwidth of exciting frequencies contained in the pulse is inversely proportional to the duration of the pulse. In order to irradiate a large spectral range, the pulse has to be as short as possible. On the other hand, large flip angles are needed for several applications and, since the flip angle depends on the product of the duration and the intensity of the pulse, large flip angles *and* large bandwidths can only be achieved by using high radiofrequency power.

5.2.2 Digitization

The built-in computer samples a certain number of data points during the FID. The number of points sampled has to be at least twice the number of oscillations in the same time period (Figure 5.4A), otherwise the sine wave is so ill-defined that a ghost frequency appears (Figure 5.4B). This gives rise to the well-known phenomenon of folding back of signals from outside the frequency window covered by the digitizer. The folding is generally minimized by using a low-pass filter. The maximum sampling rate, which is the reciprocal of the time period between two adjacent data points, depends entirely on the analog-to-digital converter, which has to be as fast as possible and with a good dynamic range (see Section 5.4.2).

If the expected frequency spread of the signals is large, not only does the pulse have to be sufficiently short, as discussed, but the sampling rate also has to be appropriate, since the large value of $\nu_0 - \nu_i$ results in a very narrow oscillation period in the FID. For example, to cover a spectral window of 100 kHz, data points have to be sampled at least every 5 μs.

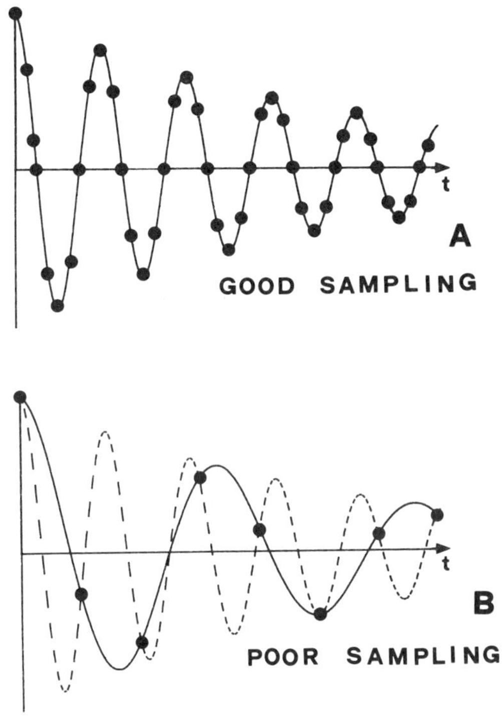

Figure 5.4 **A** Free induction decay sampled with a suitable number of data points. **B** The same free induction decay (----) sampled with an insufficient number of data points. In this case, the computer will treat the data as a ghost FID (——), giving rise to a signal at different frequency.

5.3 PULSE SEQUENCES

A 90° pulse brings the magnetization of a set of nuclei from the z axis to the xy plane, where it rotates with the Larmor frequency ν_I. If a reference system is defined as rotating about z with the same frequency, ν_I, the magnetization vector will then be fixed in the xy plane, for example, aligned along the y axis (Figure 5.5).

5.3.1 Measurements of Relaxation Times

According to the definition given in Chapter 2, T_1 is the time constant for magnetization recovery along the z axis, justifying the term longitudinal relaxation time. From the discussion so far, it is clear that the FT technique is well suited for the measurement of T_1. If a long series of 90° pulses is applied and the repetition time, τ, is not long enough to allow complete recovery of the magnetization, **M**, along the z axis, then after a few scans an equilibrium situation is reached

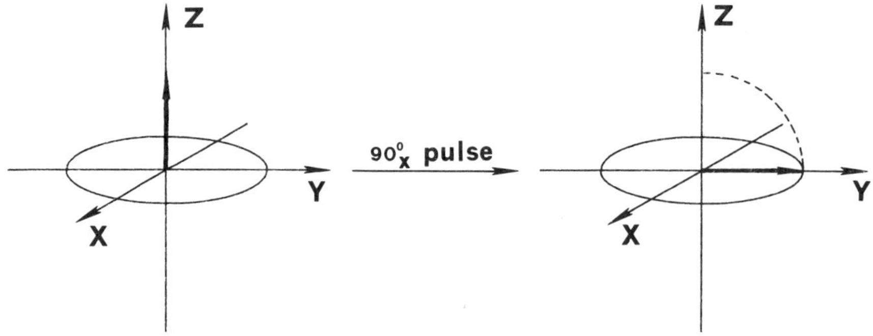

Figure 5.5 Effect of a 90°_x pulse on the macroscopic magnetization in a reference system rotating about the z axis at frequency ν_I.

in which the value of M_z will be given by [1]

$$M_\tau = M_\infty(1 - e^{-\tau/T_1}) \tag{5.1}$$

The final intensity of the Fourier-transformed signal, I_τ, will be proportional to M_τ. By repeating the same number of pulses spaced according to different values of τ, several spectra are obtained whose intensities range from approximately zero (small values of τ) to a maximum of $I_\infty(\tau \geq 5T_1)$. A two-parameter best-fitting procedure to Equation (5.1) with M_τ and M_∞ substituted by I_τ and I_∞ then gives I_∞ and the longitudinal relaxation time of the signal. This is the fastest method for obtaining T_1, and is termed progressive saturation.

A more accurate but time-consuming method is the popular inversion-recovery technique [2], which is based on the application of the following pulse sequence:

$$180^\circ_x\text{—}\tau\text{—}90^\circ_x\text{–acquisition—}D\text{—}$$

where D is a delay time chosen to be at least five times the estimated T_1 to be measured. The first 180° pulse flips the magnetization in the $-z$ direction (Figure 5.6A and B). After a time τ the magnetization has partially recovered. In Figure 5.6C, the magnetization is still negative, although smaller in absolute value, when a 90° pulse is applied. The magnetization vector on the $-y$ axis is now sampled (Figure 5.6D), giving rise to a negative Fourier-transformed signal. The delay time, D, allows complete recovery along z, and the entire sequence can be repeated N times to obtain the desired signal-to-noise ratio. The whole procedure is repeated several times with different values of τ, and the measured I_τ is fitted to the following equation:

$$I_\tau = I_\infty(1 - 2e^{-\tau/T_1}) \tag{5.2}$$

to obtain the best values of I_∞ and T_1. Figure 5.7 shows two examples of the best-fitting procedure of I_τ values, obtained in a progressive saturation and in an inversion-recovery experiment, using Equations (5.1) and (5.2), respectively.

The inversion-recovery technique is the more accurate one because the I_τ values are much less sensitive to imperfections in the shape and length of the pulse. For

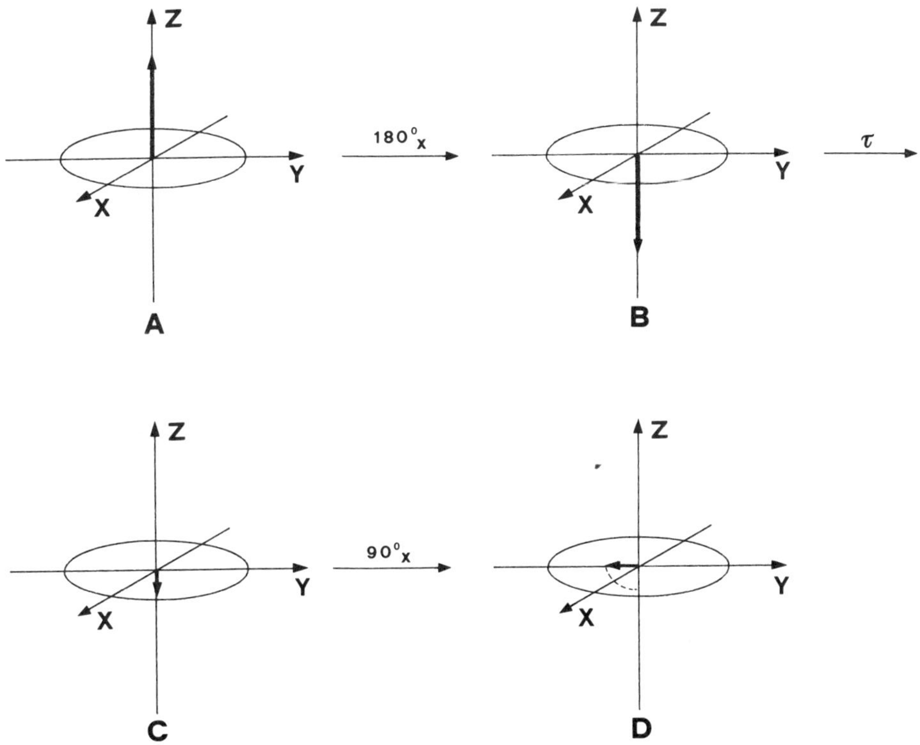

Figure 5.6 Effect of a 180°_x—τ—90°_x pulse sequence on the macroscopic magnetization in a rotating frame.

long values of T_1, however, the progressive saturation experiment is much faster, since most of the data points in Figure 5.7A can be obtained without waiting for long τ times between each scan. For example, if ten I_τ values are sampled at τ values of 2, 4, 6, 8, 10, 12, 15, 20, 30, and 50 s, using 100 scans each, and the value of T_1 is around 10 s, the progressive saturation method requires $100 \times (2 + 4 + 6 + 8 + 10 + 12 + 15 + 20 + 30 + 50) = 15700$ s whereas the inversion-recovery method would need $15700 + 10 \cdot 100 \cdot D$ s—for $D = 50$ s, that is, 65700 s (or more than 18 hours). On the other hand, very short values of T_1 cannot be measured with the progressive saturation method because τ cannot be reduced below the acquisition time (acquisition time = sampling time × data points in the FID) required for the particular experiment (see Section 5.2).

In Figures 5.5 and 5.6 the pulses are written with an index, x, to indicate that their phase is such that the magnetization vector is rotated by the appropriate angle around the x axis in the zy plane; the rotation is clockwise—from z to y on the positive quadrant. In many spectrometers, the phase of the pulse can be altered to any of the four values, x, $-x$, y, or $-y$, so that, for instance, **M** can be rotated counterclockwise about x by a pulse of phase $-x$, or can be tilted on to

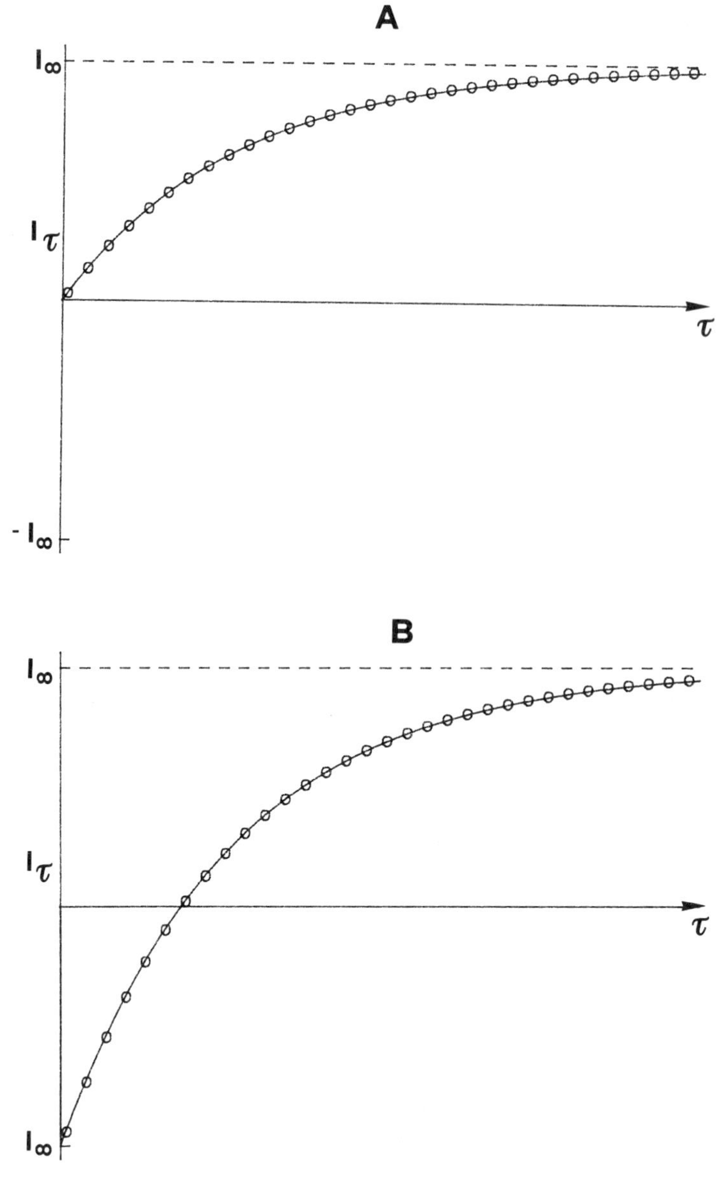

Figure 5.7 Magnetization recovery as a function of τ: **A** progressive saturation experiment; **B** inversion-recovery experiment.

the x axis by a 90°_{-y} pulse. By choosing various combinations of pulses it is possible to correct for accidental misadjustments of the pulse lengths. For example, if the inverting "180°" pulse in the inversion-recovery sequence is misadjusted and equal to, say, 170°, the inverting pulse leaves some magnetization on the y axis on the positive side, so that the resulting signal will always contain a positive contribution that sums algebraically with the detected intensity. If the first "180°" pulse is alternated between x and $-x$

$$180^\circ_x—\tau—90^\circ_x—D—180^\circ_{-x}—\tau—90^\circ_x—D$$

the spurious contribution will be alternately positive and negative, so that for an even (or large) number of scans the imperfection is cancelled. Methods have been suggested for improving a "180°_x" pulse: for example, by dividing it into two 90°_x pulses interrupted by a 180°_y pulse [3]

$$180^\circ_x \equiv \text{“}90^\circ_x\text{”}—\text{“}180^\circ_y\text{”}—\text{“}90^\circ_x\text{”}$$

Figure 5.8 shows the effect of such a composite pulse on the magnetization vector. If the actual pulse is, say, 160°, the first pulse in the sequence (80°_x) will leave **M** well above the xy plane; the central (160°_y) pulse will move it substantially below the xy plane, although somewhat out of the yz plane; and the last (80°_x) pulse will bring it much closer to the $-z$ axis than a single 160°_x pulse would have done. Results can be further improved by alternating the central pulse between y and $-y$.

For some applications it may not be convenient to irradiate all the signals in a multisignal spectrum (see also Section 5.4). In particular, pulse sequences have been developed so that T_1 can be measured for a selective single line without perturbing the equilibrium magnetization of the other nuclear sets. A particularly simple and efficient sequence consists of a train of strong pulses that are equally spaced by a

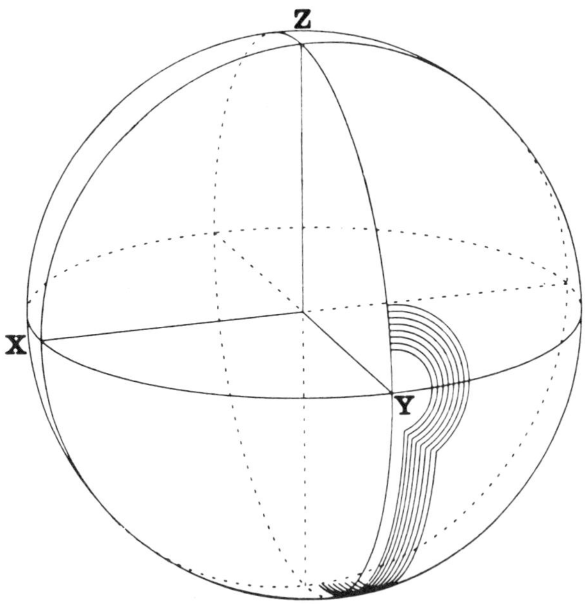

Figure 5.8 Correcting effect of a $90^\circ_x—180^\circ_y—90^\circ_x$ sequence on pulse length misadjustments. A family of trajectories has been calculated for pulse lengths between 80 and 90% of the nominal value. (Reprinted from ref. [3], with permission.)

time τ and short enough so that each pulse has a flip angle $\alpha \ll 90°$. It can be shown [4] that under these conditions only those resonances that are offset from the transmitter by $\Delta\nu = \pm n/\tau$, where n is an integer, are excited to a significant extent. The use of such a sequence also allows accurate T_1 measurements.

The transverse relaxation time, T_2, can be defined as the time constant describing the irreversible loss of phase coherence of the magnetization vectors in the xy plane. With reference to Section 5.1 and according to this definition, the FID decays exponentially with a time constant T_2. Therefore, T_2 can be directly extracted from the FID or, in the case where more than one signal is present, from the Fourier-transformed spectrum using the relation $T_2^{-1} = \pi\,\Delta\nu$ (see Section 2.1). For most applications this is a satisfactory procedure; however, for very narrow lines there is the possibility that magnetic field inhomogeneities may contribute to the line broadening so that the value of T_2 obtained from linewidth measurements is shorter than the actual T_2. If the magnetic field is not homogeneous, different regions in the sample experience slightly different magnetic fields; therefore there will be a range of Larmor frequencies, rather than a single value, and the final spectrum will be the superposition of many different signals spread out around a central ν_1 value (Figure 5.9B). Such a composite signal is of course broader than the components and its shape would probably be Gaussian rather than Lorentzian; that is, it will come from a nonexponential FID (Figure 5.9A). If the behavior of the magnetization vectors of the various components in the rotating frame is analyzed after a 90°_x pulse (Figure 5.10B), it appears that those components resonating at frequencies $\nu < \nu_1$ and $\nu > \nu_1$ move away from the y axis, in opposite directions. This gives rise to a fan of vectors centered on the y axis, which is more and more spread out with time (Figure 5.10C). Therefore, the total magnetization in the xy plane will decay not only according to the intrinsic T_2 of the signal, but also because of dephasing of the various components. This has suggested [5] the use of a 180°_x pulse applied at a time τ after the initial 90°_x pulse (Figure 5.10D). The 180°_x pulse inverts the fan on to the negative side of the y axis so that the fan tends to close up and all the vectors are refocused on $-y$ (Figure 5.10E) after the same τ previously needed to dephase them. If the FID is observed from B through E, an echo will be

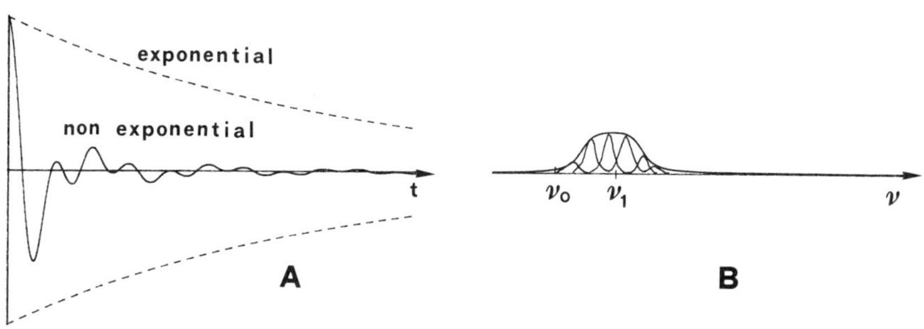

Figure 5.9 **A** Free induction decay and **B** corresponding signal in an inhomogeneously broadened sample.

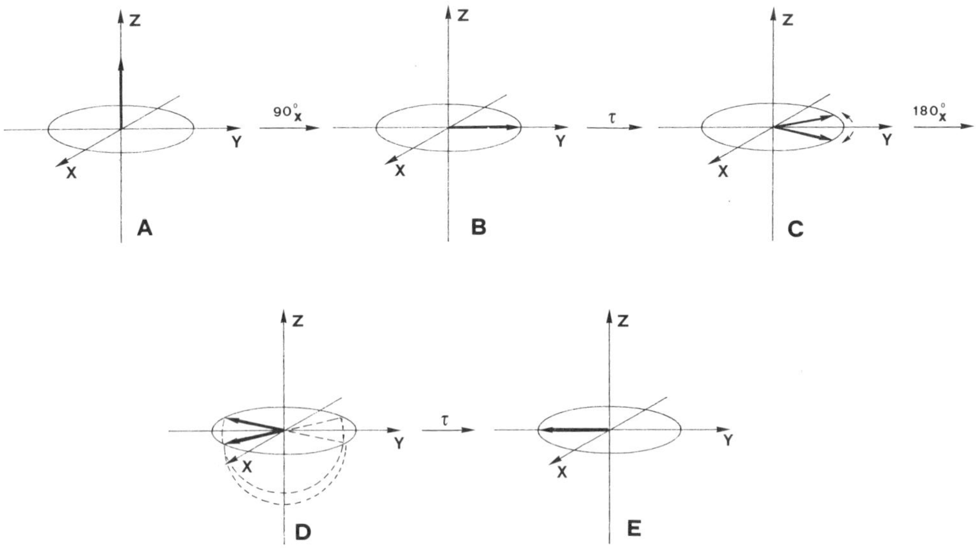

Figure 5.10 Refocusing effect of a 90°_x—τ—180°_x—τ sequence on an inhomogeneously broadened signal [5].

present after a time 2τ (Figure 5.11). The whole procedure may be repeated for different values of τ, and the heights of the echoes, h_τ, will decay exponentially with a time constant T_2.

The main drawback of the spin-echo method arises from diffusion processes in the liquid sample, since the precise refocusing of the vectors depends on the corresponding nuclei remaining in a constant magnetic field during the time of the

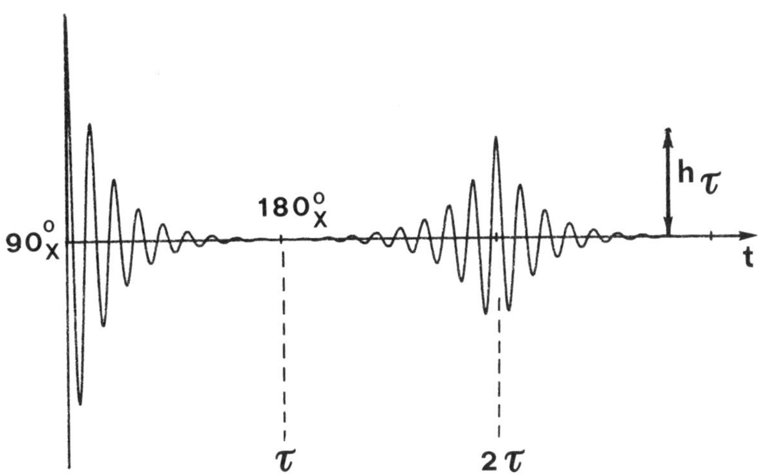

Figure 5.11 Spin echo obtained with a 90°_x—τ—180°_x—τ sequence.

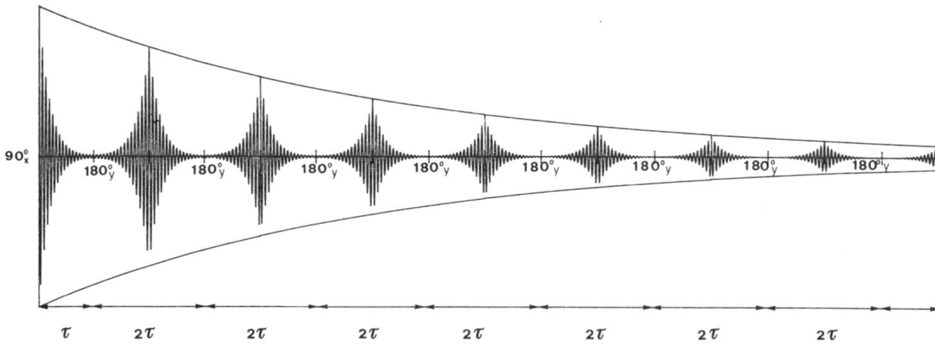

Figure 5.12 Spin-echo sequence obtained with the Meiboom-Gill modification [7].

experiment (2τ). The Carr-Purcell [6] and Meiboom-Gill [7] modifications have been developed to overcome this diffusion problem. The Meiboom-Gill modification uses the train of pulses shown in Figure 5.12; the idea is to minimize the effect of diffusion by multiple refocusing of the vectors using very short values of τ. In addition, the 180° pulses are applied along the y axis, so that refocusing always occurs on the positive side of the axis. The train consists of n inverting pulses, and the echo is detected after each inverting pulse. The height of the echo will have an exponential decay determined only by T_2.

Somewhat related to the spin-echo method is the spin-locking experiment, which allows the measurement of $T_{1\rho}$, the longitudinal relaxation time in the rotating frame. Here, a 90°_x pulse is applied immediately followed by a long, strong pulse along the y axis. In this way the spins are locked along the y axis, and any dephasing in the xy plane will be continuously corrected. After a variable time τ, during which the radiofrequency field has been on, the FID is recorded. Its initial height, or the height of the transformed signal, decays exponentially with τ with a time constant $T_{1\rho}$. During the time τ, T_2 relaxation processes reduce the intensity of the magnetization along y; since the radiofrequency pulse creates a field, B_1, along this axis, the decay is therefore a longitudinal relaxation in the rotating frame. As $T_{1\rho} = T_2$ for most liquids this experiment offers yet another way of measuring T_2 without the problems associated with inhomogeneity [32].

5.3.2 Multiple Irradiation and Two-Dimensional Spectroscopies

In the last decade, multiple irradiation and two-dimensional spectroscopies have evolved so rapidly that reviews and entire books are now dedicated to their numerous applications, see [8] and [9], for example. It is therefore not attempted here to describe the underlying theories or to cover all the experimental techniques, but simply to mention the general principles and a few experiments that will be encountered in following chapters.

Spin Decoupling. This technique is relatively old and also routinely used on CW spectrometers. If two sets of nonequivalent spins (of the same or of different nuclear species) are interacting (the interaction being described by the coupling constant J), their signals will show multiplet structure. By using a radiofrequency of suitable intensity the sample can be irradiated at the resonance frequency of one of the lines, causing saturation of that signal. The second radiofrequency (in the CW mode) is swept in the usual way, and the signal coupled to the saturated one will appear as a singlet; that is, the multiplet structure due to the coupling collapses into a single line. The decoupled line also experiences a change in intensity due to the nuclear Overhauser effect (NOE), which occurs when there is transfer of magnetization from one spin system to another. An analogous experiment may be done in the FT mode by simply using a radiofrequency pulse, instead of the second radiofrequency. Heteronuclear nonselective decoupling—for instance, ^{1}H decoupling while observing ^{13}C—is usually obtained by simultaneously irradiating all the proton resonance region, either with a pulsed radiofrequency or with a broad band CW decoupler. Depending on whether the decoupler is on all the time, only during acquisition, or only during a preparation period before acquisition, decoupled spectra with NOE or without NOE, or coupled spectra with NOE are obtained (see below).

Saturation Transfer. This technique is used to relate pairs of signals belonging to the same nuclear species in two different chemical environments (A and B) in the presence of slow chemical exchange. As seen in Section 4.2.1, two separate signals are observed under this condition. By presaturating the signal from one environment—say, B—a decrease in intensity of the other signal is obtained. This may be related to the time of chemical exchange from the other environment A through an equation of the form [10]

$$\tau_A = T_1^A \frac{M_Z^A/M_0^A - M_Z^B/M_0^B}{1 - M_Z^A/M_0^A} \tag{5.3}$$

where A denotes the observed signal, B the irradiated signal, M_Z and M_0 the magnetization observed with or without irradiation of the B signal, respectively, and T_1^A the longitudinal relaxation time of the nucleus in the A environment in the absence of exchange. If the intensity of irradiation is high enough $M_Z^B/M_0^B = 0$ and the equation is simplified. From Equation (5.3) it appears that the technique is applicable for exchange times of the order of magnitude of T_1.

Nuclear Overhauser Effect. This effect is defined as the fractional change in intensity of one NMR line when another resonance is irradiated, as a result of a change in population of the spin levels. Here, the two resonances belong to two different, noninterchanging sets of nuclei connected by some sort of coupling. The effect is measured on one signal when the other signal is irradiated, in such a way that the magnetization vectors of the coupled spin system are in an equilibrium situation (steady-state NOE). If the coupling is dipolar in origin, as is likely, for instance, among protons in a protein, the NOE depends on the inverse sixth power of the distance between the irradiated and the observed nucleus. Steady-state experiments usually provide observation of NOEs from many nuclei upon irradiation of a single line: this arises from spin diffusion [11], which countermands the structural information contained in the r^{-6} dependence. In transient NOE experiments

[11], a chosen resonance is irradiated with a selective 180° pulse. After a waiting period τ, a normal 90° observation pulse is applied, and the FID recorded and transformed. During the time τ the NOEs of the signals coupled with that selectively irradiated build up with rates proportional to the r^{-6} values of the relative signals, independently of spin diffusion processes. Alternatively [12] a weak selective pulse can be applied for a variable time τ, during which NOEs continuously build up, immediately followed by an observation pulse. If τ is long enough a steady-state NOE is approached. On the other hand, if τ is short the NOEs are truncated and it is possible to distinguish between fast-building and slow-building NOEs. This technique has been named Truncated Driven NOE [12].

Two-Dimensional Spectroscopies. A two-dimensional NMR spectrum is defined as a spectrum that is a function of two frequency variables [$S(\omega_1, \omega_2)$], and is obtained after a two-dimensional Fourier transformation of a time function $s(\tau_1, \tau_2)$.

In all two-dimensional experiments, three or four different time intervals are defined: namely, preparation, evolution, mixing (facultative), and detection periods (Figure 5.13). The experiment consists of pulses applied during the preparation and evolution periods and arranged in such a way as to focus on certain information. The signal, s, is detected during the detection period, τ_2, and the experiment is repeated for a large number of different evolution times τ_1.

Each FID is, as usual, a collection of data points characterized by two parameters (intensity and τ_2), but all the FIDs differing in τ_1 can be arranged together in a single set of data points, each characterized by an intensity and two times, τ_1 and τ_2. The resultant time function, $s(\tau_1, \tau_2)$, can then be Fourier transformed in two dimensions to give $S(\omega_1, \omega_2)$. The latter function can be displayed in several modes, of which the stacked plots (similar to the usual presentation of T_1 experiments) and the contour plots (similar to the altitude lines on a geographical map) are the most common. The physical meaning of the two frequency axes depends on the particular multipulse experiment. For example, if a spin-echo (90°_x—τ_1—180°_y—τ_1—acquisition (τ_2) sequence is applied a so-called J-resolved spectrum is obtained (Figure 5.14A). Basically, the time τ_1 only contains information on the coupling constant J, whereas τ_2 contains information on both J and the chemical shift (Figure 5.14B). A computed-aided tilting is then performed to eliminate the J component from ω_2 so that the spectrum in Figure 5.14C is obtained;

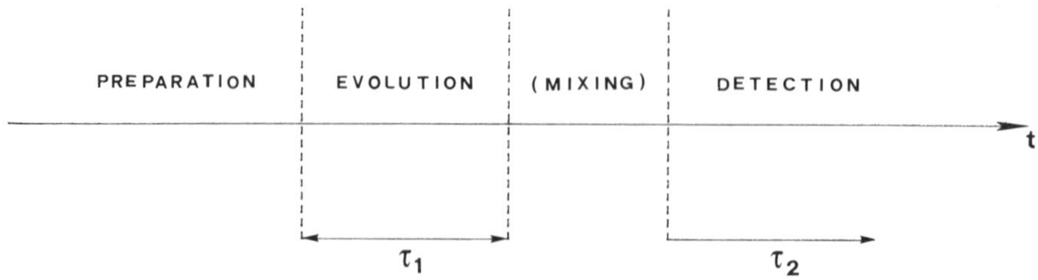

Figure 5.13 General scheme for two-dimensional experiments. (Taken from ref. [9], with permission, copyright © 1982 by Delft University Press.)

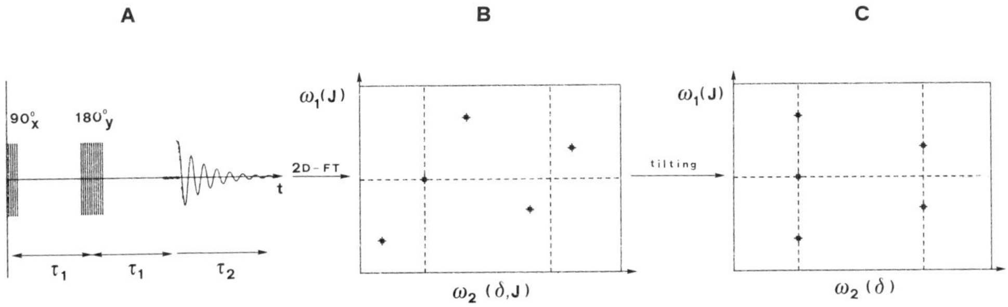

Figure 5.14 **A** Two-dimensional experiment in which a *J*-resolved spectrum is obtained for an A_2X system. **B** Spectrum after a 2D-FT as a function of τ_1 and τ_2, in which ω_1 only contains information on *J*, and ω_2 contains information on both *J* and the chemical shift, δ. **C** Computer tilting of the data points gives this final spectrum, in which the information on *J* is only retained in the ω_1 axis.

here the information on *J* and the chemical shift are localized on the two different axes. As another example, in chemical shift correlation spectroscopies, the two axes both represent chemical shift while the monodimensional spectrum lies on the diagonal (Figure 5.15). The off-diagonal signals connect pairs of signals in the

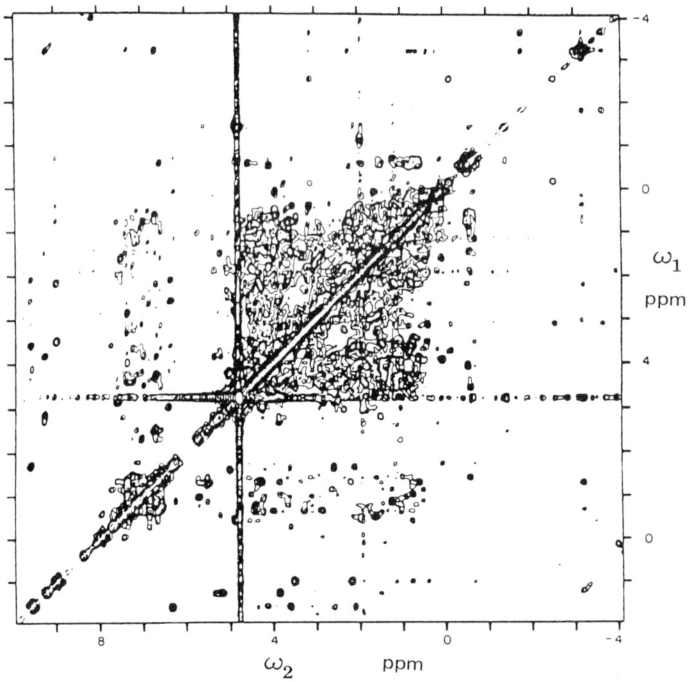

Figure 5.15 Shift correlation spectrum based on cross-relaxation of ferrocytochrome *c*. (Reprinted from ref. [13], with permission.)

monodimensional spectrum that arise from nuclei connected either by J–J coupling or dipolar coupling, which is referred to as cross-relaxation, depending on the kind of pulse sequence used. The intensity of the off-diagonal signals provides quantitative information on the magnitude of the coupling. Figure 5.15 shows an example of a shift correlation spectrum based on cross-relaxation.

5.4 SIGNAL DETECTION IN PARAMAGNETIC SYSTEMS

5.4.1 General Considerations

The effects of electron-nucleus coupling on the NMR parameters, as discussed at length in Chapters 2 and 3, can be summarized as: (1) spreading of the chemical shift range; (2) decreasing T_1; and (3) decreasing T_2 (large linewidth).

The effect on the chemical shift leads to the same kind of considerations made in Section 5.2 about the importance of using short exciting pulses and a fast analog-to-digital converter. The effects on T_1 and T_2 are opposite in nature, and may sometimes partially counterbalance each other. In fact, short T_1 times allow faster scanning and therefore higher signal-to-noise ratios for a given experiment time. On the other hand, short T_2 times imply larger linewidths, and the signal-to-noise ratio is inversely proportional to the linewidth. In the case of very large linewidths further instrumental complications may arise because the instrument has an intrinsic dead time between the end of the transmitter pulse and the switching on of the receiver. If the T_2 of the signal is of the order of such a dead time, a significant fraction of the FID may be lost. Typical dead times range from a few microseconds to several hundreds of microseconds, depending on the instrument: a dead time of 500 μs will make it very difficult to observe lines larger than $\sim$1000 Hz. Short dead times may be achieved using a suitable receiver system. In addition, it is necessary to minimize the pulse breakthrough due to pulse imperfections, and the acoustic ringing of the probehead induced by the pulse. The latter may be a serious problem, especially at low frequencies. For nuclei with a small γ_N, such as ^{14}N, this is indeed the most serious problem if the dead time has to be reduced in order to observe a very large signal. This negative effect may be reduced by using larger magnetic fields.

Proper choice of the magnetic field is a prerequisite considering all the field-dependent paramagnetic effects on T_1 and T_2 discussed in Chapter 3. In the absence of these effects the obvious choice would be the highest magnetic field available, since both resolution and sensitivity increase almost linearly with magnetic field. According to the Solomon Equations (3.3) and (3.4), both T_1^{-1} and T_2^{-1} decrease for dispersions whose position depends on the correlation time of the system: T_1^{-1} goes ultimately to zero at high magnetic field, whereas T_2^{-1} levels off at a field-independent value (Figure 3.4). Therefore it is advisable not to choose too high a field because this may result in a longer T_1 (slower scanning) with no improvement in linewidth. Anomalous behavior due to field dependence of the correlation times (Section 3.7.1) should be considered for each case.

Another general phenomenon that should be considered in paramagnetic macromolecules is the occurrence of Curie spin relaxation (Section 3.4). The line broadening introduced by this relaxation mechanism is proportional to the square of the magnetic field, and therefore becomes dramatic at very high fields. The onset of such a mechanism as the dominant one depends on the relative values of the electronic and rotational correlation times, as well as the number of unpaired electrons on the metal ion. For each particular case the optimum magnetic field, in terms of spectral resolution, will be a compromise between these different effects (see Figure 5.16, for example).

5.4.2 Detection of Isotropically Shifted Lines in Paramagnetic Macromolecules

Every attempt to obtain an FT spectrum in which lines of drastically different intensity are present, such as solute and solvent lines in diluted solutions, may result in dynamic range problems.

As seen in Section 5.4.1, the digitization process implies that the receiver signal may be only approximated by a finite series of points in the computer memory. While the reproduction quality on the x (or time) axis is limited by the speed of the analog-to-digital converter (ADC) and the memory available, that on the y (or intensity) axis is limited by the ADC dynamic range—that is, the number of points on the y axis that the computer can resolve. This number is typically 2^{12}. After a certain number of scans the summed FID will saturate the computer memory, whose dynamic range is typically between 2^{16} to 2^{24}; the acquisition software routine checks for this situation at every scan and when it detects an imminent memory overflow it divides the summed FID by two. Each subsequent FID is also divided by two before being summed in the computer memory. When the same number of scans is performed, the resulting FID is again divided by two. The next FIDs will be divided by four, and the whole scaling process can be repeated at length. However, if the sample is characterized by both an intense and a weak signal, the FID of the weak signal will only represent a small perturbation of the total signal. Upon digitization, the y variation of the small component may result in only a few digits of resolution in the ADC: therefore, after a few scalings, the new digitized signals will no longer keep trace of the small component, and further accumulation will be almost meaningless. The problem of dynamic range may be more or less severe depending on the particular instrument, but rigorous situations always require the use of additional special acquisition techniques [14], which may be software routines or pulse sequences. Software routines include Block Averaging, in which the FIDs are never scaled down but Fourier-transformed after a certain number of scans and the resulting spectra are further averaged, and Double Precision Acquisition, in which two consecutive memory words are used to define any value of the FID on the y axis, doubling the number of bits available and drastically increasing the dynamic range. Both methods require about double the memory capacity to obtain the same spectral resolution.

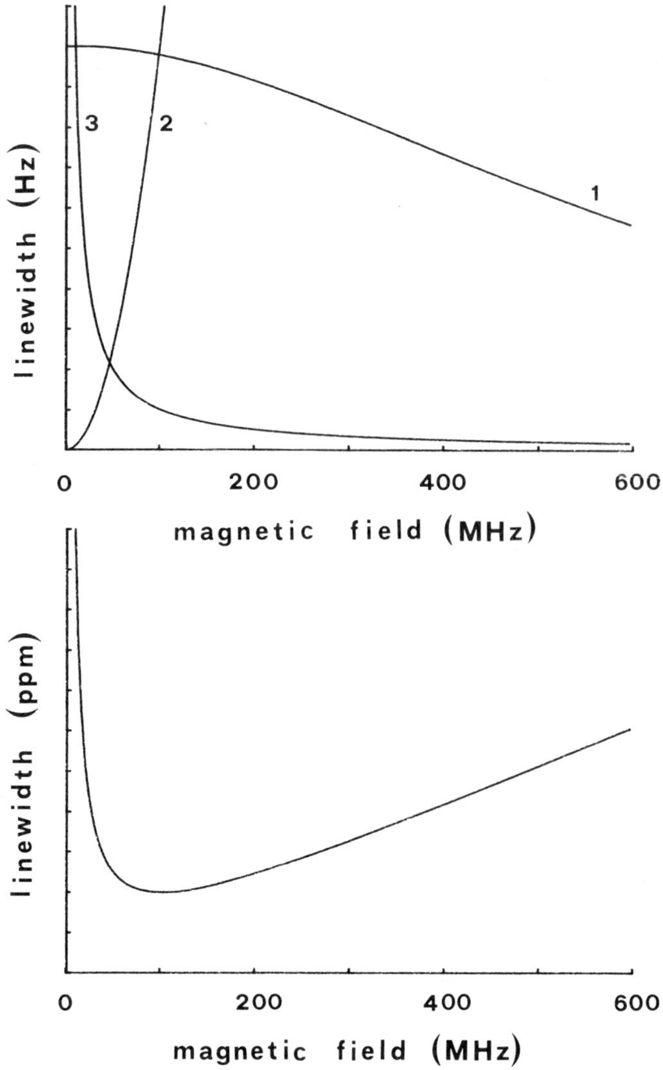

Figure 5.16 Upper diagram: field dependence of the linewidth, $\Delta\nu$ (Hz); (1) dipolar contribution according to the Solomon equation and (2) Curie spin relaxation contribution. Curve 3 shows the field dependence of the ratio between the total linewidth expressed in ppm and that expressed in hertz ($\Delta\nu$ (ppm)/$\Delta\nu$ (Hz)); the ordinate is in an arbitrary scale. Lower diagram: field dependence of the total linewidth (ppm). The pattern can be viewed as obtained by multiplying the function corresponding to curve 3 by the sum of the functions corresponding to curves 1 and 2.

Solvent-suppression pulse sequence methods usually achieve saturation of the most intense line, in order to gain dynamic range in the accumulation of the spectrum. This can be done simply by applying a fixed radiofrequency, as from a decoupler, to the solvent signal. Alternatively [15], a presaturation pulse can be applied with the carrier frequency set close to the resonance to be suppressed (Figure 5.17); this pulse is made long enough to excite only a narrow range of frequencies, including the solvent signal, and the power is reduced accordingly to obtain a 90° pulse. The magnetization due to the solvent is now tilted in the *xy* plane, and a second, short, observation pulse can be applied to detect the signals in a wide spectral range.

When saturation techniques are applied to suppress the water signal, saturation transfer (Section 5.3.2) may cause quenching of isotropically shifted signals from exchangeable protons. This phenomenon can be used if information on the exchange rate of such protons is desired, or can be avoided by using solvent suppression techniques not based on saturation of the water signal (see next paragraph).

The Redfield Sequence. As shown in Figure 5.1, a pulse of length τ excites a frequency bandwidth proportional to $1/\tau$; the excitation profile centered at ν_0 goes to zero at approximately $+1/\tau$ and $-1/\tau$, becoming negative outside this range, and dies out rapidly as it moves further out from the central range. A good way to suppress a line, provided all the other lines are on the same side, is therefore to set the carrier at about $1/\tau$ frequency units from it. Then the pulse power is adjusted to obtain a 90° pulse, or whichever value is needed. Redfield [16] proposed a modification of this experiment which makes the excitation profile smoother around zero, and so the setting of the carrier frequency is less critical. The Redfield sequence is obtained by dividing the pulse in a train of phase-alternated pulses, as shown in Figure 5.18A. The corresponding excitation profile is shown in Figure 5.18B. The value of τ is chosen in such a way as to set the flip angle at values sensibly lower than 90° to further reduce the slope around the zero point.

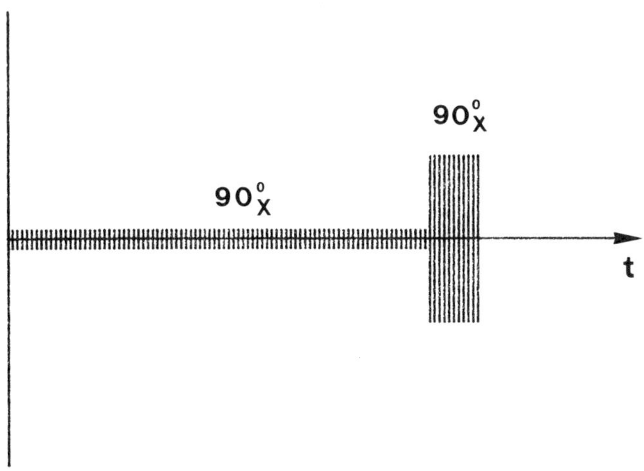

Figure 5.17 Solvent-suppressing pulse sequence. The first weak, long 90° pulse saturates only a narrow range of resonances centered at the solvent position; the second is a normal detection pulse [15].

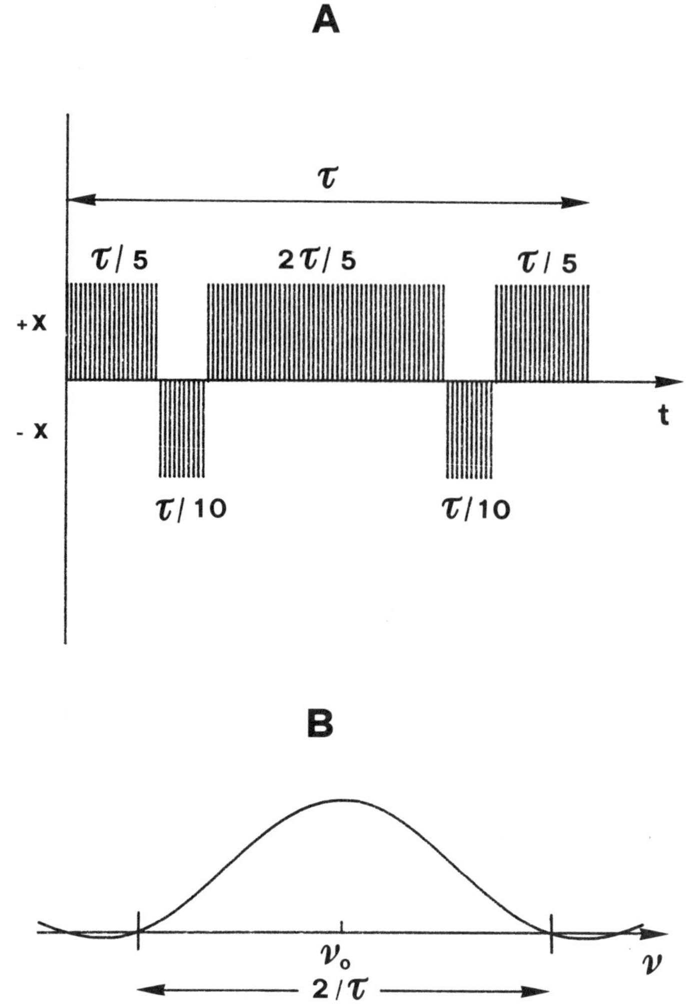

Figure 5.18 Redfield sequence for achieving a smoother profile of the excitation band around $-1/\tau$ and $+1/\tau$. Setting the solvent line on one of the two zeroes greatly suppresses its intensity [16]. For clarity, only the positive side of the pulses at $+x$ and the negative side of the pulses at $-x$ are represented in **A**.

The Modified DEFT Sequence. As mentioned before, short T_1 times of paramagnetically shifted nuclei allow fast scanning of the spectrum, and this by itself also results in partial suppression of signals with longer T_1 times. A much better way to exploit the difference in longitudinal relaxation times between the paramagnetic and the undesired diamagnetic signal in a macromolecule has recently been developed [17] as a modification of the Driven Equilibrium Fourier Transform

(DEFT) sequence, which was originally devised [18] for the opposite purpose. The DEFT sequence is shown in Figure 5.19.

In the original DEFT sequence [18], the FID is sampled after the first 90° pulse. The signals with long T_1 do not recover appreciable magnetization along the z axis during the time τ, so that the 180° pulse substantially reverses the magnetization within the xy plane. After a second τ delay the magnetization is refocused in the $-y$ direction, and the final 90° pulse forces it again towards its equilibrium value along the z axis. The system is then ready to be sampled again, avoiding long waiting times. In the modified DEFT experiment (Figure 5.20) [17], the FID is sampled after the *last* 90° pulse: in this way the slowly relaxing signals are suppressed, while the fast-relaxing ones spontaneously return to equilibrium shortly after each pulse, provided that τ is chosen so that $T_{1\,\mathrm{long}} \gg \tau \gg T_{1\,\mathrm{short}}$.

The WEFT and Super WEFT Sequences. Another way of exploiting the difference in T_1 between paramagnetically shifted nuclei and solvent nuclei is to use a conventional T_1 sequence, 180°_x—τ—90°_x—D—, with τ values chosen to give zero intensity of the solvent signal. This is the WEFT (Water Eliminated Fourier Transform) sequence [19], which has been widely used. A recent modification of WEFT, super WEFT [20], employs much shorter τ values and involves repetition of several sequences spaced by short D values in order to reach a steady-state situation before acquiring the FID. It has been shown that, if τ is much shorter than the solvent T_1 but longer (up to five times) than the T_1 values of the signals of interest, efficient elimination of the solvent line is achieved.

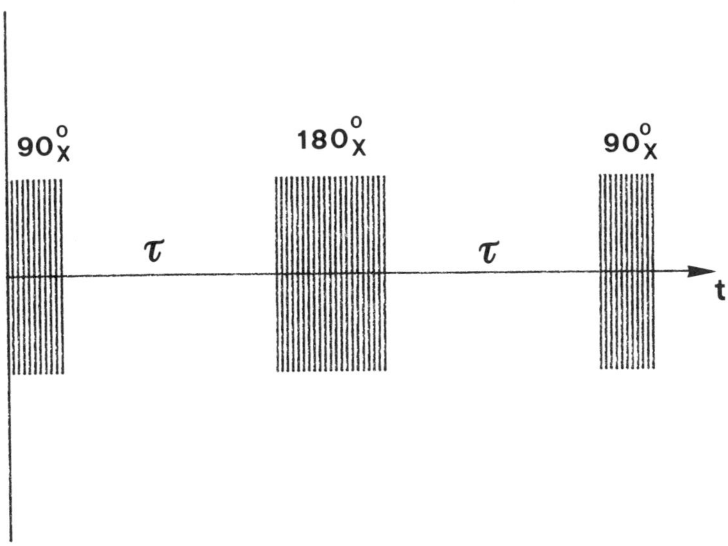

Figure 5.19 DEFT sequence [18].

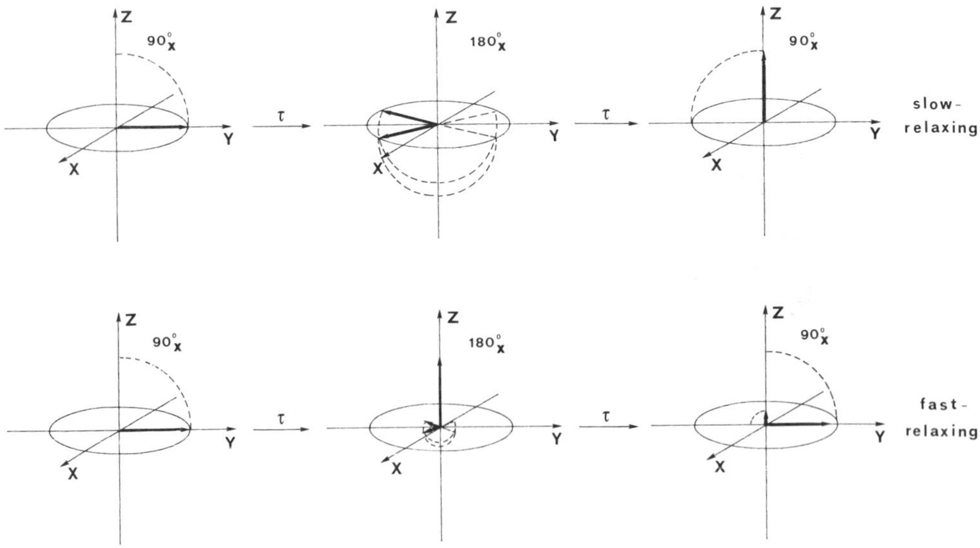

Figure 5.20 Behavior of the macroscopic magnetization for slowly relaxing and fast-relaxing signals during a modified DEFT sequence [17].

5.4.3 Measurement of Relaxation Times in Paramagnetic Systems

In principle, all the techniques described in Section 5.3 can be applied as such to paramagnetic systems; however, in the case of macromolecular samples some modifications are normally used to match the requirements of the sample itself. In general, the relaxation times of paramagnetically relaxed signals are short, so that short τ times are needed. However, some routine spectrometers are limited in the choice of τ values by the software, and the shortest τ available is typically 1 ms—this value does not allow precise measurements of T_1 values of the order of milliseconds or less. For spectrometers that do not have this limitation, the inversion-recovery sequence works quite well when used in conjunction with all the phase-shift improvements discussed in Section 5.3. There is no practical advantage in using other less accurate methods, since waiting $5T_1$ after each pulse is not a problem for fast-relaxing systems; often the acquisition time itself is longer than $5T_1$. It should be remembered that if a multisignal spectrum is analyzed the excitation band must be broad enough to cover all the signals with true 180° and 90° pulses. This is sometimes also a problem in diamagnetic heteronuclear systems if the spreading of the signals is large, and may be a dramatic problem in paramagnetic systems unless very short pulse lengths and very high pulse power can be used. If this is not the case, the spectrum should be divided into subsections and T_1 measured separately for each subsection by placing the carrier close to the signals of interest. The use of a composite 180° pulse [3] is also convenient.

Additional problems may arise when using solvent-suppression techniques. Application of the presaturation pulse requires that the transmitter be set close to the water resonance throughout the experiments, and therefore unreliable results may be obtained for signals far removed from the carrier. Furthermore, use of the Redfield sequence should be avoided because it amplifies intensity distortions. When the pulse is short enough to ensure a uniform irradiation throughout the spectrum, the best solvent-suppression method allowing measurement of T_1 with reasonable accuracy is perhaps the modified DEFT sequence. As discussed, the time τ (Figure 5.19) is chosen to be $T_{1\,\text{long}} \gg \tau \gg T_{1\,\text{short}}$ so that the paramagnetic signals are completely relaxed while the diamagnetic signals are not. Of course, by reducing τ to the order of $T_{1\,\text{short}}$ or less, the paramagnetic signals also start to be quenched. In the absence of pulse imperfections all the signal intensities are zero for τ tending to zero. Therefore by varying τ a set of measurements can be performed which are analogous to a progressive saturation experiment. The following equation [17] relates the signal intensity to T_1 and the time τ:

$$I_\tau = I_\infty[1 - 2e^{-\tau/T_1} + (1 - a)e^{-2\tau/T_1}] \tag{5.4}$$

where $a = \cos\theta$ and θ is the flip angle of the first pulse (Figure 5.19). The parameter a represents possible deviations from the 90° flip angle of the first pulse and can be used as a third parameter, together with I_∞ and T_1, for the best-fitting procedure.

Compared to T_1, the measurement of T_2 on paramagnetically broadened signals does not require the application of particular pulse techniques, since the effect of field inhomogeneity on broad lines is usually negligible. The equation $T_2^{-1} = \pi\Delta\nu$ fits for most cases, the only requirement being that the line is Lorentzian. If it is not, strictly speaking, T_2 would not fit its definition of a time constant for an exponential process.

More sophisticated pulse techniques of the type described in Section 5.3.2 are not (or not yet) routinely applied to paramagnetic species. The main limitation, which will be hard to overcome, is intrinsic in the short T_1 of these systems. Saturation transfer and NOE experiments may be difficult to perform because they require saturation of a paramagnetically relaxed line and the time scale of the experiments is much shorter. Furthermore, the range of τ_M values that can be investigated through saturation transfer experiments is narrower, being of the order of T_1. Exceptions to this are investigations of slow equilibria between diamagnetic and paramagnetic species and NOE studies on some paramagnetic heme proteins (see Chapter 7). Two-dimensional spectroscopies have not yet been applied, to the authors' knowledge, to paramagnetic systems. Again, the complicated pulse sequences involved generally contain preparation, evolution, and mixing periods during which little T_1 relaxation should occur: if this is not the case, the final FIDs would have no memory of the preceding perturbations [33].

5.5 NMRD

The information content of nuclear longitudinal relaxation measurements in both paramagnetic and diamagnetic systems can be greatly increased by performing such measurements as a function of the magnetic field. For paramagnetic species, the reason is apparent from the functional form of the equations discussed in Chapter 3 and from the relevant experimental data, which will be presented in the following chapters. The field dependence of a relaxation rate is called relaxation dispersion: hereafter nuclear magnetic relaxation dispersion will be abbreviated as NMRD. In principle, NMRD would be helpful for any chemical system, but practical limitations, as will be shown, restrict its use, with a few exceptions, to water protons. Despite the potential of NMRD, no commercial instrument is available for performing routine NMRD experiments, and most researchers have to overcome this problem by somehow adapting a multinuclear pulsed NMR system to the purpose.

5.5.1 Changing the Magnetic Field

The basic instrument must contain: (1) a variable field magnet (usually an electromagnet); (2) a frequency synthesizer, and a broadband transmitter and receiver; and (3) one (or more) broadband tunable probes covering a large frequency range. A variable field external lock or, at least, a reasonably good flux stabilizer for the magnetic field are also desirable although not strictly necessary (see below). Once these requirements are met, the experiment can be done by first measuring the water proton T_1 at the regular proton frequency—100 MHz for a 2.35 T magnet, for example—with a usual pulse sequence—such as that used in an inversion-recovery experiment, for example. Then the electromagnet current is reduced to bring the field to a lower value (for instance, 1.88 T, corresponding to 80 MHz resonance frequency), the synthesizer is set to 80 MHz, the transmitter and receiver are retuned accordingly (changing the probehead and matching boxes if necessary), the pulse durations are readjusted, and the T_1 measurement is repeated. Using this procedure, it is possible to go as low as is permitted by the lowest frequency probe (typically 2 to 4 MHz); but sensitivity problems will ultimately arise. (Of course, the set of data may be implemented by measurements at 200, 220, 270, 300, 360, and 400 MHz, each on a different commercial instrument, if available.) Except at 100 MHz, the standard lock system is useless, be it an external ^{1}H lock or an internal ^{2}H lock, since both operate at a fixed frequency and once the field has been lowered the lock signal remains far off-resonance. This problem may be partly overcome by choosing magnetic field values at which an easily observable nucleus resonates at the same frequency at which ^{2}H resonates when the field is set at 2.35 T; in this way the ^{2}H lock channel may be used. Table 5.1 shows the various combinations of magnetic fields and lock nuclei for a 100 MHz commercial instrument [21]. Alternatively, by using an additional frequency synthesizer and an especially built tunable external lock probe [22], the proton signal of an external lock can be followed over the entire field range.

If none of these instrumental systems is used the field will be unlocked, and care should be taken to keep it as stable as possible. Long-term stability is not usually required, as the water signal can be observed with a single pulse, but short-term field variations during the sampling of the FID (typically from 0.1 to

Table 5.1 Magnetic Field Values and Corresponding ^1H Larmor Frequencies at Which the Standard ^2H Lock Circuit of a 100-MHz Instrument Can Be Used Employing Different Nuclei as Lock Substances in a Coaxial Tube. [21]

Magnetic field (T)	^1H Larmor frequency (MHz)	Lock nucleus resonating at 15.351 MHz
2.3488	100.000	^2H
1.6255	69.205	^{113}Cd
1.3631	58.036	^{23}Na
1.1238	47.846	^{11}B
0.9278	39.500	^7Li
0.8907	37.921	^{31}P
0.6248	26.601	^{205}Tl
0.3833	16.317	^{19}F

10 s) will be reflected in the signal lineshape. Since a T_1 experiment requires reliable signal intensities to follow the recovery of magnetization on the z axis, it is better to use the areas of the peaks rather than their heights; alternatively, the initial intensity of the FID can be used directly in the diode-detection mode, instead of the phase-detection mode. In the diode-detection mode, several transients can be summed up to average out the noise without the worry of field fluctuations, which would alter the shape of the FID in the phase-detection mode. Each research group solves the problem according to its particular instrumental setup; it should be clear, however, that the final result is hardly a classical high resolution spectrum, because the field homogeneity of the magnet is in general optimized only for the usual high field working conditions. For this reason, and for the dramatic decrease in sensitivity at low fields, the experiments are usually limited to single and intense resonance lines, which in general are solvent lines.

5.5.2 Pulsing the Magnetic Field

Besides the complications inherent with the type of measurements just discussed, and their being rather tedious, the main disadvantage of NMRD is the rather high lower limit of practically accessible fields. This problem has been overcome by using a completely different approach, developed in the 1950s by Redfield (see, for example, ref. [23]), and pursued later to a high degree of refinement and reliability by Koenig and Brown [24]. Their approach uses a solenoid without any iron core to generate a magnetic field. In this way the field can be turned on and off, or changed from one value to any other, rapidly; typical rise or fall times for the field are of the order of one millisecond per MHz. The solenoid is such that the highest magnetic field obtainable is 1.2 T (50 MHz for protons); it is cooled with liquid nitrogen to dissipate the heat produced in the copper wire by the high current required. Although the field cannot be kept on at the maximum value for more than about one or two minutes, this is usually enough for most purposes, as will be seen later.

In Chapter 2, T_1 was defined as the time constant for the first-order process in which the Boltzmann equilibrium of the macroscopic magnetization along the z axis is reached from *any* nonequilibrium situation. Thus, the onset of the equilibrium when a sample is placed into a magnetic field or after the magnetization has been reversed in a 180°—τ—90° experiment, for instance, follows the same rate law (Figure 5.21).

Now, suppose a sample is taken from a strong magnetic field and rapidly put into a weak one: the magnetization will decay according to Figure 5.22A and the time constant required to bring it to the equilibrium value at that field is the same as that for a sample that started at zero field (Figure 5.22B). T_1 will then be the time constant for the field at which the magnetization evolves towards equilibrium; that is, the weak field.

If data points could be taken along the curve in Figure 5.22A, this would measure the T_1 at low field by sampling magnetization values of the order of magnitude of those obtained in the high field magnet. This can, in fact, be accomplished by using the previously described solenoid, and a transmitter and receiver system tuned at a single frequency value (4 MHz in the following example). The procedure is as follows (Figure 5.23):

1. The magnetic field is turned on and set to a value corresponding to, say, 45 MHz (Figure 5.23A).

2. The field is maintained for a time of the order of $5T_{1(45)}$ (where $T_{1(45)}$ is the T_1 at 45 MHz) to allow the magnetization on the z axis to build up substantially (Figure 5.23B).

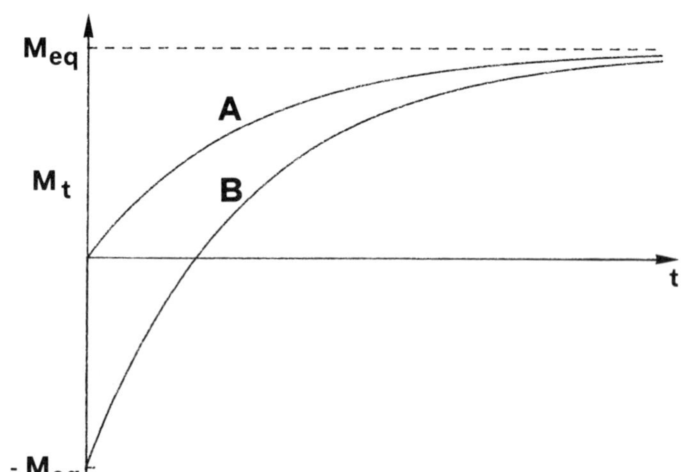

Figure 5.21 Macroscopic magnetization as a function of time when the magnetic field is turned on (**A**) or after a 180° pulse (**B**).

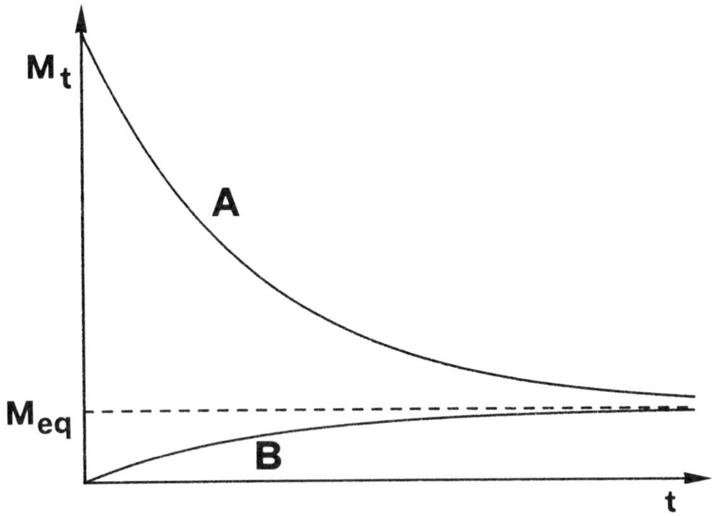

Figure 5.22 The time constant for reaching the equilibrium value of the macroscopic magnetization at a certain field is the same starting from a higher field (**A**) or from zero field (**B**).

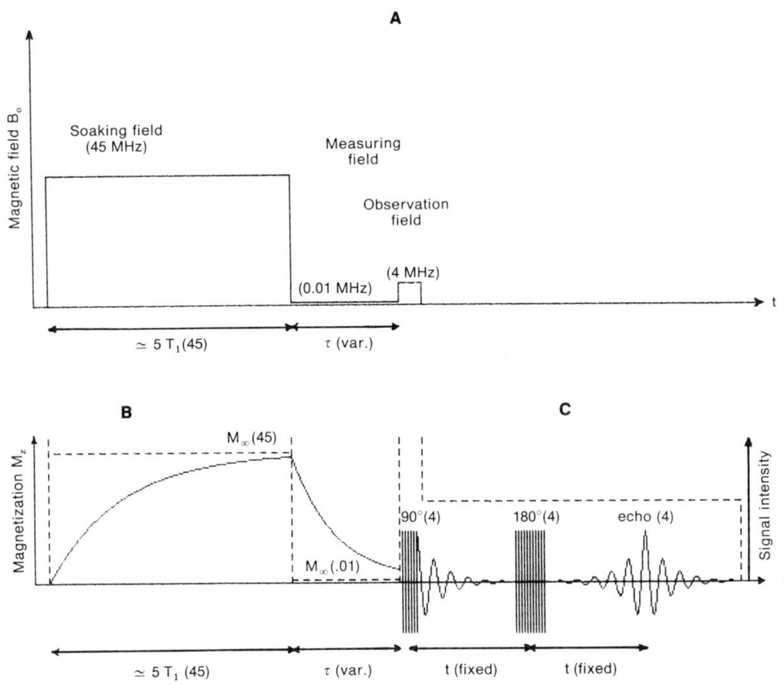

Figure 5.23 Pulsed field-water proton T_1 measurements [23, 24]. Sequence for measuring T_1 at fields lower than 22.5 MHz.

3. The field is then lowered to a value (as low as 0.01 MHz, as in this example) corresponding to the desired Larmor frequency and kept at this value for a time τ, during which M_z decays towards its new equilibrium value, M_∞.

4. The field is then raised to a value corresponding to a fixed observation frequency (4 MHz here).

5. As soon as the field is set to 4 MHz a 90°—t—180° pulse sequence is applied (Figure 5.23C), with t of the order of milliseconds, and the receiver turned on. After a time, t, an echo will appear (of the type shown in Figure 5.11). The height of the echo is proportional to the value of M_z at the time τ, if the time elapsed during the 90°—t—180°—t sequence and the raise time of the magnetic field from 0.01 to 4 MHz are negligible compared to $T_{1\,(4)}$.

6. The field is then turned off to let the magnetization dissipate (and the system cool down!), and the whole procedure is repeated using a different value of τ. T_1 can then be calculated for that particular magnetic field from the heights of the echoes, as a function of τ.

In summary, T_1 at 0.01 MHz is measured at 4 MHz by taking advantage of the sensitivity obtainable at 45 MHz. Of course, the same kind of measurement can be made for any field value, provided it is not too close to 45 MHz, as the observed variation in **M** would then be too small. (It should be remembered that the observation field is always 4 MHz, so only one transmitter and receiver system is needed, and there is no need to retune the coil, either.)

When measuring relaxation rates at fields higher than 22.5 MHz, there is a larger change in magnetization when starting from zero magnetic field and going up as compared to starting from 45 MHz and going down. Therefore, for fields higher than 22.5 MHz the procedure is simplified as follows (Figure 5.24): (1) the magnetic field is turned on and set to the value of interest (for instance, 30 MHz) for a time (Figure 5.24A) τ, during which magnetization starts building up according to $T_{1(30)}$ (Figure 5.24B); (2) the field is switched to the 4 MHz observation value (Figure 5.24C); and step (5) is executed. Again, everything is repeated for different values of τ.

The instrument currently in use at the IBM, T. J. Watson Research Center of Yorktown Heights (NY) can measure smoothly from 0.01 to 50 MHz, has a completely automatized data-processing system, and a reproducibility within 1% over the entire field range.

5.6 MEASUREMENTS OF MAGNETIC SUSCEPTIBILITY IN SOLUTION

The absolute chemical shift of a nucleus in solution is determined by its chemical environment (Chapter 2) and also, to a small extent, by the bulk magnetic susceptibility of the sample. As the effect of the magnetic susceptibility on the chemical shift is the same for all the nuclei, when chemical shifts are reported relative to an internal reference signal the magnetic effect is therefore cancelled. However,

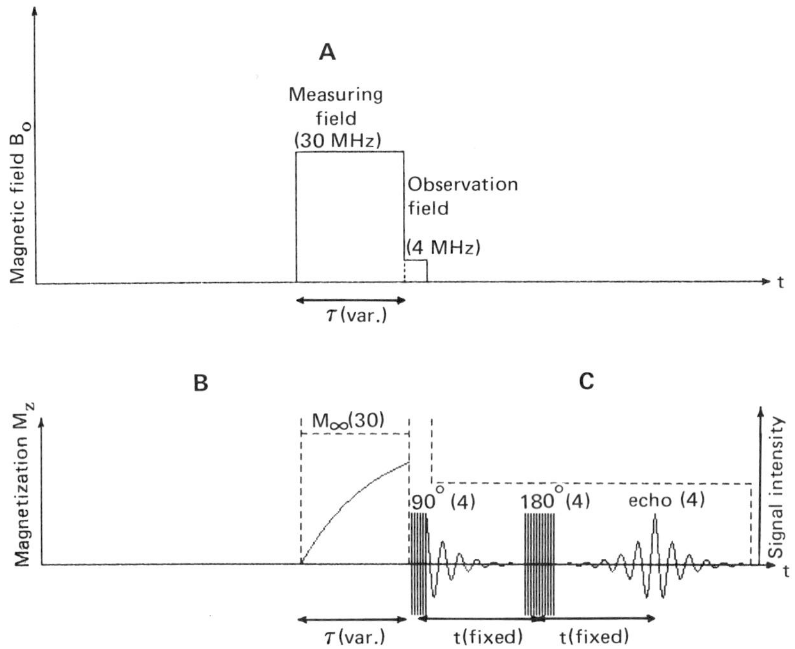

Figure 5.24 Pulsed field-water proton T_1 measurements [23, 24]. Sequence for measuring T_1 at fields higher than 22.5 MHz.

it is possible to measure the magnetic susceptibility of a solute molecule through an experimental procedure often referred to as the Evans method [25, 26].

The contribution to the absolute chemical shift from the magnetic susceptibility of a sample is given by the following equation [27]:

$$\left(\frac{\Delta\nu}{\nu_0}\right)_\chi = m\left(\frac{1}{3} - \alpha\right)\left[\chi - \chi_0\left(1 + \frac{\rho_0 - \rho_s}{m}\right)\right] \tag{5.5}$$

where $\Delta\nu$ is the chemical shift in hertz, ν_0 is the spectrometer frequency, m is the mass of solute per unit volume, α is a demagnetization factor, χ is the mass susceptibility of the solute, χ_0 is the mass susceptibility of the solvent, and ρ_0 and ρ_s are the densities of the solvent and the solution, respectively.

The mass susceptibility of a paramagnetic solute, χ, can be further defined in terms of a paramagnetic and diamagnetic term:

$$\chi = \chi^P - \chi^D$$

where the opposite contribution of the diamagnetic term is taken account of by the minus sign. The demagnetization factor, α, depends on the geometry of the sample and its orientation relative to the external magnetic field. For a spherical sample, $\alpha = \frac{1}{3}$ and the effect on the chemical shift vanishes. For cylindrical samples, $\alpha = \frac{1}{2}$ if the magnetic field is perpendicular to the cylinder's axis (as in most electromagnets) and $\alpha = 0$ if the magnetic field is parallel to the cylinder's axis (as is usual in superconducting magnets). Note that the quantity $(\frac{1}{3} - \alpha)$ is therefore equal to $-\frac{1}{6}$ in the former case and $\frac{1}{3}$ in the latter; that is, opposite in sign and double in magnitude. Therefore, a high field cryomagnet would be preferable because of α and higher resolution.

The experimental setup for measuring the paramagnetic susceptibility of a metalloprotein is as follows:

1. To a solution of a paramagnetic metalloprotein of known concentration (typically in the millimolar range), a small amount of an inert probe substance is added, such as 2,2-dimethyl-2-silapentane-5-sulphonate (DSS), dioxane, or a tetramethylammonium salt.

2. The solution is placed in a coaxial NMR tube, with the solvent and probe substance in the outer tube (Figure 5.25), and the signals of the probe substance are recorded. The signals will appear in different positions owing to the different bulk magnetic susceptibility of the solutions.

3. The measurement is then repeated for a corresponding diamagnetic protein (apoprotein, or a protein substituted with a different, diamagnetic, metal), which provides a shift of the probe signal due only to its diamagnetic susceptibility.

4. If the concentrations are the same, the difference in $\Delta\nu/\nu_0$ gives χ^P directly:

$$\Delta\left(\frac{\Delta\nu}{\nu_0}\right) = m\left(\frac{1}{3} - \alpha\right)\chi^P \tag{5.6}$$

Note that the terms containing the solvent susceptibility cancel because the two solutions have the same density, ρ_s. Equation (5.6) can also be rewritten as

$$\Delta\left(\frac{\Delta\nu}{\nu_0}\right) = 1000\text{M}\left(\frac{1}{3} - \alpha\right)\chi_M^P \tag{5.7}$$

where 1000M is the concentration of the protein in mol m^{-3} and χ_M^P is the paramagnetic contribution to the molar susceptibility, which is directly related to the effective magnetic moment of the paramagnetic center (Sections 1.3 and 1.4):

$$\mu_{\text{eff}}^2 = \frac{3kT}{N_A\mu_0}\chi_M^P \tag{5.8}$$

Typical values of χ_M^P range from 2×10^{-8} to 3×10^{-7} m^3 mol^{-1}, so that for a millimolar solution in a superconducting magnet (1000M = 1, $\alpha = 0$), $\Delta(\Delta\nu/\nu_0)$ goes from 0.007 to 0.1 ppm, that is from 2 to 30 Hz on a 300 MHz instrument.

Possible sources of errors may arise if the probe is not completely inert. In fact, even a very weak interaction of the probe with the paramagnetic center may cause specific effects on the chemical shift, which may substantially alter the measurement, given the relatively small effects being measured. The best approach is to repeat the measurements using different probes. Another source of error lies in the fact that very large macromolecules have a sizeable χ^D term, whose absolute value may be of the same order of magnitude or even larger then χ^P.

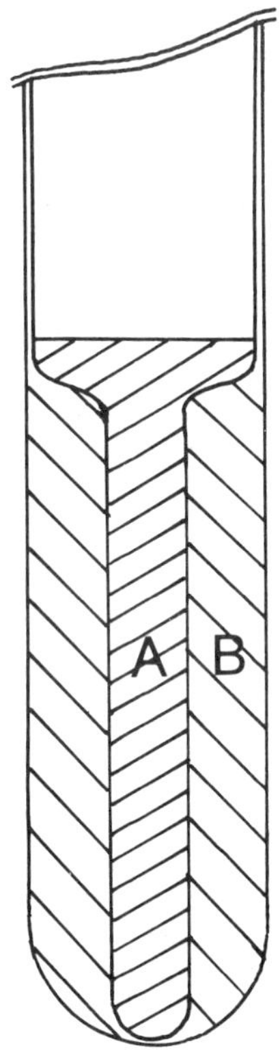

Figure 5.25 Coaxial NMR tubes [25] for the measurement of magnetic susceptibility of a paramagnetic metalloprotein. Solution **A** contains the paramagnetic metalloprotein and an inert probe substance. Solution **B** contains the probe substance in the same solvent. The measurement is then repeated substituting the paramagnetic metalloprotein in **A** with a diamagnetic analog of the same concentration.

5.7 NMR OF ORIENTED MICROCRYSTALS

NMR of solid samples is widely applied to the investigation of organic materials, taking advantage of magic angle spinning (MAS) and, for heteronuclei, cross-polarization (CP) techniques [28]. The need for MAS arises because of the presence of anisotropic contributions to the chemical shifts, which average in solution but do not in solid, nonrotating samples, thus giving rise to a wide distribution of shift values that reflect the random orientation of the crystals. To be efficient, MAS has to be fast enough to average out the chemical shift anisotropy; this is the main requirement limiting the use of solid state NMR on paramagnetic samples, since the anisotropies introduced by the paramagnetic center are generally too large to be efficiently averaged out by MAS. One way to overcome the problem is to use a large single crystal as the sample; studies of this type have been successfully performed on small inorganic complexes [29]. The difficulty in obtaining crystals of suitable size, coupled to the intrinsic difficulties in observing paramagnetically shifted signals in macromolecules, have up to now prevented MAS investigations on paramagnetic metalloproteins. Recently, however, encouraging results have been obtained [30] by exploiting a physical phenomenon that is sometimes observed in EPR experiments [31], and which has never been related to possible

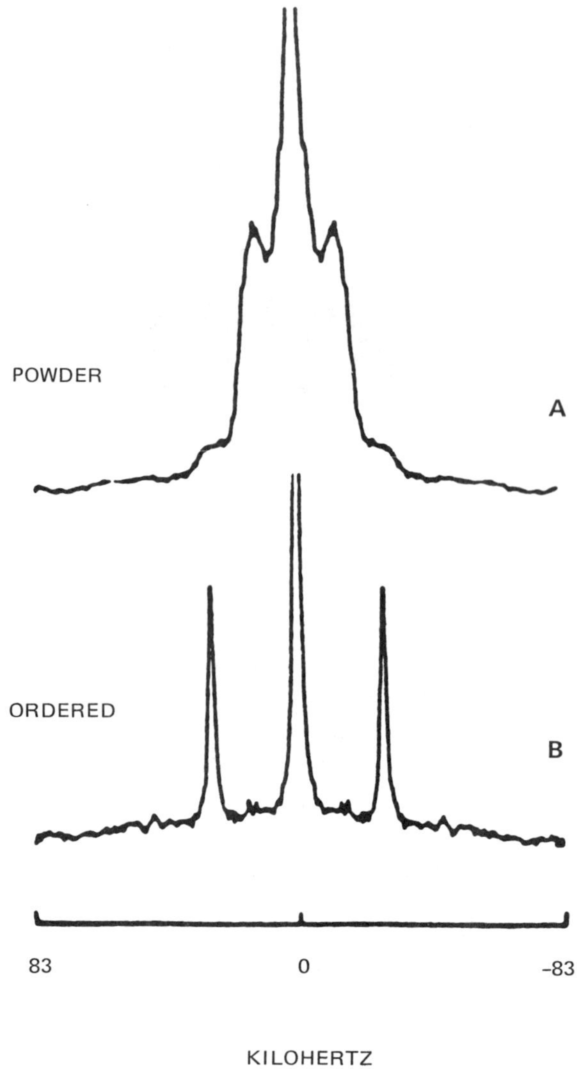

Figure 5.26 55.3 MHz ^{2}H NMR spectra of met-aquamyoglobin crystals specifically deuterated at the methyl groups of Met 55 and Met 131. **A** Randomly oriented crystals. **B** Magnetically ordered crystals. (Reprinted from ref. [30], with permission; copyright © 1980 by American Chemical Society.)

applications in NMR. That is, when a suspension of microcrystals of a paramagnetic substance is placed in an intense magnetic field, the crystals tend to orient along the field and deposit in an ordered fashion. The efficiency of the ordering depends on the magnetic anisotropy of the crystals, which may not be the same as that for the single molecules. The crystals, in fact, align their axis of maximum magnetic susceptibility along the magnetic field. In principle, the crystal is free to rotate along this axis even if the symmetry is not axial, because there is no magnetic driving force in the xy plane, and no ordering is expected in these directions. However, in most cases, single crystals appear as plates or needles. In practice, they tend to deposit according to their geometry so that a two- or three-dimensional ordering is obtained. In the latter case, the result is equivalent to a single crystal.

The average size of the crystals is also important: if they are too small, Brownian motions will tend to oppose the ordering effects; if they are too large, they will have a large inertial momentum, and gravity will drive them down before the orientation is complete.

As an example [30], Figure 5.26 shows the ^{2}H NMR spectrum of met-aquamyoglobin (high spin iron(III), $S = \frac{5}{2}$) specifically deuterated at the methyl groups of Met 55 and Met 131. Spectrum A is obtained with randomly oriented crystals, spectrum B with magnetically ordered crystals. The lines are much sharper in the case of spectrum B, and the known orientation of the crystals in the magnetic field and of the molecules in the crystal allows structural information about the orientation of the methyl groups to be extracted from the measured quadrupole splitting of the lines.

REFERENCES

1. Freeman, R., and Hill, H. D. W. *J. Chem. Phys.* **54** (1971) 3367.
2. Vold, R. L., Waugh, J. S., Klein, M. P., and Phelps, D. E. *J. Chem. Phys.* **48** (1968) 3831.
3. Leviti, M. H., and Freeman, R. *J. Magn. Reson.* **33** (1979) 473.
4. Morris, G. A., and Freeman, R. *J. Magn. Reson.* **29** (1978) 433.
5. Hahn, E. L. *Phys. Rev.* **80** (1950) 580.
6. Carr, H. Y., and Purcell, E. M. *Phys. Rev.* **94** (1954) 630.
7. Meiboom, S., and Gill, D. *Rev. Sci. Instrum.* **29** (1958) 688.
8. Keller, R. M., and Wüthrich, K. In *Biological Magnetic Resonance*, L. J. Berliner and J. Reuben (eds.), Plenum Press, New York, 1981, Volume 3, p. 1.
9. Bax, A. *Two-Dimensional Nuclear Magnetic Resonance in Liquids*, Reidel, Dordrecht, 1982.
10. Gupta, R. K., and Redfield, A. G. *Biochem. Biophys. Res. Commun.* **41** (1970) 273.
11. Gordon, S. L., and Wüthrich, K. *J. Am. Chem. Soc.* **100** (1978) 7094.
12. Wagner, G., and Wüthrich, K. *J. Magn. Reson.* **33** (1979) 675.
13. Baumann, R., Kumar, A., Ernst, R. R., and Wüthrich, K. *J. Magn. Reson.* **44** (1981) 76.
14. Redfield, A. G. *Methods Enzymol.* **49** (1978) 253.
15. Campbell, I. D., Dobson, C., and Ratcliffe, R. G. *J. Magn. Reson.* **27** (1977) 455.
16. Redfield, A. G., Kunz, S. D., and Ralph, E. K. *J. Magn. Reson.* **19** (1975) 114.
17. Hochmann, J., and Kellerhals, H. *J. Magn. Reson.* **38** (1980) 23.
18. Becker, E. D., Ferretti, J. A., and Farrar, T. C. *J. Am. Chem. Soc.* **91** (1969) 7784.
19. Patt, S. L., and Sykes, B. D. *J. Chem. Phys.* **56** (1972) 3182.
20. Inubushi, T., and Becker, E. D. *J. Magn. Reson.* **51** (1983) 128.

21. Mildvan, A. S., and Gupta, R. K. *Methods Enzymol.* **49** (1978) 322.
22. Cosgrove, T., Jarvis, R. A., and Sissons, D. S. *J. Magn. Reson.* **45** (1981) 254.
23. Anderson, A. G., and Redfield, A. G. *Phys. Rev.* **116** (1959) 583.
24. Hallenga, K., and Koenig, S. H. *Biochemistry* **15** (1976) 4255; Brown, R. D., Brewer, C. F., and Koenig, S. H., *Biochemistry* **16** (1977) 3883.
25. Evans, D. F. *J. Chem. Soc.* (1959) 2003.
26. Phillips, W. D., and Poe, M. *Methods Enzymol.* **24** (1972) 304.
27. Dickinson, W. C. *Phys. Rev.* **81** (1951) 717.
28. Haeberlen, U. *High Resolution NMR in Solids. Selective Averaging*, Academic Press, New York, 1976.
29. Reuveni, A., and McGarvey, B. R. *J. Magn. Reson.* **36** (1979) 7.
30. Oldfield, E., and Rothgeb, T. M. *J. Am. Chem. Soc.* **102** (1980) 3635.
31. Nilges, M. J., and Belford, R. L. *J. Magn. Reson.* **35** (1979) 259.
32. Note added in proof. The theory for the dependence of $T_{1\rho}$ on the radiofrequency field, B_1, has been recently developed for paramagnetic systems (Chopra, S., McClung, R. E. D., and Jordan, R. B. *J. Magn. Reson.* **59** (1984) 361.
33. Note added in proof. A report appeared on 2D experiments on a paramagnetic heme protein (Santos, H., Turner, D. L., Xavier, A. V., and LeGall, J. *J. Magn. Reson.* **59** (1984) 177.

CHAPTER 6

Copper

6.1 COPPER SYSTEMS

Copper is often encountered in studies of biological relevance and is essential to living systems. Most of its properties relate to the redox properties of the Cu^{2+}/Cu^{+} couple although the Cu^{3+}/Cu^{2+} couple has also been invoked.

The redox properties are very sensitive to the donor atoms and geometry of the complexes. Copper(I) is a d^{10} ion and is always diamagnetic. Copper(III) is a d^8 ion and in principle could be either high spin ($S = 1$) or low spin ($S = 0$); however, the three positive charges result in it exhibiting diamagnetism in all known inorganic compounds except K_3CuF_6. Thus, the only ion of relevance to the present discussion is copper(II), which is always paramagnetic unless antiferromagnetically coupled to neighboring copper(II) ions.

Copper(II), owing to its paramagnetic properties, has also been used as a probe; that is, as a substitute for the native metal in several metalloproteins and in many zinc enzymes, although the resulting derivatives do not generally have any of the catalytic activity of the native proteins.

6.2 THE ELECTRONIC PROPERTIES OF COPPER(II)

Copper(II) has a $3d^9$ electronic configuration, which gives rise to a ^2D free ion ground term (see Section 1.2). The tenfold degeneracy of this term is reduced upon coordination of the ion. In principle, pure octahedral and tetrahedral symmetries can never be observed because Jahn-Teller distortions tend to remove the orbital degeneracy of the ground state. Furthermore, in natural systems, the metal and its donor atoms give rise to an array far from any idealized stereochemistry, although it may still be useful to refer to one of them (Figure 6.1) in order to predict certain properties of the chromophore. The separation of the electronic energy levels depends on the coordination number and stereochemistry, as well as on the nature of the ligands; however, the ground state Kramers doublet is

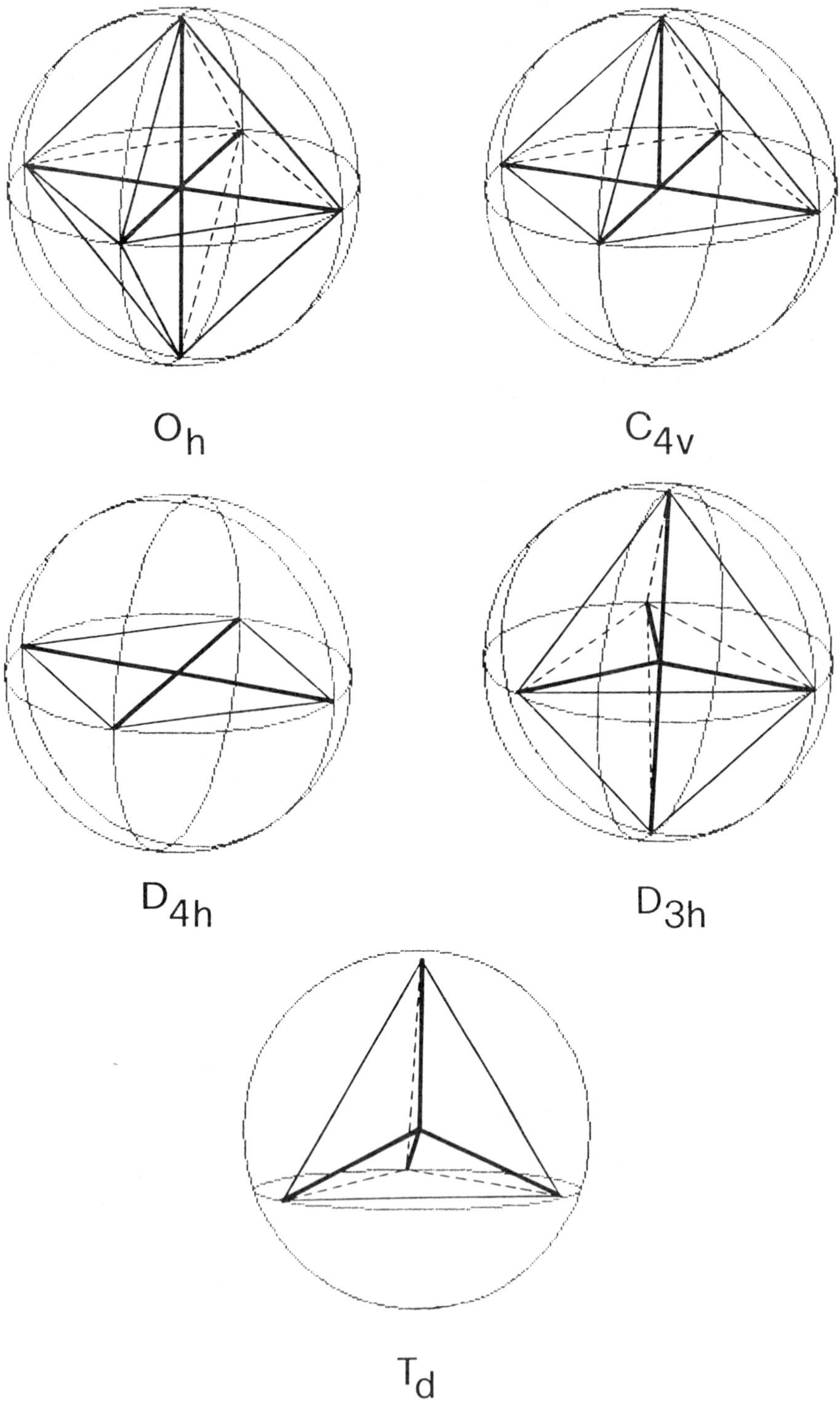

Figure 6.1 High symmetry stereochemistries in coordination compounds and their corresponding group-theoretical labels.

always well isolated from the excited states. As outlined in Section 3.7.1, when the excited levels are far from the ground state, the electronic relaxation mechanisms are relatively inefficient. Copper(II) complexes therefore have relatively sharp ESR signals, and it is often possible to record these spectra at room temperature.

The natural isotopic mixture of copper is composed of two isotopes, ^{63}Cu (69%) and ^{65}Cu (31%), both with nuclear spin $I = 3/2$ and similar magnetogyric ratios (Table 1.1); the single $-\frac{1}{2} \rightarrow +\frac{1}{2}$ electron spin transition expected for copper(II) (Figure 6.2) is therefore split into a quartet due to hyperfine coupling with

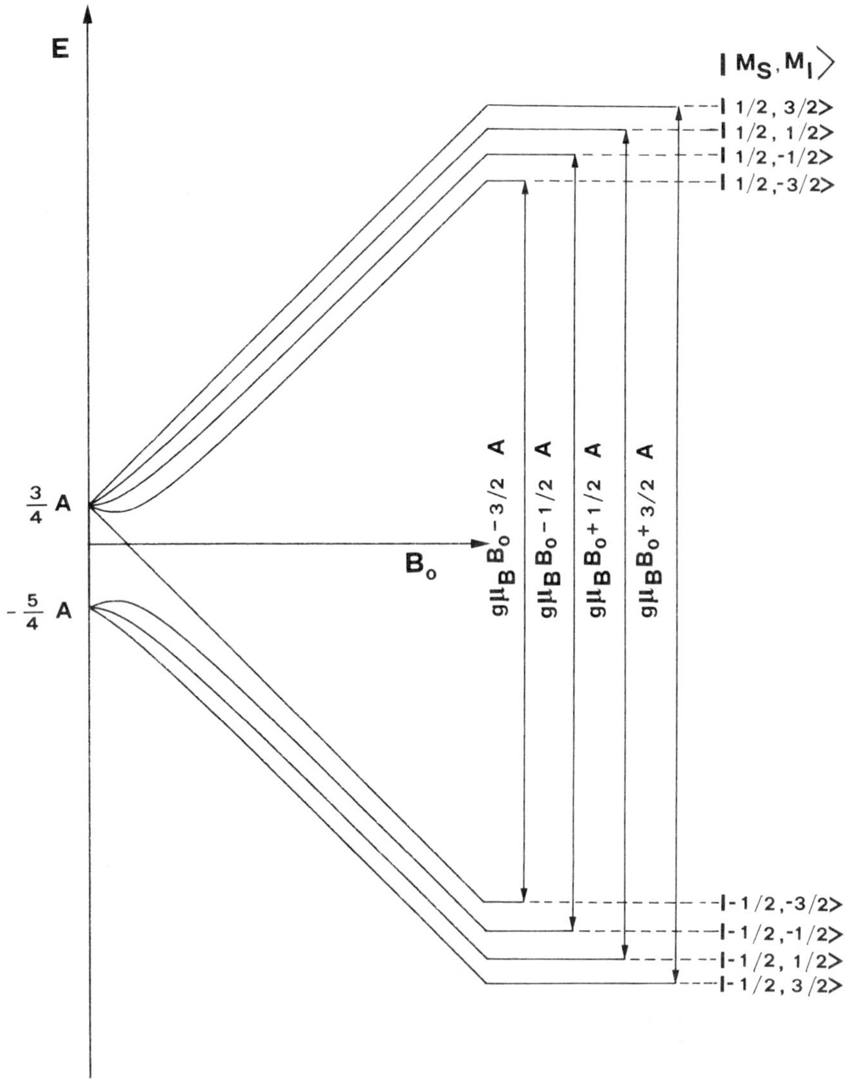

Figure 6.2 Electronic and nuclear spin levels of a system with $S = \frac{1}{2}$, $I = \frac{3}{2}$ in the presence of isotropic hyperfine coupling, as a function of the magnetic field, $\mathbf{B}_0$. The ESR transitions and spin wavefunctions in the high field limit are shown on the right.

the nuclear magnetic moments. From linewidth measurements of a variety of copper proteins, the electronic transverse relaxation time, T_{2e}, has been estimated to be around 10^{-9} s at X-band fields. As pointed out in Section 3.7.1, T_{2e} values of this order of magnitude are indicative of large paramagnetic effects on the nuclear relaxation rates in macromolecules with long rotational times (τ_r for a medium-sized protein is $\sim 10^{-8}$ s). Under these conditions, the correlation time for the electron-nucleus interaction will be determined by the relatively long electronic relaxation time. The situation is different in small complexes, such as copper(II)-amino acid systems, where the rotational time is shorter ($\sim 10^{-10}$ s); in such cases, the paramagnetic effects will be reduced by about one order of magnitude or more.

It is generally extremely difficult to observe well-resolved NMR spectra from nuclei of copper(II) complexes, and, indeed, very few examples are reported in the literature [1]. Isotropically shifted signals in copper-containing proteins are not usually detected; however, the fact that some resonances in a copper-substituted protein are broadened beyond detection may sometimes be useful in assigning the metal ligand residues, by comparison with copper(I) systems. Most of the applications of NMR to copper in biochemistry therefore deal with systems in which an exogenous ligand exchanges with the bulk solution, which usually contains the ligand in large excess. In such cases, the paramagnetic effect is averaged over a large number of nuclei (see Section 4.2) and a signal is readily observed. As a matter of fact, the high nuclear-relaxing properties of copper are often an advantage when the ligand is in very large excess, as in the case of water, the solvent itself. Together with manganese(II), gadolinium(III), and a few other metal ions, copper(II) is a good nuclear-relaxing agent.

6.3 THE TYPES OF COPPER(II) IN SYSTEMS OF BIOLOGICAL RELEVANCE

The foregoing considerations mainly apply to a broad class of copper(II) proteins and enzymes, either natural or artificially substituted with copper, customarily referred to as type 2. There are two more classes of copper(II), which are termed type 1 and type 3 [2–4]. The three classes of copper(II) are characterized by their different spectroscopic properties. Both type 1 and type 2 copper(II) ions are found isolated in proteins or well separated from each other. Type 3 copper is characterized by pairs of copper(II) ions lying in close proximity; this arrangement causes magnetic coupling between the two unpaired electrons, and the resulting dimer is not detectable by ESR. Some well-characterized copper proteins and enzymes containing type 3 binuclear copper centers are reported in Table 6.1. The binuclear copper clusters appear to be utilized for either oxygen transport (hemocyanin) or substrate hydroxylation (tyrosinase); that is, for functions analogous to those displayed by some mononuclear iron(II) centers, such as in hemoglobin or cytochrome P-450 (see Section 7.2).

Table 6.1 Some Naturally Occurring Copper(II) Proteins and Enzymes

Protein or Enzyme	MW	Subunits	Cu(II)(Type)/ Subunit	Other Metals/ Subunit	Function	Copper Donors
Aminooxidase (from pig plasma)	180,000	2	1(2)	—	Redox catalyst	2N(His), 2O(?), 2O(H_2O)
Superoxide dismutase	32,000	2	1(2)	1Zn	$2O_2^- + 2H^+ \rightleftharpoons O_2 + H_2O_2$	4N(His), O(H_2O)
Tyrosinase (from *Neurospora crassa*)	46,000	1	2(3)	—	Monoxygenase	2[2N(His), O(H_2O), X/2, O_2^{2-}/2][a]
Hemocyanin	2.3×10^6	10 × 5	2(3)	—	Oxygen transport	2[3N(His), O/2(Tyr)(?), O_2^{2-}/2] or 2[2N(His), O(H_2O), O/2(Tyr), O_2^{2-}/2][a]
Galactose oxidase	68,000	1	1(2)	—	Redox catalyst	2N(His), O(H_2O), O(Tyr)?, S(Cys)(?)
Ceruloplasmin	132,000	1	2(1) 1(2) 4(3)	—	Redox catalyst	2N(His), S(Cys), S(Met) 3–4N(His) 3–4N(His), X/2(?)[a]
Azurin	14,000	1	1(1)	—	Electron transport	2N(His), S(Cys), S(Met)
Plastocyanin	10,500	1	1(1)	—	Electron transport	2N(His), S(Cys), S(Met)
Stellacyanin	20,000	1	1(1)	—	Electron transport	2N(His), S(Cys), X
Laccase	64,000–140,000	1	1(1) 1(2) 2(3)	—	Redox catalyst	2N(His), S(Cys), S(Met) 3N(His), O(H_2O) 3N(His), O/2(Tyr)[a]
Cytochrome oxidase	120,000–200,000	9–13 (nonequivalent)	1(1)/holoenzyme 1(2)/holoenzyme	2Fe/holoenzyme	Redox catalyst	N(His), N(?), 2S(Cys) 2N(?), 2S(?), X?

[a] Unidentified ligands are indicated as X; ligands, L, bridging two metal ions are indicated as L/2.

Type 1 copper is found in the so-called blue proteins, which may or may not contain other types of copper as well. Two characteristic features of this class of proteins are: (1) a very intense absorption in the visible region, which is responsible for the deep blue color of the chromophore; and (2) an unusually small value of $A_{\parallel}$, the parallel component of the hyperfine coupling tensor. The first feature has been explained in terms of low energy ligand-to-metal charge transfer bands arising from the presence of sulfur ligands in the coordination sphere [5]. The second feature has recently been accounted for as the concurrence of several effects: (1) pseudotetrahedral environment of the copper ion [6, 7]; (2) large covalency of the coordination bond [8, 9]; and (3) large ligand spin-orbit coupling constants of the sulfur donor atoms [10]. Pseudotetrahedral geometry is known to reduce $A_{\parallel}$, compared to tetragonal symmetries, owing to the reduction of the contact component and the increased magnetic anisotropy; covalent bonds and ligand spin-orbit coupling are both known to further reduce the dipolar interaction between the electronic and nuclear magnetic moments [62]. Among the blue copper proteins (Table 6.1), there are some small proteins (MW 10,000 to 20,000) containing a single mononuclear copper center. Their function is to perform rapid electron transfer by changing the oxidation state of the metal—oxygen is not involved in the process.

Type 2 copper is present in enzymes that have several different functions (Table 6.1). Although the electronic properties of type 2 copper are to some extent more usual than those of type 1 and type 3 copper, a variety of stereochemistries have been proposed for type 2 copper on the basis of detailed ESR and optical spectroscopic investigations. As will be shown in this chapter, this class of copper enzymes is the most thoroughly investigated using NMR.

The copper center in most enzymes artificially substituted with copper [11] is also of type 2 (Table 6.2), the exception being copper(II) liver alcohol dehydrogenase [12], which may be considered the only example of an artificial blue copper protein.

Another important group of copper proteins is represented by the oxidases (Table 6.1), some of which contain all three types of copper, reflecting their functional complexity. Copper-containing oxidases catalyze the oxidation of biological substrates by molecular oxygen through a four-electron transfer process that ultimately leads to the formation of water.

In addition to the proteins and enzymes listed in Table 6.1, copper(II) in organisms is also present in a variety of other chemical environments. As a trace element essential for life, copper has to be assimilated from the diet at the intestinal level; it is rapidly transferred from the intestinal mucosa to the blood and transported to the cells where specific copper-containing enzymes are synthesized. The medium in which the copper(II) ion is transported is essentially an aqueous one, rich in good coordinating agents like amino acids and oligopeptides, in addition to proteins like albumin, which may be specifically involved in copper transport. As for other transition elements, the behavior of copper(II) ions in blood can be described by equilibria involving the hexaaqua ion and a number of complexes in which one or more water molecules have been replaced by nitrogen,

Table 6.2 Some Copper(II)-Substituted Proteins and Enzymes

Protein or Enzyme	MW	Subunits	Native Metal	Type of Copper	Donor Atoms
Carbonic anhydrase	30,000	1	Zn	2	3N(His), 2O(H_2O)
Carboxypeptidase A	34,500	1	Zn	2	2N(His), O(Glu), O(H_2O)
Thermolysin	34,500	1	Zn	2	2N(His), O(Glu), O(H_2O)
Alkaline phosphatase	94,000	2	Zn(catalytic sites)	2	[3N(His), O(H_2O)](?)
Transferrin	81,000	2	Fe	2	2N(His), 2O(Tyr), O(HCO_3^-), O(H_2O)(?)
Liver alcohol dehydrogenase	80,000	2	Zn(catalytic sites)	1	N(His), 2S(Cys), O(H_2O)

oxygen, or sulfur donors from the various species in solution. NMR techniques have been of considerable help in shedding light on the behavior of copper in water solutions that simulate the blood plasma environment.

6.4 COPPER-WATER INTERACTIONS IN THE $Cu(OH_2)_6^{2+}$ ION

The study of water-paramagnetic ion interactions in metalloproteins gains significantly from the combined use of ^1H, ^2H, and ^{17}O relaxation measurements because these nuclei not only differ in magnetogyric ratios but also in their relative weight of dipolar and contact contributions, and thus provide a means of multiple fitting of the data to the relaxation theories. Moreover, their exchange rates may be different, and in some cases a more complete kinetic picture can be obtained from the exchange mechanisms operative in the hydration sphere. The main factor limiting the application of this type of investigation to ^2H and ^{17}O nuclei arises from their intrinsic relaxation rates, which are dominated by quadrupolar interactions [13, 14] and are much higher than those observed for protons. A very strong paramagnetic interaction is therefore required for appreciable relaxation rate enhancements, with respect to diamagnetic solutions, to be observed. From this point of view, the transverse relaxation of ^{17}O is favored, since the coupling constant for the contact interaction is larger than for deuteron (see Section 3.6); indeed, for the latter, no measurements of paramagnetic effects due to copper(II) are available in the literature.

6.4.1 ^1H NMR Studies

The longitudinal relaxation rate, T_1^{-1}, of a water proton in pure water is around $0.28\ s^{-1}$ at room temperature. As commented in Section 3.1, this value is determined mainly by dipolar interactions with nearby protons, and is modulated by all the molecular motions that take place in the fluid. Slightly higher values ($\sim 0.4\ s^{-1}$) are observed if the liquid contains dissolved molecular oxygen, which is paramagnetic, in equilibrium with air. The value of $0.28\ s^{-1}$ is independent of the magnetic field and only slightly dependent on temperature. If a copper(II) salt is dissolved in, for example, 1mM concentration, the value of the longitudinal relaxation rate is increased to about $1\ s^{-1}$ in a magnetic field corresponding to 100 MHz proton Larmor frequency. At lower fields, the effect is greater, reaching $2.5\ s^{-1}$ in the low field limit (≤ 1 MHz). An analogous increase is observed in T_2^{-1}.

The increase in the water proton relaxation rate is due to the interaction with the unpaired electron in the hexaaqua copper complex. Thorough studies have been performed on the frequency and temperature dependence of such rates in order to obtain a deeper insight into the mechanisms controlling the electron-nucleus coupling in this simple system.

Figure 6.3 shows water proton relaxation data obtained over a wide range of frequencies (0.1 to 100 MHz) at three different temperatures [15]. Several observations may be made even before analyzing the data in detail:

1. A single dispersion region is apparent corresponding to $\omega_s\tau_c \cong 1$, centered around 10 MHz. A proton Larmor frequency of 10 MHz corresponds to an electron resonance frequency ω_s (see Section 3.1.1) of

$$\omega_s = 658\omega_I = 658 \times 2\pi \times 10^7 = 4.13 \times 10^{10} \text{ rad s}^{-1}$$

which gives rise to a τ_c value of 2 to 3×10^{-11} s, which is of the right order of magnitude for the rotational correlation time of the complex. As has been seen, the electronic relaxation times for copper(II) are expected to be much longer, and therefore do not contribute to the correlation time in small complexes.

2. There is a rather pronounced temperature dependence. A fivefold decrease in the paramagnetic effect is observed between 5 and 65 °C. This decrease of the paramagnetic effect with temperature indicates that the observed rates are not controlled by exchange (see Section 4.2.3). Rather, the temperature dependence arises from variations in τ_c, as shown by the shift of the inflection point of the dispersion; the rotational time is in fact expected to be strongly temperature dependent (Section 3.9).

The transverse relaxation rate, T_2^{-1}, is slightly larger than T_1^{-1}, the difference being more pronounced at high field. According to the Solomon equations (see Section 3.1.1), $T_2^{-1} = T_1^{-1}$ at low field and $T_2^{-1} = \frac{7}{6} T_1^{-1}$ at high field; the data

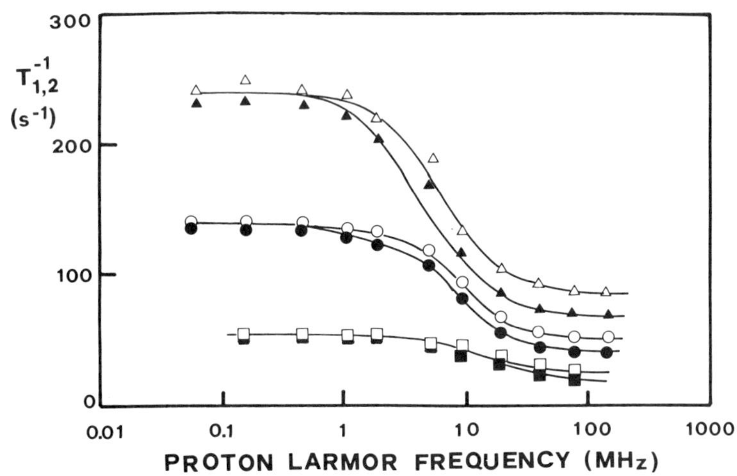

Figure 6.3 T_2^{-1} (open symbols) and T_1^{-1} (filled symbols) of water protons in 6.65×10^{-2}M Cu^{2+} solutions as a function of the proton Larmor frequency at 5 °C (△), 25 °C (○), and 65 °C (□) [15]. The continuous lines represent a fitting of the data according to the Solomon equations.

are therefore in qualitative agreement with such behavior. The fitting [15] of the data to the Solomon equations is reported in Figure 6.3. The calculated value of τ_c is 3.0×10^{-11} s at 25 °C. Furthermore, assuming that the measured effect is entirely due to the presence of water in the first coordination sphere, and that there are twelve equivalent protons of this kind, the metal-proton distance is estimated to be 280 pm, which is in agreement with the expected 200 pm metal-oxygen distance. The main conclusion to be drawn from such an analysis is that contact interactions do not seem to play an appreciable role as a relaxation mechanism, although the correlation time for the latter is longer because it does not depend on the rotational reorientation of the complex (Section 3.2). Thus, contact contributions in all the water-copper(II) interactions encountered in biological macromolecules may be confidently neglected because the dipolar interactions will surely be more effective due to the increased rotational times of such systems.

6.4.2 ^{17}O NMR Studies

The 8 MHz ^{17}O linewidth is increased from 80 to ~3000 Hz in a 0.04M copper(II) solution; the signal is also isotropically shifted (see Section 2.3.2). By taking into account the ^{17}O magnetogyric ratio and the estimated rotational correlation time, τ_r, of the $Cu(OH_2)_6^{2+}$ complex ($\sim 3 \times 10^{-11}$ s), it can be shown that such a large line broadening cannot be accounted for by a dipolar interaction [16]; the contact term must be dominant in this case, which is possible since its correlation time is not reduced by the rapid rotation of the complex (Section 3.2). From the isotropic shift values, the coupling constant A_c/h can be estimated to be 5.0×10^7 Hz [16]; if the latter parameter is used to calculate τ_e from the contact contribution to nuclear relaxation (Equation 3.16), the electronic correlation time is determined to be about one order of magnitude shorter than the value of T_{1e} obtained from ESR measurements.

This observation has led to the proposal that the relevant correlation time for the contact interaction in this case is the rearrangement time for the tetragonal Jahn-Teller distortion of the complex. This may be true if the coupling constant in the axial position of the tetragonally elongated complex is vanishingly small. This has been demonstrated [17] in an investigation of the copper complexes $Cu(en)(OH_2)_4^{2+}$ and $Cu(en)_2(OH_2)_2^{2+}$, where en is ethylenediamine. The contact interaction, as measured from both isotropic shifts and line broadening, was found to be reduced by a factor of two on passing from the hexaaqua complex to $Cu(en)(OH_2)_4^{2+}$, and was abolished in $Cu(en)_2(OH_2)_2^{2+}$. Thus, the interaction was only due to the water molecules in the equatorial plane, the number of which passes from 4 to 2 to 0 on addition of the chelating ligand ethylenediamine. The coupling constant in the axial position was estimated to be more than two orders of magnitude smaller than in the equatorial plane. It has been argued [18], however, that Jahn-Teller distortions cannot provide an efficient interruption of the contact coupling, and therefore cannot determine the correlation time. If so, another explanation is needed for the observed short correlation time.

6.5 DETECTION OF COORDINATED WATER IN NATIVE COPPER PROTEINS AND COPPER-SUBSTITUTED PROTEINS

6.5.1 ^{1}H NMR Studies

Water proton relaxation measurements have been performed on a number of copper(II)-containing proteins. Any time a sizeable relaxation rate enhancement is observed on passing from the diamagnetic system—reduced copper(I) protein, apoprotein, or protein substituted with a diamagnetic metal—to the copper(II)-containing system, such an effect can be ascribed to some kind of paramagnetic interaction. Without any commitment as to the kind of relaxation mechanism or chemical situation involved, such an effect can be used as a titration reporter (see Section 4.4). For example, the number of copper(II) ions capable of binding to apoalkaline phosphatase can be estimated to be at least two by simply titrating copper(II) into the apoenzyme and observing a break in the measured relaxation enhancement [19].

When the aim is to determine whether or not water is present in the first coordination sphere of the metal ion, a more sophisticated approach is applied. Table 6.3 reports the results of such studies for several copper proteins investigated through ^{1}H NMR at variable magnetic fields—these will be discussed in

Table 6.3 G^a and τ_c Values Obtained from Magnetic Field Dependence of Water T_1^{-1} Values in Solutions of Copper(II)-Containing Proteins and Enzymes at Room Temperature^b

Protein or Enzyme	G (pm^{-6})	τ_c (s)	References
CuBCA II (pH 5.5)^c	3.9×10^{-15}	1.9×10^{-9}	[20]
CuBCA II + N_3^- ^{c}	4.7×10^{-15}	2.6×10^{-9}	[20]
CuBCA II + $C_2O_4^{2-}$ ^{d}	0.93×10^{-15}	3.1×10^{-9}	[21]
CuBCA II + PTS^d	0.52×10^{-15}	2.8×10^{-9}	[21]
CuHCA I (pH 7.2)^d	3.9×10^{-15}	2.6×10^{-9}	[22]
Cu_2TRN^c	0.72×10^{-15}	5.7×10^{-9}	[23]
Cu_2AP^d	3.0×10^{-15}	3.0×10^{-9}	[24]
Cu_2Mg_4AP + 2P_i^d	1.0×10^{-15}	3.0×10^{-9}	[24]
Cu_2Zn_2SOD^c	1.7×10^{-15}	1.8×10^{-9}	[25]
Benzylamine oxidase^d	5.5×10^{-15}	7.0×10^{-9}	[26]
Laccase^d	3.7×10^{-15}	2.7×10^{-9}	[27]

^{a} G is defined in Equation (6.1).

^{b} Abbreviations: BCA II: bovine carbonic anhydrase II; PTS: *p*-toluenesulfonamide; HCA I: human carbonic anhydrase I; TRN: transferrin; AP: alkaline phosphatase; P_i: inorganic phosphate; SOD: superoxide dismutase.

^{c} Fit of the complete dispersion.

^{d} Fit of the high-field region.

more detail later. In most cases, temperature dependence studies have also been performed in order to discriminate between slow and fast exchange of the protons with the bulk solution. In the range of magnetic fields corresponding to proton Larmor frequencies of 4 to 100 MHz, a sensible T_1^{-1} dependence has been observed in several cases (Figure 6.4)*. The presence of a dispersion between 10 and 100 MHz would give a τ_c of 10^{-8} to 10^{-9} s for $\omega_I\tau_c \simeq 1$. Such value is consistent with the theoretical estimates and the approximate values of the electronic relaxation times of copper(II) from ESR measurements. By fitting the high field data to the Solomon equation, information can be obtained on the presence of water in the first coordination sphere. Before discussing this point, it is worthy to point out that the τ_c values obtained from such a fitting are remarkably close to each other (Table 6.3) and relatively close to the ESR estimate.

The T_1^{-1} field dependence at low magnetic fields is far from that predicted by the Solomon equation. It can be seen from Figure 6.4 that: (1) although there is evidence of a low field dispersion, attributable to $\omega_s\tau_c \sim 1$, the drop of the T_1^{-1} values is far less than the predicted value of $\frac{7}{10}$; (2) the inflection point is at a higher frequency than expected from the $\omega_s/\omega_I = 658$ ratio; and (3) there is a more or less pronounced raise in the T_1^{-1} values at intermediate fields, between the $\omega_s\tau_c \simeq 1$ and $\omega_I\tau_c \simeq 1$ regions. Although some of the experiments date back to

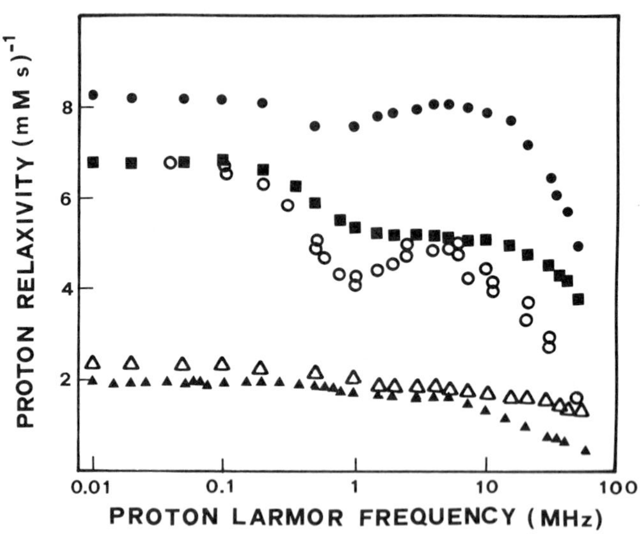

Figure 6.4 Water proton NMRD data for copper(II) protein solutions: (■) copper(II) bovine carbonic anhydrase II at 25 °C [22]; (●) its adduct with N_3^- [22]; (△) its adduct with HCO_3^- [22]; (○) copper(II) superoxide dismutase at 0 °C [28]; (▲) copper(II) transferrin at 25 °C [23, 29].

* The NMRD data in Figure 6.4 are expressed in relaxivity, that is, water proton relaxation enhancements per millimolar metalloprotein solution.

1972 [28, 29], a completely satisfactory theoretical analysis of such a behavior has only recently been achieved: the electron-metal nucleus hyperfine coupling accounts for several of these properties [30]. Particularly, its effects are expected to vanish when: (1) $g\mu_B B_0$ is larger than the hyperfine coupling energy (see Figure 6.2) and (2) the electron-metal nucleus hyperfine coupling constant A is smaller than $\hbar\tau_c^{-1}$. The former condition indicates that the Solomon equation is valid at high field, that is, on the $\omega_I\tau_c$ dispersion; the latter condition accounts for the Solomon equation being valid in fast rotating systems. When $A > g\mu_B B_0$, $\hbar\tau_c^{-1}$, the experimental data deviate from the Solomon equation (Figure 6.5) [31]. In particular, when Cu^{2+} ions are dissolved in ethyleneglycol and the temperature is decreased, τ_c^{-1} ($= \tau_r^{-1}$) decreases, the $\omega_s\tau_c$ dispersion shifts to higher field, and the overall pattern becomes similar to that of a macromolecule [31].

A fitting using the $A_{\parallel}$ and $A_{\perp}$ values obtained from EPR measurements is shown in Figure 6.6 for the water proton NMRD data of superoxide dismutase at various temperatures [25]. The best fit parameters for the 25 °C data are shown in Table 6.3, together with those for copper(II) substituted bovine carbonic anhydrase II [20] and transferrin [23] obtained from analogous treatments.

The Solomon equation in its basic form contains the two unknowns r and τ_c. This form of the equation only holds when considering a single proton interacting with the paramagnetic center. In this case, a two-parameter fit will directly give τ_c and the distance of such a proton from the metal ion. In real cases, however, the number of interacting protons is unknown and, in principle, each of them may be at a different distance from the metal. The term $1/r^6$ in the Solomon equation

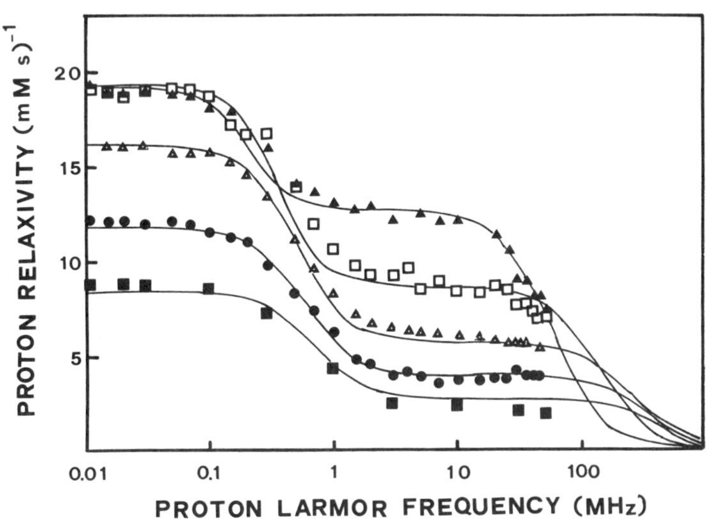

Figure 6.5 Solvent proton NMRD data for copper(II) ions in ethyleneglycol solutions at −9, 5, 15, 25, and 39 °C in order of decreasing relaxivity at 10 MHz. τ_r ranges from about 3×10^{-9} s at −9 °C to 4×10^{-10} s at 39 °C [31].

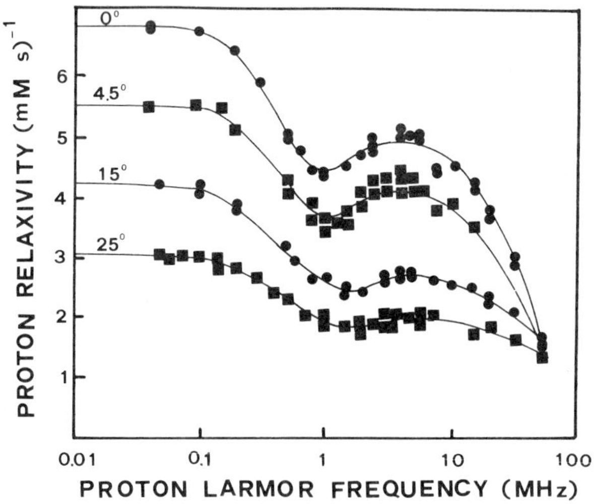

Figure 6.6 Water proton NMRD data for superoxide dismutase solutions at various temperatures [28]. The solid lines are best-fit curves obtained with the inclusion of the effect of hyperfine coupling with the metal nucleus [25].

should therefore be substituted by a parameter

$$G = \sum_i \frac{n_i}{r_i^6} \tag{6.1}$$

where n_i is the number of protons at distance r_i and $\sum_i n_i$ is the total number of protons. Obviously, there is no way of breaking G down and obtaining information on the single protonated species. In practice, either the expected G value ($G_{\text{calc}}^{n=2}$) for a single coordinated water molecule ($n = 2$ protons at $r = 280$ pm) is calculated and compared to the value obtained from a best-fitting procedure [21]; or the ratio $G_{\text{exp}}/G_{\text{calc}}^{n=2}$ is expressed as the hydration number q [32–34]; or it is assumed that two interacting protons are at the same distance, and r is obtained by a computer-fitting procedure [35] and compared with the expected value of 280 pm for a coordinated water molecule. The procedure used is a matter of taste. However, expressing the results in terms of q relies too much on the accuracy of the numerical result, suggesting that it is possible to actually count the coordinated water molecules and discriminate between, say, one or two water molecules in the absence of other spectroscopic information. Actually, $q = 1$ could either mean one water molecule with a metal-oxygen distance of 195 pm or two molecules with a metal-oxygen distance of 235 pm, which is not uncommon in the case of copper complexes. For the very same reason, it would be hard to discriminate, on the basis of the q value, between a regularly coordinated water molecule or a hydroxide ion, provided that the proton belonging to the latter species can undergo fast exchange with the bulk water (see Section 9.2.2). It is somewhat easier to assess whether or not at least one exchangeable proton is present in the first coordination sphere: if not, any paramagnetic effect would be due to a second sphere and outer-sphere relaxation (see Section 4.5), and such effects can be estimated to be around one-fifth of that due to the first coordination sphere.

Table 6.3 shows the G values obtained for several copper proteins, in addition to the τ_c values obtained from best-fitting procedures. The copper ion is of type 2 in all cases, although laccase also contains type 3 copper, which is diamagnetic, and type 1 copper, which does not significantly contribute to the proton relaxation enhancement (see below). Bearing in mind the aforementioned limitations on the interpretation of G and recalling that one regularly coordinated water molecule would give a G value of 4×10^{-15} pm^{-6}, it can be safely concluded that water is present in the coordination sphere of copper carbonic anhydrase and its adduct with N_3^- [20], laccase [27], and alkaline phosphatase at low pH [19, 24]. The ligands oxalate and *p*-toluenesulfonamide remove water from coordination in bovine carbonic anhydrase [21], as probably does inorganic phosphate in alkaline phosphatase [24]. Of course, a much more detailed picture could be obtained for each particular case by using this information in conjunction with the bulk spectroscopic knowledge of the enzyme. For example, the copper ion in carbonic anhydrase would be five coordinated [22, 36, 37] with three histidine nitrogens and two water molecules, one of which would be semicoordinated and scarcely contributing to the overall G value.

The lower G value observed in superoxide dismutase [25, 28, 38] is explained by an axial position of the coordinated water [25, 39], with a Cu—O bond of 240 pm. An increase in T_{1p}^{-1} and possibly G at high pH has been observed [40], and will be discussed in Section 6.5.2.

The low G value of copper transferrin is not fully understood [23]; however, there is general agreement that the metal coordination sphere contains two histidine nitrogens, two tyrosinate oxygens, and an additional oxygen from the synergistic bicarbonate (carbonate?) ion. In addition, all spectroscopic studies so far indicate six coordination [41]; therefore, a solvent molecule is expected to be present as a sixth ligand. If this is indeed the case then a very slowly exchanging water molecule (or hydroxide ion) is consistent with the experimental data. The observed G value would be due to second sphere effects.

Type 1 copper has also been investigated through water proton relaxation measurements: complete field dependence data are available for ceruloplasmin [29], azurin [29], and liver alcohol dehydrogenase substituted with copper at the catalytic sites [42]. Ceruloplasmin contains seven copper ions [43], four of which are type 3, two are inequivalent type 1, and one is type 2. The effect on water proton relaxation is unusually low and almost field independent [29]. There are two possible reasons for this: (1) inaccessibility of the solvent molecules to the copper ions; and (2) electronic relaxation times shorter than the usual values observed for type 2 copper proteins. However, the situation is complicated by the fact that the copper(II) ions are nonequivalent, so more precise information cannot be obtained.

Azurin is a smaller protein containing a single blue copper center. Data [29] show a frequency dependence that differs from that observed for type 2 copper proteins (Figure 6.7). First, the overall relaxation enhancement is smaller; and

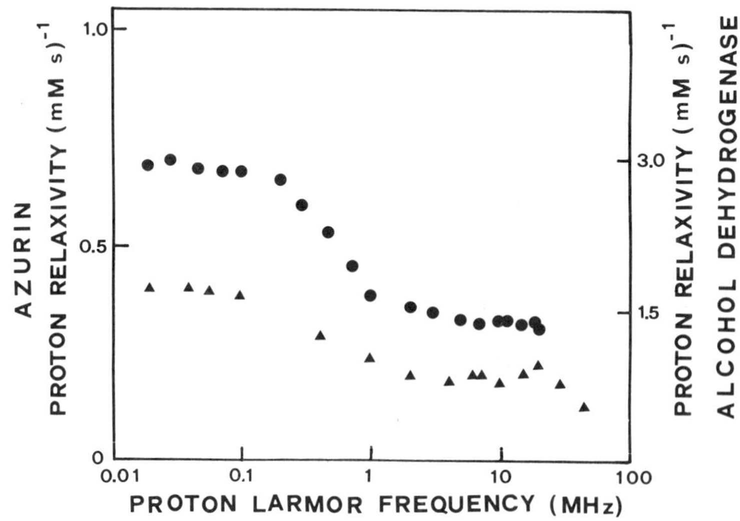

Figure 6.7 Water proton NMRD data for: (●) copper(II) liver alcohol dehydrogenase at 5 °C [42]; (▲) copper(II) azurin at 6.5 °C [29].

second, the low field dispersion is more pronounced than the high field one. The correlation time has been estimated around 6×10^{-10} s [29], which could be consistent with a τ_s value for a type 1 copper. If the data are analyzed in terms of outer-sphere relaxation (Section 4.5), the above value would represent the diffusional correlation time, τ_D, and a copper-proton distance of 500 pm is obtained. Crystallographic studies have then confirmed that the copper in azurin is in a pseudotetrahedral environment, which is inaccessible to the solvent [44].

The shape of the field dependence of copper(II)-substituted liver alcohol dehydrogenase [42] is very similar to that of azurin, but the relaxation enhancement is five to ten times larger; again, only a single major dispersion is observed around 0.45 MHz (Figure 6.7). This would give a τ_c of either 5.5×10^{-10} s ($\omega_s\tau_c \sim 1$) or 4×10^{-7} s ($\omega_I\tau_c \sim 1$). The latter value is discarded because of the limit set by the rotational time of the molecule, which is almost one order of magnitude faster. Analysis of the data in terms of the basic form of the Solomon equation is satisfactory, yielding a G value typical of coordinated water. Note that the value of 5.5×10^{-10} s for τ_c is shorter than the corresponding value for type 2 copper, which is again in keeping with the expectations for a pseudotetrahedral blue copper protein containing sulfur ligands in the coordination sphere.

6.5.2 ^{17}O NMR Studies

A few copper enzymes have been recently investigated through water ^{17}O NMR: namely, laccase [27], superoxide dismutase [45], and copper(II)-substituted carbonic anhydrase [46]. In all cases, T_1 and T_2 have been measured at a single frequency (8 MHz). T_1 has been found not to be affected by the presence of the protein, while T_2 is sizeably reduced, except in laccase [27]. In the case of superoxide dismutase, the paramagnetic contribution to the line broadening was resolved by subtracting the linewidth observed in the presence of the reduced copper(I) protein, and was found to be sizeable. A complete pH dependence has been performed with superoxide dismutase [45] with the aim of investigating the ionization that causes the ^{1}H T_1^{-1} values to increase by about 50% at alkaline pH. The response of ^{17}O was much more dramatic, T_2^{-1} increasing at high pH by about one order of magnitude. As mentioned in Section 6.5.1, water in superoxide dismutase is thought to be bound in the axial position of a somewhat distorted square pyramidal copper(II) chromophore [47]; since the paramagnetic interaction with ^{17}O is likely to be contact in origin, such a dramatic increase could be due to the oxygen becoming more equatorial. It has been proposed that one of the four histidine ligands is replaced by an OH^- ion in the equatorial plane, possibly leaving a water molecule in the axial position [45], which is consistent with a variety of experimental data on the binding of anions (see Section 6.6).

A pH dependence study has also been performed with copper(II) carbonic anhydrase [46]. Its paramagnetic effect is substantial over the entire pH range investigated, although it is almost completely abolished by some metal ligands and enhanced by others. The results of the pH dependence study are consistent with the proposal of two coordination positions from ^{1}H NMR data. From the temperature dependence of the ^{17}O linewidth, the exchange rate appears to be fast on the NMR time scale; a lower limit is estimated to be 10^7 s^{-1} [46]. According to the suggestions of many authors, the coordinated water in metal-substituted carbonic anhydrases ionizes to hydroxide with a pK_a in the range 6 to 9; the ^{17}O NMR data at high pH indicate that, if this is the case, the hydroxide group is also capable of exchanging rapidly with the bulk solvent. A further discussion on the exchange properties of coordinated hydroxide as revealed by NMR will be given in Section 9.2.2.

6.6 LIGAND AND SUBSTRATE BINDING TO COPPER(II) PROTEINS

The foregoing observations on the relaxation behavior of copper(II) proteins with respect to coordinated water permit a more sound interpretation of the data obtained on ligand and substrate binding to such systems. With a few exceptions, field-dependent measurements are not feasible on nuclei other than water protons; therefore the correlation time obtained from water proton relaxation has to be used to analyze the system under investigation. Of course, this is somewhat arbitrary unless both water and the ligand are simultaneously present in the coordination sphere; however, the relative constancy of the copper(II) electronic relaxation times, at least among the type 2 copper proteins, may suggest that no major change occurs in the electronic relaxation time upon substitution of water with the investigated ligand.

Relaxation mechanisms of ligand nuclei may be more complex than for water protons. It has been shown that the longitudinal relaxation rate enhancement for water protons is entirely due to dipole–dipole interactions; this also holds for ligand protons, although the possibility of ligand-centered dipolar contributions should be taken into account (see Sections 3.1.2 and 6.7). When dealing with heteronuclei, ligand-centered dipolar contributions are more important; moreover, due to the smaller magnetogyric ratios, the relative weight of the dipolar interaction is reduced. On the other hand, the contribution of the contact term may be sizeable for nuclei like ^{13}C and ^{31}P, which are usually connected to the metal ion through an oxygen bridge (carboxylate or phosphate groups), and becomes much more dramatic for directly bound nuclei, such as ^{13}C in CN^-, ^{19}F, and $^{14,15}N$. An example of this type of effect was seen in the water ^{17}O NMR studies discussed in Section 6.5.2.

Owing to the importance of ^{19}F NMR investigations, both from the point of view of their biochemical relevance and the theoretical implications, the copper(II)–fluoride interaction will be analyzed in some detail.

The contribution of the contact term to the overall coupling is expected to be large for ^{19}F, as was the case for ^{17}O. In CuF^+, the contact coupling constant, $A_c/\hbar$, has been estimated [48] to be of the order of 10^9 rad s^{-1} (Table 6.4). In a macrocyclic copper complex that can only bind one fluoride ion in the axial position, $A_c/\hbar$ is about 30 times smaller [39] than that in CuF^+ (Table 6.4). ^{19}F relaxation data, like ^{17}O NMR, should in principle provide a way of discriminating between axial and equatorial coordination. From the data reported in Table 6.4, it would seem that fluoride binds to copper as an equatorial ligand in superoxide dismutase, the $A_c/\hbar$ values being much closer to those observed for CuF^+ than for $CuTMCF^+$. In galactose oxidase, the values of $A_c/\hbar$ have been found to be smaller than those in CuF^+ [49], but only by a factor of four. For this reason, the authors favor axial coordination [49, 51]. It should be recalled that the coupling in the axial position is expected to be small only if the metal orbital involved is mainly d_z^2 in nature: deviations from tetragonal symmetries may introduce substantial unpaired spin density, even in an "axial" position.

An interesting application of ^{19}F relaxation data has been proposed to monitor the presence of submicromolar concentrations of superoxide dismutase [52]. The extremely high relaxing capability of copper(II) towards fluoride in this enzyme [39, 50] seems to be unmatched by any other paramagnetic substance usually

Table 6.4 $A_c/\hbar$ Values for $^{19}F^-$ Coordinated to Copper(II) Ions

Complex	$A_c/\hbar$ (rad s^{-1})	References
CuF^+(aq)	1.0×10^9	[48]
$CuTMCF^+$	3.2×10^7	[39]
$GOaseF^-$	2.6×10^8	[49]
$SODF^-$	$>4.0 \times 10^{8a}$	[39, 50]

^{a} Lower-limit value, T_{2p}^{-1} being in the exchange-controlled region [47].
Abbreviations: TMC: 1,4,8,11-tetramethyl-1,4,8,11-tetraazacyclotetradecane; GOase: galactose oxidase; SOD: superoxide dismutase.

present in biological preparations; this property has also been recently extended to monitor O_2^- flows [53] by exploiting the reduction of copper that takes place during the catalytic cycle of superoxide dismutase. This is, indeed, a very sophisticated analytical application of paramagnetic NMR.

Ligand proton relaxation measurements are quite common in investigations of interactions of ligands with paramagnetic metalloproteins. Among the many studies, the interaction of methanol, as substrate, with copper(II)-substituted alcohol dehydrogenase is noteworthy. In this study, the frequency dependence of the longitudinal relaxation times of the CH_3OD protons in D_2O was investigated. The data have shown a weak direct interaction of the alcohol with the metal center [42].

The foregoing arguments on the higher nuclear relaxation observed in equatorial, with respect to axial, coordination sites of copper(II) have been applied in a systematic study of anion binding to superoxide dismutase [39, 54]. The data obtained on ^{14}N, ^{15}N, and ^{13}C for the anions cyanate, azide, and thiocyanate are shown in Table 6.5 together with the data for $^{19}F^-$. A comparison of this data with the relaxation rates observed in the axial position of the copper macrocycle $CuTMC^{2+}$ enables the enzyme data to be interpreted in terms of equatorial coordination [39, 54].

The study of the interaction between the substrates $^{13}CO_2$ and $H^{13}CO_3^-$ and copper(II)-substituted carbonic anhydrase (Figure 6.8) will now be described in some detail, since this provides an opportunity to discuss ligand exchange problems associated with relaxation experiments. A water solution containing a suitable amount of both $^{13}CO_2$ and $H^{13}CO_3^-$ (pH 6.4) gives rise to two distinct ^{13}C signals separated by $\sim$40 ppm. At the experimental 20 MHz ^{13}C resonance frequency the separation is therefore 800 Hz. The interconversion rate should be larger than $2\pi 800/\sqrt{2} \simeq 3{,}500\ s^{-1}$ (see Section 4.2.1) to obtain an averaged

Table 6.5 Nuclear Relaxation Parameters for Anions Bound to Copper(II) in Superoxide Dismutase (SOD) and Cu(1,4,8,11-Tetramethyl-1,4,8,11-Tetraaza cyclotetradecane) $(ClO_4)_2$ (TMC) [39]

	SOD		TMC	
Ligand	$(fT_{1p})^{-1}$, (s^{-1})	$(fT_{2p})^{-1}$, (s^{-1})	$(fT_{1p})^{-1}$, (s^{-1})	$(fT_{2p})^{-1}$, (s^{-1})
NCS^- (^{13}C)	1.0×10^4	5.0×10^5	2.7×10^2	2.2×10^4
(^{14}N)		1.6×10^7		
NCO^- (^{13}C)	5.3×10^3	5.6×10^5	2.8×10^2	6.9×10^4
(^{14}N)		5.0×10^6		
N_3^- (terminal ^{15}N)		1.1×10^7		7.3×10^4
(terminal ^{14}N)		7.0×10^6		
(central ^{14}N)		2.1×10^5		
F^- (^{19}F)	5.2×10^6	2.1×10^7	2.5×10^4	6.7×10^5

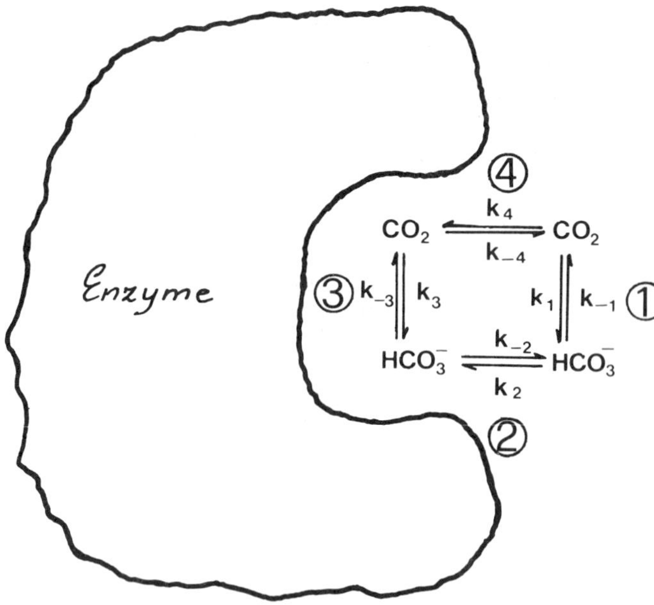

Figure 6.8 Equilibria involved in the CO_2/HCO_3^-/carbonic anhydrase system.

signal, while for the present system this rate is as low as 0.1 s^{-1}; this is why an enzyme is required to catalyze the reaction in biological systems. Upon addition of carbonic anhydrase, the lifetimes of both the $^{13}CO_2$ and $H^{13}CO_3^-$ species are strongly reduced and only one signal is observed [55].

The relevant equilibria for the CO_2/HCO_3^-/carbonic anhydrase system are reported in Figure 6.8, from which it is clear that equilibria 2, 3, and 4 all have to be fast for the enzyme to be efficient. On the other hand, if the aim is to obtain information on the binding sites of each species by using a paramagnetic metal in the active site, then the equilibria 2 and 4 should be fast while 3 should be slow on an NMR time scale. It should be noted that the copper derivative has been chosen here [22, 56] because it is almost inactive and is an efficient nuclear-relaxing agent, both of which are favorable properties. Indeed, two separate signals are observed, the relaxation times of which are strikingly different and shorter than in the corresponding diamagnetic solution (Table 6.6). Therefore, at least one of the three equilibria involving the enzyme is much slower than 3,500 s^{-1}, and both species are affected, to different extents, by the paramagnetic center. The temperature dependence of the relaxation rates suggests that $H^{13}CO_3^-$ is in the exchange-controlled region: k_{-2} is estimated to be 5×10^2 s^{-1} for the human derivative [22] and 5×10^3 s^{-1} for the bovine one [56]—these values also represent the lower limits for T_{1M}^{-1} and T_{2M}^{-1}. By using these values and the copper relaxation times obtained from water proton relaxation data, an upper limit for the metal-carbon distance is obtained, which is only consistent with direct coordination of the bicarbonate ion.

Table 6.6 ^{13}C NMR Relaxation Parameters for the System $^{13}CO_2/H^{13}CO_3^-$ Interacting with Copper(II)-Substituted Bovine (BCA II) and Human (HCA I) Carbonic Anhydrases [22, 56]

Enzyme	Substrate	pH	T(°C)	T_1^{-1} (s^{-1})	T_2^{-1} (s^{-1})
apoBCA II (1.3×10^{-3}M)	HCO_3^- (7.7×10^{-2}M)	6.9	13	0.71	[a]
	CO_2(2.5×10^{-2}M)			0.25	3.3
CuBCA II (1.1×10^{-3}M)	HCO_3^- (7.0×10^{-2}M)	6.6	13	77	77
	CO_2 (4.5×10^{-2}M)			1.7	5.0
apoHCA I (2.3×10^{-3}M)	HCO_3^- (7.2×10^{-2}M)	6.7	4	0.18	4.9
	CO_2 (2.9×10^{-2}M)			0.06	
apoHCA I (2.3×10^{-3}M)	HCO_3^- (7.2×10^{-2}M)	6.0	13	0.11	4.4
	CO_2 (1.5×10^{-1}M)			0.06	3.6
CuHCA I (2.0×10^{-3}M)	CO_2 (9.6×10^{-2}M)	6.3	4	1.1	
CuHCA I (2.0×10^{-3}M)	HCO_3^- (8.6×10^{-2}M)	6.3	13	10.1	13.4
	CO_2 (9.6×10^{-2}M)			1.1	4.9
CuHCA I (2.0×10^{-3}M)	HCO_3^- (6.8×10^{-2}M)	6.9	31	38	38

[a] Linewidth smaller than 1 Hz.

The exchange of the CO_2 molecule is not controlled by its dissociation rate from the enzyme binding site, k_{-4} [22]. If the paramagnetic effect is entirely due to the coupling with the paramagnetic center, a binding distance of 600 to 700 pm from the metal is obtained, if it is assumed that CO_2 saturates the binding site. However, the paramagnetic effect could be partly due to equilibrium 3; that is, the bulk CO_2 could be relaxed by the product of a hydrolyzed HCO_3^- ion from the active site. In order to estimate the rate of such a residual enzymatic process, k_{-3}, the following reasoning is applied: (1) since both k_2 and k_{-4} are fast, the result of the catalytic reaction 3 would be to introduce an exchange broadening term in the CO_2 T_2^{-1} values while the T_1^{-1} values of CO_2 cannot be affected by exchange, and thus are entirely due to paramagnetic effects; (2) the experimental T_1/T_2 ratios are equal to those expected from the Solomon equations; (3) from the experimental uncertainty in the T_2^{-1} of CO_2, an upper limit for k_{-3} is estimated to be 10 s^{-1}, which is much less than the experimental $^{13}CO_2$ relaxation rates. Therefore, a residual catalytic process, if any, does not appreciably affect the relaxation parameters of the $^{13}CO_2$ species.

In summary, a careful consideration of all equilibria involved in the system [22] enables several conclusions to be drawn from the ^{13}C NMR data: (1) the binding site of HCO_3^- in carbonic anhydrase is the metal ion; (2) the dissociation rate of HCO_3^- in the copper derivative is much slower than in the active enzyme; (3) the binding site of CO_2 lies in the active site at about 600 to 700 pm from the metal; (4) the turnover number for copper(II)-substituted carbonic anhydrase is not higher than 10 s^{-1}—that is, it is four or five orders of magnitude smaller than in the native enzyme.

The copper(II)-substituted carbonic anhydrase system also provides a good example of ^{31}P NMR measurements. Here, the interaction of the phosphate buffer, $H_2PO_4^-/HPO_4^{2-}$, with the enzyme is investigated. Only one ^{31}P NMR signal is observed in this case because the interconversion between the two acid-base species is fast on the ^{31}P chemical shift time scale. Large paramagnetic effects on both T_1 and T_2 have demonstrated [57] a direct binding to the metal, and their pH dependence indicates that the predominant interaction is between the $H_2PO_4^-$ species and the low pH form of the enzyme.

6.7 RELEVANT INFORMATION FROM NMR STUDIES OF SMALL COPPER COMPLEXES OF BIOLOGICAL INTEREST

Up till now the discussion has been restricted to the well-defined field of copper proteins; that is, macromolecules where at least one copper(II) ion is tightly bound and possibly capable of interacting with solvent and solute molecules from the bulk solution. The copper protein is thus a rigid system with a well-defined long rotational correlation time, so that the relevant correlation time for the paramagnetic interaction with ligand nuclei is often determined by the electronic relaxation time. Furthermore, it is not necessary to consider the chemical exchange of the copper ion with the bulk solution because the rate of this process, if present, is extremely slow. The only important exchange parameters are those of the ligand of interest, the situation being more complicated only in the presence of two chemical species in equilibrium. Owing to the high affinity of copper for the metal coordination sites in proteins there are no copper-activated enzymes; the most representative paramagnetic metal in this class of enzymes is manganese, which will be discussed in Chapter 8. Nevertheless, the interaction of copper(II) with biological molecules is worthy of investigation in order to further understand the mechanism by which it is transported in tissues, the role it plays in genetics, its possible involvement in redox processes, and its role in redox enzymes and enzyme inhibition in general. A large amount of NMR work has been devoted to the clarification of the interaction of copper(II) with two main classes of small and middle-sized molecules: nucleic acid components, such as polynucleotides, nucleotides, and nucleotide bases, and protein components, such as oligopeptides and amino acids. The interest is mainly focused on the identification of ligand donor atoms in order to abstract information on the coordination behavior in more complex copper-containing systems, such as nucleic acids and proteins. However, such studies sometimes only result in a better knowledge of the coordination compound *per se*, from a purely inorganic point of view.

Owing to the relative ease of obtaining well-resolved proton NMR spectra of simple molecules like nucleotides and oligopeptides, the investigation of their interaction with paramagnetic metal ions dates back more than two decades. In the first studies, analysis of the line broadening of proton resonances was used to propose the donor atoms involved in copper coordination under exchange conditions; water proton relaxation was later employed to obtain qualitative information on the displacement of water molecules from the hexaaqua ion as a consequence of coordination of the molecule of interest. In more recent years, nuclei such as ^{13}C, ^{14}N, ^{15}N, and ^{31}P have been investigated in order to obtain a more detailed picture of the structures of adducts in solution. Rather than giving examples of the many studies existing in the literature, the peculiarities and difficulties of these studies as compared to the applications of NMR discussed in the previous sections will be emphasized.

These types of systems may appear simple, but in most cases they do not involve a single chemical species. There will in general be a number of different complexes in solution, and their relative concentration will vary with that of the metal and ligand, and the pH. The ligands in each particular complex may undergo fast or slow exchange on an NMR time scale, and the temperature dependence of their relaxation rates will not only depend on the exchange regime operative, but also on variations of the equilibrium distribution of all the species involved. Since most of these complexes have relatively low molecular weights, their rotational correlation times are of the order of 10^{-10} s^{-1}—that is, shorter than the copper(II) electronic relaxation time. The dipolar terms of the paramagnetic relaxation rates are therefore dependent on the rotational correlation times, which are expected to differ for each particular species. For example, suppose that on addition of increasing amounts of a potentially multidentate ligand, L, to a copper(II) salt in solution, the number of coordinated water molecules is progressively decreased, owing to the formation of CuL, CuL_2, and CuL_3 species. The water proton relaxation should then decrease as a result of the decreasing number of coordinated water molecules, but should also increase because the rotational correlation time will increase on passing from Cu to CuL, CuL_2, and CuL_3. Actually, the water dependence of ^1H T_1^{-1} on ligand concentration is seldom monotonic. As a result, it will be difficult to guess the hydration number of the intermediate species on the basis of ^1H T_1^{-1} measurements at a single frequency.

If it is assumed, for simplicity, that a single copper complex exists in solution with a ligand that is in fast exchange on the NMR time scale, then at a ligand-to-metal ratio suitable for the NMR spectrum to be recorded, the relaxation parameters of the ligand nuclei should give information on the identity of the donor atoms. Although this is qualitatively true, warnings have been made [58] against attempting to map the ligand arrangement around the metal ion. In fact, unlike the situation encountered in studying copper proteins, in this case the correlation times for the dipolar and contact interactions are not the same; the former is dominated by the rotational time whereas the latter is dominated by the copper electronic relaxation time, which is one to two orders of magnitude longer. Indeed, the contact interaction may partially contribute to the T_1^{-1} values and be the dominant term in T_2^{-1} [58]. In addition, ligand-centered dipolar relaxation may also be sizeable, especially for heteronuclei.

As mentioned before, in real cases the situation is further complicated by the simultaneous presence of more than one complex, with which the investigated ligand in excess is in chemical exchange. It is often assumed that, if only one of the complexes is the dominant species, small fractions of other complexes may be neglected. The existence of one dominant species may be verified by other spectroscopic techniques, like ESR, taking advantage of a different time scale. However, a fundamental breakdown may invalidate this assumption, as has been recently pointed out [59]. If the dominant species, CuL, contains the ligand behaving as a multidentate, the exchange rate of the latter is likely to be extremely low and no paramagnetic contribution from it will be observed in the bulk ligand. On the other hand, a small fraction of copper complexes may exist in which a second ligand molecule behaves as a monodentate, possibly because only one coordination position is left by the presence of the first ligand. In this case, the main contribution to the ligand relaxation will come from the second ligand molecule, and the chemical conclusions will be drawn on a species present only as a minor component. Some striking examples of this behavior have been reported [59].

From the foregoing comments it is clear that these types of studies require the collection of a vast amount of data on each single system before any definitive conclusions can be drawn: several different nuclei should be investigated, through both T_1 and T_2 measurements; concentration, temperature, pH, and possibly frequency dependence of the relaxation parameters should be obtained; parameters such as the rotational correlation times should be checked when possible through independent techniques like EPR lineshape analysis at variable field [60]; and finally the data should be completely analyzed in terms of species distribution.

In conclusion, the available data have allowed some progress in the understanding of metal-polypeptide and metal-nucleotide (nucleoside) interactions. However, future efforts should be especially directed toward those systems, the understanding of which leads to the elucidation of biological processes.

6.8 THE OBSERVATION OF ISOTROPICALLY SHIFTED ^{1}H NMR SIGNALS IN A COPPER(II) PROTEIN

The ^{1}H NMR signals of a copper(II) complex can be observed if the electronic relaxation times of the ion are decreased at least two orders of magnitude through magnetic coupling with a fast relaxing metal ion (see Section 3.8). A typical example is provided by copper-zinc superoxide dismutase (see scheme in Figure 6.9).

If zinc(II) is substituted with cobalt(II), the two metal ions are magnetically coupled and the electronic relaxation time of copper is reduced and approaches that of cobalt(II) (Section 3.8). Under these circumstances ^{1}H NMR signals are observed for ligands coordinated to both metal ions (Figure 6.9) [61]. A tentative assignment is proposed in Figure 6.9 based on the idea that nuclear relaxation rate enhancements are smaller for protons of ligands attached to copper(II) with $S = \frac{1}{2}$ than for protons of ligands attached to cobalt(II) with $S = \frac{3}{2}$. The observation of NMR spectra under the above conditions opens to NMR the investigation of the active site of superoxide dismutase.

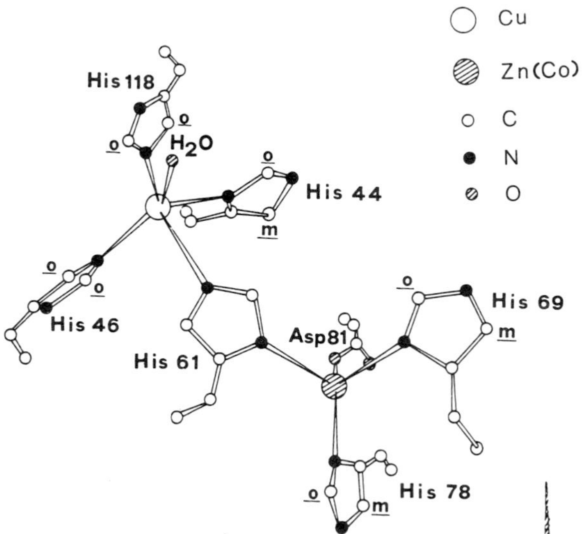

Signal	Linewidth (Hz)	T_1 (ms)	Proposed assignment
A	430	1.1	o-H(Co),o-H(Cu)
B	115	4.1	NH(Cu)
C	240		NH(Cu)
D	295	3.1	m-H(Co)
E	295	3.1	m-H(Co)
F	380		NH(Co)
G	210	2.8	o-H(Cu)
H	280	1.7	o-H(Cu)
I	530	1.4	o-H(Co)
J'	530	1.6	o-H(Co)
J	310		NH(Co)
K	105	4.5	NH(Cu)
L	190	4.2	m-H(Cu)
M	220	2.5	o-H(Cu)
N	220	2.5	o-H(Cu)
O	280	2.4	o-H(Cu)
P	480	1.2	o-H(Cu),m-H(Co)
Q	330	2.2	β-CH_2
R	330	2.2	β-CH_2

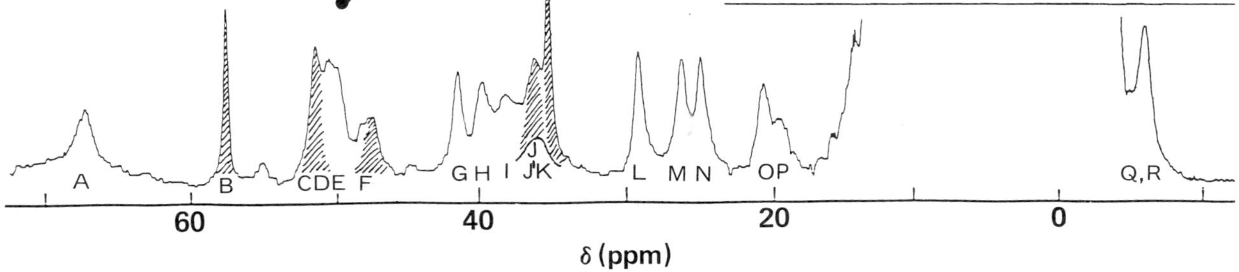

Figure 6.9 300 MHz ^1H NMR spectra of Cu_2Co_2 superoxide dismutase. The dashed signals belong to histidine NH protons; *o* and *m* on histidine rings refer to *ortho-like* and *meta-like* positions with respect to the coordinating nitrogen. T_1 values of the signals and a tentative assignment are also shown. Chemical shifts are in ppm from TMS [61].

REFERENCES

1. Bertini, I., Dei, A., and Scozzafava, A. *Inorg. Chem.* **14** (1975) 1526.
2. Reinhammar, B. In *Advances in Inorganic Biochemistry*, G. L. Eichhorn and L. G. Marzilli (eds.), Elsevier North Holland, Amsterdam, 1979, Volume 1, p. 91.
3. Malkin, R., and Malmström, B. G. In *Advances in Enzymology*, F. F. Nord (ed.), Wiley, New York, 1970, Volume 33, p. 177.
4. Malmström, B. G., Andreasson, L.-E., and Reinhammar, B. In *The Enzymes*, P. D. Boyer (ed.), Academic Press, New York, Third Edition, 1975, Volume XII, p. 507.
5. Gray, H. B., and Solomon, E. I. In *Metal Ions in Biology*, T. G. Spiro (ed.), Wiley, New York, 1981, Volume 3, p. 1.
6. Yokoi, H., and Addison, W. *Inorg. Chem.* **16** (1977) 1341.
7. Bertini, I., Canti, G., Grassi, R., and Scozzafava, A. *Inorg. Chem.* **19** (1980) 2198.
8. Malmström, B. G., and Vänngård, T. *J. Mol. Biol.* **2** (1960) 118.
9. Peisach, J., and Blumberg, W. E. *Arch. Biochem. Biophys.* **165** (1974) 691.
10. Bencini, A., Gatteschi, D., and Zanchini, C. *J. Am. Chem. Soc.* **102** (1980) 5234.
11. Bertini, I., and Scozzafava, A. In *Metal Ions in Biological Systems*, H. Sigel (ed.), Marcel Dekker, New York, 1981, Volume 12, p. 31.
12. Maret, W., Dietrich, H., Ruf, H.-H., and Zeppezauer, M. *J. Inorg. Biochem.* **12** (1980) 241.
13. Connick, R. E., and Wüthrich, K. *J. Chem. Phys.* **51** (1969) 4506.
14. Meiboom, S. *J. Chem. Phys.* **34** (1961) 375.
15. Hausser, R., and Noack, F. *Z. Physik* **182** (1964) 93.
16. Lewis, W. B., Alei, M. Jr., and Morgan, L. O. *J. Chem. Phys.* **44** (1966) 2409.
17. Lewis, W. B., Alei, M. Jr., and Morgan, L. O. *J. Chem. Phys.* **45** (1966) 4003.
18. Hertz, H. G. In *Water: A Comprehensive Treatise*, F. Franks (ed.), Plenum Press, New York, 1973, Volume 3, p. 301.
19. Zukin, R. S., and Hollis, D. P. *J. Biol. Chem.* **250** (1975) 835.
20. Bertini, I., and Luchinat, C. In *Biochemical and Inorganic Aspects of Copper Coordination Chemistry*, K. D. Karlin and J. Zubieta (eds.), Adenine Press, 1985.
21. Bertini, I., Canti, G., and Luchinat, C. *Inorg. Chim. Acta* **56** (1981) 99.
22. Bertini, I., Borghi, E., Canti, G., and Luchinat, C. *J. Inorg. Biochem.* **18** (1983) 221.
23. Bertini, I., Briganti, F., Koenig, S. H., and Luchinat, C. *Biochemistry*, in press.
24. Bertini, I., Luchinat, C., Maldotti, A., Scozzafava, A., and Traverso, O. *Inorg. Chim. Acta* **78** (1983) 19.
25. Bertini, I., Briganti, F., Luchinat, C., Mancini M., and Spina G. *J. Magn. Reson.* **63** (1985) 41.
26. Barker, R., Boden, N., Cayley, G., Charlton, S. C., Henson, R., Holmes, M. C., Kelly, I. D., and Knowles, P. F. *Biochem. J.* **177** (1979) 289.
27. Goldberg, M., Vuk-Pavlović, S., and Pecht, I. *Biochemistry* **19** (1980) 5181.
28. Gaber, B. P., Brown, R. D., Koenig, S. H., and Fee, J. A. *Biochim. Biophys. Acta* **271** (1972) 1.
29. Koenig, S. H., and Brown, R. D. *Ann. N.Y. Acad. Sci. U.S.A.* **222** (1973) 752.
30. Bertini, I., Luchinat, C., Mancini, M., and Spina, G. In *Magneto-Structural Correlations in Exchange Coupled Systems*, R. D. Willett, D. Gatteschi, and O. Kahn (eds.), Reidel, Dordrecht, 1985, p. 421.
31. Banci, L., Bertini, I., and Luchinat, C. *Inorg. Chim. Acta* **100** (1985) 173.
32. Burton, D. R., Forsén, S., Karlström, G., and Dwek, R. A. In *Progress in NMR Spectroscopy*, J. W. Emsley, J. Feeney, and L. H. Sutcliffe (eds.), Pergamon Press, Oxford, 1980, Volume 13, p. 1.

33. Mildvan, A. S., and Gupta, R. K. In *Methods in Enzymology*, C. H. W. Hirs and S. N. Timasheff (eds.), Academic Press, New York, 1978, Volume XLIX, p. 322.
34. Lanir, A., Gradstayn, S., and Navon, G. *Biochemistry* **14** (1975) 242.
35. Koenig, S. H., Brown, R. D. III, and Lindstrom, T. R. *Biophys. J.* **34** (1981) 397.
36. Bertini, I., and Luchinat, C. In *Metal Ions in Biological Systems*, H. Sigel (ed.), Marcel Dekker, New York, 1983, Volume 15, p. 101.
37. Bertini, I., Lanini, G., and Luchinat, C. *J. Mol. Cat.* **23** (1984) 133.
38. Koenig, S. H., Brown, R. D., and Brewer, C. F. *Proc. Natl. Acad. Sci. U.S.A.* **70** (1973) 475.
39. Bertini, I., Borghi, E., Luchinat, C., and Scozzafava, A. *J. Am. Chem. Soc.* **103** (1981) 7779.
40. Terenzi, M., Rigo, A., Franconi, C., Mondovì, B., Calabrese, L., and Rotilio, G. *Biochim. Biophys. Acta* **351** (1974) 230.
41. Chasteen, N. D. *Adv. Inorg. Biochem.* **5** (1983) 202.
42. Andersson, I., Maret, W., Zeppezauer, M., Brown, R. D., and Koenig, S. H. *Biochemistry* **20** (1981) 3424.
43. Reinhammar, B., and Malmström, B. G. In *Metal Ions in Biology*, T. G. Spiro (ed.), Wiley, New York, 1981, Volume 3, p. 109.
44. Adman, E. T., Stenkamp, R. E., Sicker, L. C., Jensen, L. H. *J. Mol. Biol.* **123** (1978) 35.
45. Bertini, I., Luchinat, C., and Messori, L. *Biochem. Biophys. Res. Commun.* **101** (1981) 577.
46. Bertini, I., Canti, G., and Luchinat, C. *Inorg. Chim. Acta* **56** (1981) 1.
47. Tainer, J. A., Getzoff, E. D., Beem, K. M., Richardson, J. S., and Richardson, D. C. *J. Mol. Biol.* **160** (1982) 181.
48. Eisenstadt, M., and Friedman, H. L. *J. Chem. Phys.* **48** (1968) 4445.
49. Kurland, R. J., and Marwedel, B. J. *J. Phys. Chem.* **83** (1979) 1422.
50. Viglino, P., Rigo, A., Stevanato, R., Ranieri, G. A., Rotilio, G., and Calabrese, L. *J. Magn. Reson.* **34** (1979) 265.
51. Marwedel, B. J., Kosman, D. J., Bereman, R. D., and Kurland, R. J. *J. Am. Chem. Soc.* **103** (1981) 2842.
52. Rigo, A., Ugo, P., Viglino, P., and Rotilio, G. *FEBS Lett.* **132** (1981) 78.
53. Rigo, A., Viglino, P., Argese, E., Terenzi, M., and Rotilio, G. *J. Biol. Chem.* **254** (1979) 1759.
54. Bertini, I., Luchinat, C., and Scozzafava, A. *J. Am. Chem. Soc.* **102** (1980) 7349.
55. Simonsson, I., Jonsson, B.-H., and Lindskog, S. *Eur. J. Biochem.* **93** (1979) 409.
56. Bertini, I., Borghi, E., and Luchinat, C. *J. Am. Chem. Soc.* **101** (1979) 7069.
57. Bertini, I., Luchinat, C., and Scozzafava, A. *FEBS Lett.* **93** (1978) 251.
58. Espersen, W. G., and Martin, R. B. *J. Am. Chem. Soc.* **98** (1976) 40.
59. Kuroda, Y., and Aiba, H. *J. Am. Chem. Soc.* **101** (1979) 6837.
60. Basosi, R., Antholine, W. E., Froncisz, W., and Hyde, J. S., *J. Chem. Phys.* **81** (1984) 4849.
61. Bertini, I., Lanini, G., Luchinat, C., Messori, L., Monnanni, R., and Scozzafava, A. *J. Am. Chem. Soc.* **107** (1985) 4931.
62. Note added in proof. A comprehensive treatment of the electronic properties of blue copper proteins is given in: Penfield, K. W., Gewirth, A. A., and Solomon, E. I. *J. Am. Chem. Soc.* **107** (1985) 4519.

CHAPTER 7

Iron

7.1 ELECTRONIC PROPERTIES OF IRON(II) AND IRON(III)

Iron is the most abundant transition metal in the earth's crust; iron compounds are known with oxidation numbers ranging from +2 to +6, the most stable being by far the +2 and +3 states. Its abundance, and the easy availability of at least two different oxidation states, are probably the reasons why iron is also the most abundant transition metal in living organisms, and essential to every form of life. Its presence in living organisms is mainly linked to redox processes involving the +2 and +3 oxidation states. Higher oxidation states may also be reached, at least formally, as intermediates in reactions involving more than one electron, but only iron(II) and iron(III) give rise to stable biological compounds.

Iron(II) has a $3d^6$ electronic configuration, which gives rise to a ^5D free ion ground term (see Section 1.2). Tetrahedral or pseudotetrahedral complexes are always high spin ($S = 2$); octahedral or pseudooctahedral complexes may have either high spin ($S = 2$) or low spin ($S = 0$) electronic configurations depending on the ligand field strength. Temperature-dependent spin equilibria are also observed in intermediate cases. Iron(II) complexes with two unpaired electrons ($S = 1$) may also be observed when the deviation from octahedral symmetry, O_h, is large: if there is a large separation in the energy of the two orbitals derived from the e_g degenerate set in O_h symmetry, the six electrons may adopt a high spin configuration among the lowest four orbitals, giving rise to an $S = 1$ triplet ground state of either ^3E_g or $^3A_{2g}$ symmetry.

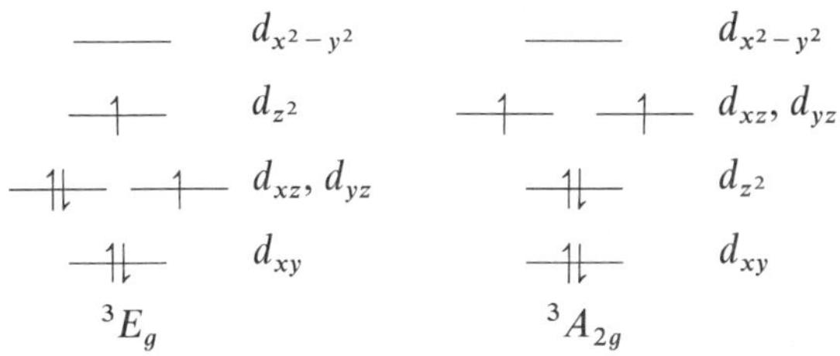

High spin iron(II) complexes that can be referred to as pseudooctahedral are expected to have several energy levels (originating from the $^5T_{2g}$ ground level in O_h symmetry) relatively close to the ground level; as discussed in Section 3.7 the presence of low-lying excited states favors electronic relaxation processes, which are extremely short in this case (10^{-12} to 10^{-13} s). Similar considerations hold for pseudotetrahedral complexes (^5E ground state). Consequently, the nuclear-relaxing efficiency is low and, qualitatively, it may be anticipated that the NMR spectra of iron(II) compounds will be well resolved. For complexes with an intermediate spin ($S = 1$) there are also closely spaced energy levels originating from a $^3T_{1g}$ ground term in the ideal O_h symmetry, or from $^3A_{2g}$ and ^3E_g levels in D_{4h} symmetry, and the observed electronic relaxation times are again short. Low spin iron(II) is diamagnetic and will not be considered further here.

Iron(III) is a d^5 ion with a ^6S free ion ground term. In complexes it gives rise to high ($S = \frac{5}{2}$), low ($S = \frac{1}{2}$), and intermediate ($S = \frac{3}{2}$) spin states. The electronic properties of high spin iron(III) are expected to be similar to those of manganese(II) (see Chapter 8). Having a $^6A_{1g}$ ground state, high spin iron(III) has no first-order zero field splitting. However, spin-orbit coupling with an excited $^4A_{2g}$ level arising from the ^4P free ion term (giving a $^4T_{1g}$ term in O_h symmetry that is further split by lowering the symmetry in $^4A_{2g}$ and ^4E_g levels) is quite efficient in this case owing to the small sextuplet-quadruplet energy separation; this situation is responsible for a relatively large second-order zero field splitting which introduces considerable magnetic anisotropy and reduces the electronic relaxation times well below the values (10^{-8} to 10^{-9} s) observed for manganese(II). The relaxation times for high spin iron(III) complexes can be as low as 10^{-11} s; the NMR signals are expected to be relatively broad, but still observable, provided that the signals are sufficiently spread out by the isotropic shift.

Low spin iron(III) is by far the most thoroughly investigated ion using high resolution NMR: its ground state configuration, $^2T_2$, grants sufficiently short electronic relaxation times. For slowly rotating proteins, the mechanism that limits the resolution of spectra is Curie spin relaxation (see Section 3.4).

7.2 IRON IN BIOLOGICAL SYSTEMS

Besides being the most abundant transition metal in living organisms, iron is also present in a large variety of proteins and enzymes. It is difficult to classify iron-containing systems according to the spectroscopic properties of the iron encountered and, moreover, such a classification would only loosely correspond to the biological functions of the systems. However, at least two classes of complexes may be identified with some confidence: namely, heme proteins and iron-sulfur proteins. As heme proteins are of particular interest for both their NMR implications and biological relevance, they deserve a large share of this chapter.

7.2.1 Heme-Containing Biological Compounds

About 70% of the iron in mammals is present as porphyrin complexes, or hemes, which have the general formula shown in Figure 7.1. The key feature of porphyrin

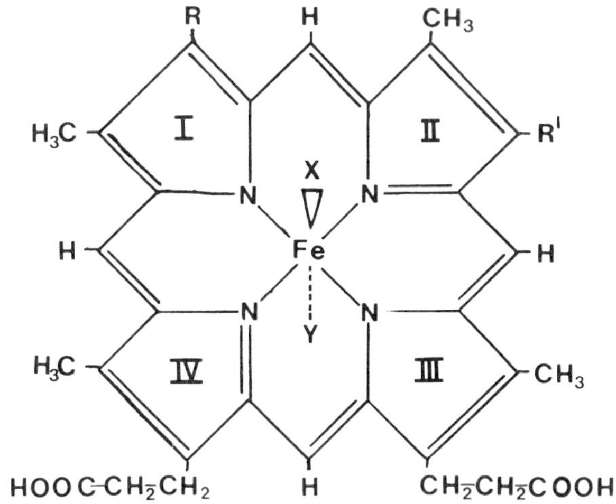

Figure 7.1 General formula for a heme moiety present in natural systems. R, R′, X, and Y groups are reported in Table 7.1.

complexes is the presence of a relatively rigid macrocyclic tetradentate ligand which is substantially planar and allows easy access of monodentate ligands to both the axial coordination positions.

Heme groups are found in a variety of proteins and enzymes. They are held in the host polypeptide by coordinative bonds of amino acid residues in one or both of the axial positions of the iron complex, and may be further linked to the protein by covalent bonds through one or both of the R and R′ groups. The identity of the R and R′ groups, and the ligands in the axial positions, give rise to complexes with different biological functions and spectroscopic properties. This feature has allowed nature to employ heme iron in a variety of biological functions and possibly to trigger them at the molecular level through movements or changes of the axial donor(s). The characteristics of some heme-containing biological compounds are summarized in Table 7.1. The axial ligands, the oxidation number, and the spin state of iron in the compounds listed in Table 7.1 are those encountered in the resting state and are subject to changes during redox processes and/or binding of substrates and inhibitors. All of the reported heme-containing compounds are in fact involved in redox reactions *in vivo*; possible exceptions are myoglobin and hemoglobin, which are oxygen-carrier proteins.

One of the fundamental tasks in living organisms is the transport of oxygen from the respiratory organs to tissues in which metabolic processes are occurring. In most animals, this function is performed by hemoglobin and myoglobin: hemoglobin carries oxygen from the lungs to the tissues via the circulatory system, and myoglobin binds the cellular oxygen and releases it when it is required for metabolic processes. Myoglobin (Table 7.1) is a small monomeric protein with a single hemin group. Hemoglobin, on the other hand, is a tetramer with two pairs of identical subunits labelled α and β; each subunit binds a dioxygen molecule. An important characteristic of hemoglobin is its ability to bind oxygen in a cooperative way;

Table 7.1 Some Selected Heme-Containing Biological Compounds

Protein or Enzyme	MW	Subunits	Heme Groups	Heme Metal	Spin State	Other Metals	Function	R	R′	X	Y
Myoglobin (deoxy)	17,500	1	1	Fe(II)	High	—	Oxygen carrier	$-CH{=}CH_2$	$-CH{=}CH_2$	N(His)	H_2O(?)
Hemoglobin (deoxy)	64,500	4	4	Fe(II)	High	—	Oxygen carrier	$-CH{=}CH_2$	$-CH{=}CH_2$	N(His)	H_2O(?)
Cytochrome *c*	12,500	1	1	Fe(II)/Fe(III)	Low	—	Electron carrier	$-CH(CH_3)-S(Cys)$	$-CH(CH_3)-S(Cys)$	N(His)	S(Met)
Cytochrome b_2, cytochrome b_5	235,000 12,000	4 1	4 1	Fe(II)/Fe(III)	Low	—	Electron carrier	$-CH{=}CH_2$	$-CH{=}CH_2$	N(His)	N(His)
Cytochrome *a* } cytochrome Cytochrome a_3 } oxidase	120,000	7	{1 {1	Fe(II)/Fe(III) Fe(II)/Fe(III)	Low Low	1Cu 1Cu	Redox enzyme	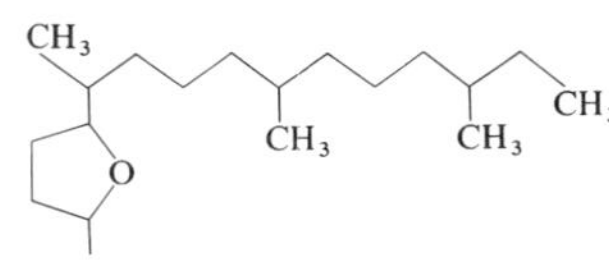	$-CH{=}CH_2$	N(His) N(His)	N(His) (?)
Cytochrome P450 (monooxygenases)	45,000	1	1	Fe(III)	Low	—	Redox enzyme	$-CH{=}CH_2$	$-CH-CH_2$	S(Cys)(?)	(?)
Cytochrome *c*′	24,000	2	2	Fe(III)	High	—	Electron carrier	$-CH(CH_3)-S(Cys)$	$-CH(CH_3)-S(Cys)$	N(His)	(?)
Peroxidase	40,000	1	1	Fe(III)	High	—	Redox enzyme	$-CH{=}CH_2$	$-CH{=}CH_2$	N(His)	H_2O(?)
Catalase	220,000–250,000	4	4	Fe(III)	High	—	Redox enzyme	$-CH{=}CH_2$	$-CH{=}CH_2$	O(Tyr)	H_2O(?)

that is, binding of the first molecule increases the affinity of the other hemes for oxygen, as does the second and third. As a consequence, the binding curve for oxygen, as a function of the partial pressure of oxygen, is steeper for hemoglobin than for myoglobin. Thus, hemoglobin binds oxygen more easily in the lungs and releases it to myoglobin in the tissues where the oxygen pressure is much lower. In both hemoglobin and myoglobin, the iron(II) ion is coordinated by a histidyl imidazole and possibly by a water molecule in the two axial positions. Water, if present, is reversibly substituted by an oxygen molecule. Both proteins can be irreversibly oxidized to met-hemoglobin and met-myoglobin, which contain high spin iron(III).

Cytochromes are electron transport proteins that often have low molecular weight. They are composed of a single polypeptide chain with one hemin group. The redox function of these proteins is performed by the heme iron which during a reversible one-electron reaction passes from iron(III) to iron(II). Although cytochromes are relatively small molecules, an understanding of their detailed biochemical functions is not easy, mainly because they differ from one organism to another and from tissue to tissue in the same species. A broad type of classification divides them into cytochromes b, c_1, and c; the latter reacts with the enzyme cytochrome oxidase which contains two distinct types of hemes named a and a_3. The reduction potential of the Fe^{3+}/Fe^{2+} couple increases in the order cyt $b <$ cyt $c_1 <$ cyt $c <$ cytochrome oxidase. Together, the cytochromes constitute part of the electron transport chain in mammalians, which commences with the catabolic oxidation of succinate or NADH by dehydrogenases; the latter enzymes are then reoxidized by a quinone derivative called coenzyme Q. In this process, coenzyme Q passes two electrons to two cyt b molecules, which are then transferred to cyt c_1, cyt c, and finally to cytochrome oxidase which reduces molecular oxygen to water.

In cyt b, the heme is not covalently linked to the protein, whereas in cyt c and c_1 the R and R′ groups are linked to cysteine residues (Table 7.1). The axial ligands are two histidines in cyt b, and a histidine and a methionine residue in cyt c. A common feature of all the cytochromes seems to be the inaccessibility of the iron ion to solute molecules. Electron transfer may thus occur through the partially exposed heme ring or by outer-sphere mechanisms.

In addition to the a and a_3 heme groups, cytochrome oxidase contains two copper ions, one of type 1 and the other of type 2. The latter is closely connected with the a_3 heme moiety; one of the copper ligands probably bridges it to the heme iron through axial ligation.

Despite their name, cytochromes P-450 are enzymes that catalyze the hydroxylation of C—H bonds, and therefore act as monoxygenases. They are similar to the other cytochromes in that they are single chain proteins with a single heme, but they have a higher molecular weight (Table 7.1). In the resting state, the iron ion is low spin iron(III); one of the axial ligands is a cysteine sulfur.

Heme proteins also include dioxygenases, such as tryptophan-2,3-dioxygenase, which catalyzes the reaction of tryptophan with molecular oxygen to give a product, formylkynurenine, in which both oxygen atoms are incorporated. The enzyme is composed of four subunits, resulting in a relatively high overall molecular weight, and two heme groups. The isolated enzyme contains mostly high spin iron(III), but there is increasing evidence that the iron has to be reduced to high spin iron(II) to be able to bind oxygen (analogously to hemoglobin and myoglobin) in the presence of substrate. Little information is available concerning the nature of the R and R′ groups, and the axial ligands.

Peroxidases and catalases perform the following reaction on peroxide groups:

$$ROOH + AH_2 \rightleftharpoons ROH + A + H_2O$$

In the case of catalase, $ROOH = AH_2 = H_2O_2$ and the reaction is the dismutation of hydrogen peroxide. Both enzymes contain high spin iron(III), which is believed to be the starting species in the catalytic cycle. Several intermediate species, compound I, II, and III, can be identified in the catalytic pathway, each characterized by a different state of the heme group. It has been proposed that compound II contains iron(IV), compound I, formerly iron(V), while compound III is a dioxygen adduct of iron(II). Peroxidase has an histidyl group as one of the axial ligands, whereas the X-ray structure of catalase has identified a tyrosyl oxygen in one of the two axial positions. In both enzymes, the sixth coordination position may be occupied by water.

7.2.2 Nonheme Iron Proteins

Owing to the very low solubility of $Fe(OH)_3$ at physiologic pH, dietary iron must be efficiently sequestered at the level of the intestinal mucosa by strong chelating agents in order to be delivered to the cells, where specific iron proteins and enzymes are synthesized, or to iron storage systems. This task is performed by low molecular weight compounds called siderochromes in fungi and bacteria and by proteins with a molecular weight around 80,000, with the general name of transferrins, in animals. Transferrins have two binding sites that are extremely efficient in coordinating iron(III) near pH 7. In order to bind iron the protein requires bicarbonate (or carbonate) as a synergistic anion. Iron(III) is at least six coordinated with a high spin configuration. Two histidyl nitrogens and two tyrosyl oxygens from the protein are probably involved in metal coordination, in addition to an oxygen of the synergistic ion and possibly a water molecule.

Uteroferrin belongs to the class of acid phosphatases. It contains two iron atoms per molecule which are probably close to each other and antiferromagnetically coupled. In analogy to transferrins, histidines and tyrosines have been proposed as metal ligands. Hemerythrin also contains two iron atoms in close proximity.

The two principal iron storage compounds in the human body are hemosiderin and ferritin. Although very little is known about hemosiderin, ferritin has been characterized in some detail. It has a shell of proteic material (24 identical subunits) surrounding a core of hydrated iron(III) oxide (FeO—OH) microcrystals with iron(III) in an octahedral environment of oxygens. The overall molecular weight is about 900,000, almost half of which is due to the core which contains about 4,300 iron atoms. The core also contains 1% phosphate.

Iron-sulfur proteins are electron transport proteins found in bacteria, plants, and animals. They are all characterized by the presence of iron and sulfur in the chromophore and rather low molecular weights (5,000 to 25,000). With the exception of rubredoxin, which contains a single iron coordinated to four cysteine sulfurs in a pseudotetrahedral arrangement, all the other iron-sulfur proteins also contain inorganic sulfur, often in a 1:1 stoichiometry with respect to iron. The established types of coordination polyhedra are summarized in Figure 7.2. Redox proteins may be isolated and studied both in the reduced and oxidized forms. In the case of rubredoxin, the two forms contain high spin iron(II) and iron(III), respectively. In 2Fe-2S types (ferredoxins), the oxidized form contains two high spin iron(III) centers that are strongly antiferromagnetically coupled; by addition of one electron to the system a Fe(II)-Fe(III) dimer is obtained that exhibits an EPR spectrum still consistent with magnetic coupling. 4Fe-4S types may have three different oxidation states, formally corresponding to $Fe^{III}_3Fe^{II}$, $Fe^{III}_2Fe^{II}_2$, and $Fe^{III}Fe^{II}_3$, respectively. For a given protein, only two states that differ by one electron are easily accessible; the first two oxidation states are characteristic of the so-called high potential iron proteins (HiPIP), whereas proteins with the second two states accessible are called 4Fe-4S ferredoxins. The extent of magnetic coupling and the observability of the EPR spectrum depends on the oxidation state and the type of protein [1]. Recently, other kinds of iron-sulfur clusters, such as 3Fe-4S (Figure 7.2), 3Fe-3S and 5Fe-5S, have been characterized.

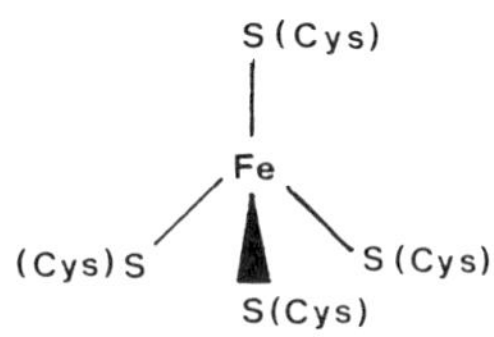

I

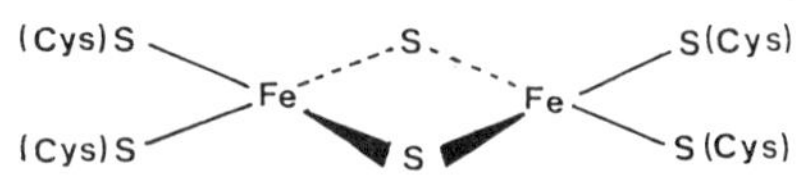

II

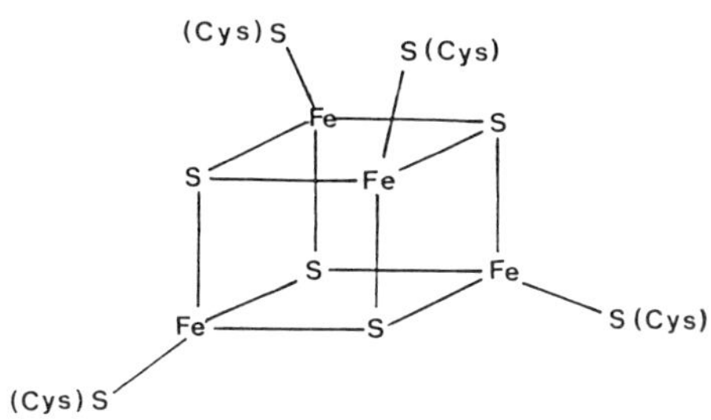

III

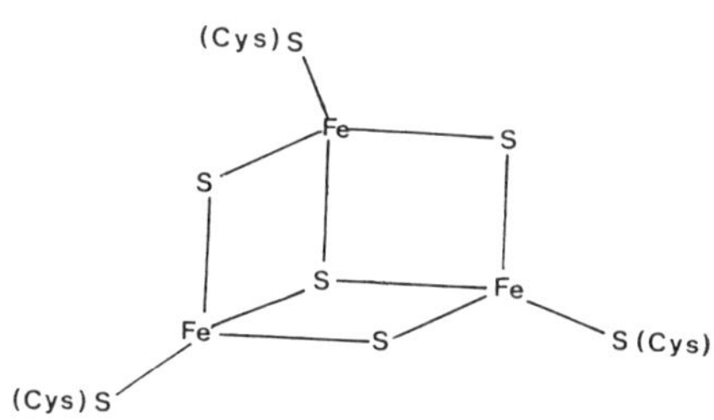

IV

Figure 7.2 Core structures in iron-sulfur proteins.

7.3 ¹H NMR STUDIES OF HEMES

NMR studies of hemoproteins owe much of their success to the vast amount of available literature, which is still growing, for isolated metalloporphyrins, M-P. Such studies have provided the basis for the interpretation, at least qualitatively, of NMR spectra of much more complicated natural systems. As already discussed in Section 7.1, the electronic relaxation times of iron with oxidation numbers +2 and +3, and in every paramagnetic spin state, are such that the NMR lines are expected to be sharp enough to allow detection of the ligand proton signals. Furthermore, the heme moieties contain a number of protons that are ideally situated to give information on the electron delocalization through NMR. For these reasons, the past decade has seen a tremendous output of NMR data on paramagnetic metalloporphyrins and heme-containing proteins, together with a progressive development in the complexity of the investigated systems, the sophistication of the experimental techniques, and the theoretical approaches. Although the nuclear-relaxing properties of heme iron have also been exploited in investigating water and ligand binding, most of the attention has been devoted to the study of the isotropic shifts of the heme protons. The criteria of spectral analysis applied to model compounds will therefore be reviewed before a discussion of the data obtained on biological compounds.

An extensive treatise on porphyrins is presented in *The Porphyrins*, edited by D. Dolphin; Volume 4 contains an excellent review by G. N. La Mar of the NMR properties of paramagnetic metalloporphyrins, updated to about 1977. In addition, Volumes 1 and 2 of the present series, dedicated to iron porphyrins, thoroughly cover the physicochemical properties of these systems. In particular, the chapter by H. M. Goff in Volume 1 deals with the NMR properties in a clear and synthetic way, reviewing data reported up to 1981.

7.3.1 Types of Iron Porphyrins

The porphyrin system is based on a central dinegative anion that is planar with D_4 rotational symmetry. The labelling of the positions is shown in the scheme below; the $\alpha\beta\gamma\delta$ positions are collectively called meso.

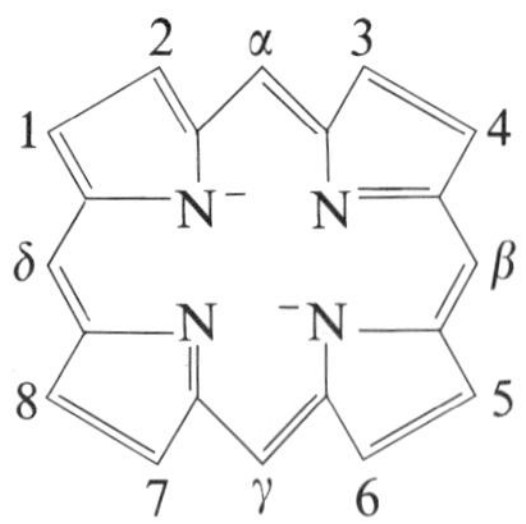

The large variety of natural and artificial porphyrins is due to the presence of different substituents at the twelve available positions.

The π molecular orbitals are responsible for the spectroscopic and redox properties of metalloporphyrins. Filled bonding π molecular orbitals of the dinegative porphinate ion have energies slightly lower than those of metal d orbitals, but relatively close; the lowest antibonding π orbital has also comparable energy. It follows therefore that the energies of the metal d_{xz} and d_{yz} orbitals are extremely dependent on the extent of metal-ligand π bonding [2].

Iron in both the $+2$ and $+3$ formal oxidation states can be coplanar with the ligand when it is four or six coordinated with two axial equivalent bases; when it is five coordinated with an axial ligand the metal ion is removed from the ligand plane by up to 60 pm [3], thus affecting the order of energy levels. Related to this is the overall spin state of the system, according to the classification presented in Section 7.1. In order to briefly discuss the most pertinent cases, metalloporphyrins will be viewed according to their oxidation state.

Iron(II). The simplest iron(II) compounds have the formula Fe(II)-P. They usually have intermediate spin with $S = 1$; however, the five-coordinated adducts have high spin with $S = 2$. Binding of two axial nitrogen bases generally causes spin pairing [4]; this also occurs when one of the two ligands is dioxygen or carbon monoxide [4, 5].

Iron(III). A general scheme of metal d orbital energies is shown below:

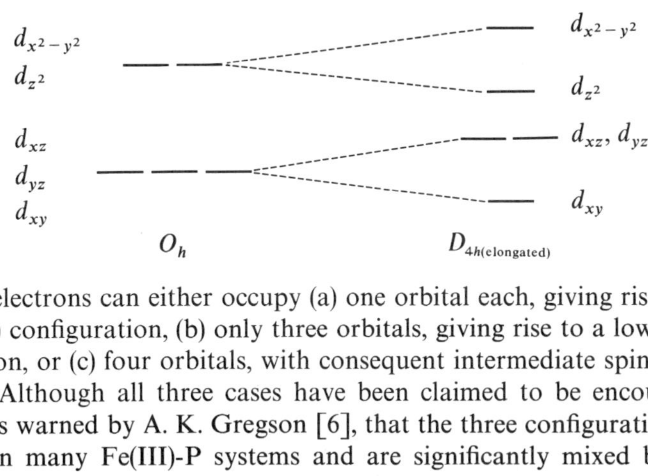

The five d electrons can either occupy (a) one orbital each, giving rise to the high spin ($S = \frac{5}{2}$) configuration, (b) only three orbitals, giving rise to a low spin ($S = \frac{1}{2}$) configuration, or (c) four orbitals, with consequent intermediate spin ($S = \frac{3}{2}$) configuration. Although all three cases have been claimed to be encountered, it is probable, as warned by A. K. Gregson [6], that the three configurations are close in energy in many Fe(III)-P systems and are significantly mixed by spin-orbit coupling. Fortunately, spin-orbit coupling effects are sensitive to very small variations in ligand field parameters, which in turn are sensitive to temperature. Consequently, when the mixing is strong, the magnetic moments and therefore the isotropic shifts are dramatically temperature dependent [6].

In general, the high spin state is associated with coordination of a single weak-field anionic ligand, such as F^-, Cl^-, or Br^- [4], or bis ligation of two weak-field ligands, such as H_2O [7] or tetramethylene sulfoxide [8]. Admixed spin states have been proposed for very weak-field anionic complexes such as ClO_4^- and $C(CN)_3^-$ [9].

Low spin complexes occur when two strong-field donors—imidazole, cyanide, or pyridine—bind the metal ion or when cyanide is one of the two ligands.

Higher Oxidation States. One- or two-electron oxidation products of iron(III) porphyrins are known. One-electron oxidation products can be (1) low spin ($S = \frac{1}{2}$) iron(III) π cation radicals [10]; (2) low spin ($S = 1$) iron(IV) derivatives containing the ferryl, Fe=O, group [11, 12, 152]; or (3) high spin ($S = \frac{5}{2}$) iron(III) π cation radicals [13, 14]. Two-electron oxidation products are usually iron(IV) π cation radicals [15].

The Porphyrin Ligands. A large variety of synthetic porphyrins, substituted at every position, has been studied in order to investigate the factors affecting the isotropic shift patterns and the physicochemical properties. Natural porphyrins have low symmetry; they contain four methyl groups, two propionic acid groups, and two other groups R and R′ (see Figure 7.1). Heme derivatives are bound to proteins through axial donation and hydrophobic interactions; in addition, covalent bonds via the R groups may often be present. Heme-containing molecules with both types of bonding have been isolated and reinserted into the protein. Differences in isotropic shifts have provided useful information on the structure of the heme within the protein.

For a synthetic porphyrin with C_4 symmetry, the z axis is perpendicular to the porphyrin plane and the x and y axes are along the Fe—N bonds. The $d_{x^2-y^2}$ orbital is therefore strongly σ antibonding and the pair of degenerate orbitals, d_{xz} and d_{yz}, of e symmetry are capable of π bonding. Thus, unpaired spin density may reside in a σ molecular orbital or in two sets of ligand π molecular orbitals that are considered to overlap with the metal orbitals, that is, ligand $3e_\pi$ and $4e_\pi$—these represent the highest filled and lowest empty molecular orbitals, respectively. Single occupancy of the $3e_\pi$ molecular orbital may result from ligand-to-metal charge transfer, and vice versa for single occupancy of the $4e_\pi$ molecular orbital. The former molecular orbital places no spin density on the meso position whereas the latter accounts for large spin density on the same position. Both molecular orbitals provide some spin density at the pyrrole positions.

Information on the spin density pattern can be obtained from both ^1H and ^{13}C NMR measurements. Spin density analysis would require simultaneous discussion of both kinds of spectra; however, owing to the greater feasibility of ^1H NMR spectra, the spectra of all hemes will be discussed according to the nucleus of the NMR experiment and not to the spin density pattern.

7.3.2 Electronic Relaxation Properties and NMR Spectra

^1H NMR spectra have been recorded for all of the aforementioned species. Sharp signals have been shown by low spin iron(III), and high and intermediate spin iron(II). For all of these, low-lying excited states provide efficient electronic relaxation pathways. In the case of high spin iron(III), there are no excited states with the same spin multiplicity; therefore, electron relaxation is less efficient. For example, T_{1e} for low spin Fe(III)-P is around 2×10^{-12} s while for high spin Fe(III)-P it is in the range 10^{-10} to 10^{-11} s. La Mar and Walker [16] have proposed that in high spin Fe(III)-P the modulation of zero field splitting (Section 3.7.1) is the predominant mechanism allowing the electron to relax. The $^6A_1$ splits into three Kramers doublets as follows:

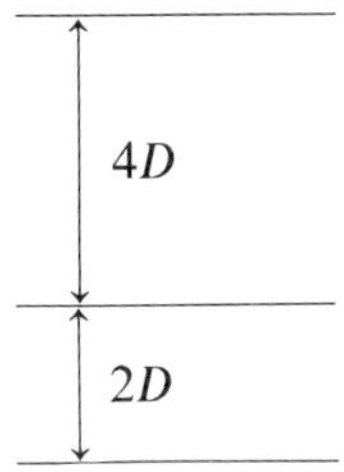

where D is of the order of 10 cm^{-1}. The available D values are consistent with the relationship $T_{1e}^{-1} \propto D^2$ (see Equation 3.27).

A similar mechanism has been proposed to hold for high spin Mn(III)-P [16] (Section 8.8). Fe(II)-P in both $S = 1$ and $S = 2$ states has a $T_{1e} \simeq 10^{-12}$ to 10^{-13} s. Fe(IV)-P ($S = 1$) is somewhat similar to Fe(II)-P and therefore provides sharp signals [11, 17]. Short T_{1e} values are also observed for the systems Fe(III)-π cation-P and Fe(IV)-π cation-P [13–15], probably as a result of the coupling between the spin of the radical and the high spin iron(III) or iron(IV) systems. Coupling between the two systems leads the overall T_{1e} to be equal to or smaller than the shorter T_{1e} (Section 3.8). Similarly, the coupling between the iron centers in μ-oxo P-dimers causes an appreciable decrease in the electronic relaxation times.

The ^{1}H NMR spectra of high spin *meso*-tetraphenylporphyrin iron(III) chloride, TPPFeCl, and its low spin *bis*-imidazole adduct, $TPPFeIm_2^+$, are shown in Figure 7.3 together with the spectra of high spin TPPFe(II)2MeIm and intermediate spin TPPFe(II). The signals are well spread with that due to the pyrrole protons dramatically moving with the oxidation state and spin state of iron. Assignment of the signals is generally performed by analyzing the linewidth pattern, since the dipolar mechanism is the dominant relaxation mechanism. T_1 measurements would also allow a safe check of the assignment [24]. The spectra will now be discussed according to the oxidation state of the metal ion.

7.3.3 High Spin Iron(III)

High spin iron(III) has one unpaired electron in each of the five d orbitals, and so can give rise to both σ and π contact shifts. Furthermore, zero field splitting results in relatively large magnetic anisotropy which is not reflected in the g anisotropy. Typical D values range from a few wave numbers to about 15 cm^{-1}, depending on the heme moiety and the axial ligand. In particular, the latter gives

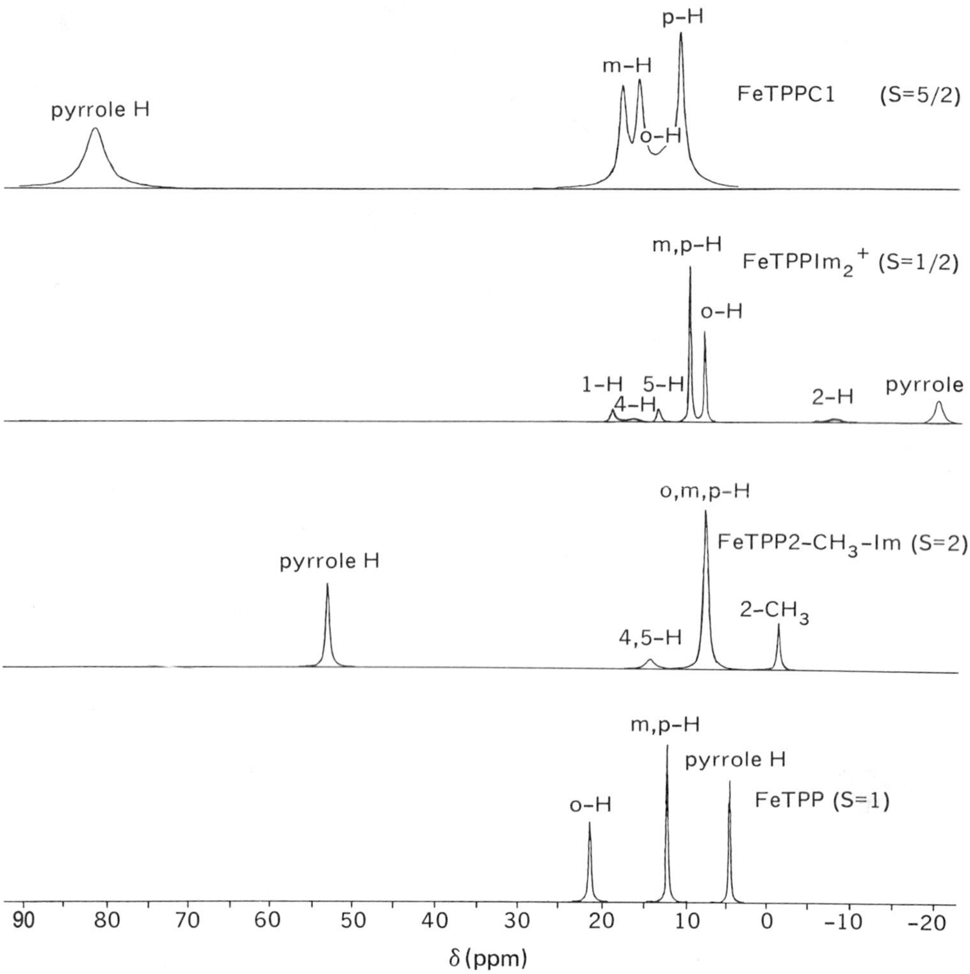

Figure 7.3 ^{1}H NMR spectra of *meso*-tetraphenylporphyrin (TPP) iron complexes [18–23]. Chemical shifts are in ppm from TMS.

rise to increasing D values in the order $I^- \gtrsim Br^- > Cl^- > NCS^- > F^-$. Sizeable dipolar shifts are thus expected. The magnetic susceptibility in such cases depends on T^{-2} at temperatures for which $kT \gg D$; therefore, the dipolar shifts also follow a T^{-2} dependence, whereas the contact shifts essentially follow a T^{-1} dependence (Sections 2.3 and 2.4). La Mar *et al.* have separated contact and dipolar contributions by considering the temperature dependence of the isotropic shifts [25]. Anisotropic magnetic susceptibility data are also available for some Fe(III)-PX compounds, where X is Cl^-, Br^-, I^-, or NCS^-; from these the dipolar shifts have been calculated using Equation (2.18). The contact and dipolar contributions for FeTPPBr [26] are shown in Table 7.2. The splitting of the ortho and meta

Table 7.2 Dipolar and Contact Shifts for FeTPPBr at 293 K Calculated from Single Crystal Magnetic Susceptibility Data [26]

Protons	$\frac{3\cos^2\theta - 1}{r^3}$ ($cm^{-3} \times 10^{-21}$)	Dipolar Shift (ppm)	Contact Shift (ppm)
Pyrrole H	−6.62	10.94	62.38
Meso *o*-H	−4.85	8.02	−6.54
Meso *o*-H′	−1.81	2.99	−5.65
Meso *m*-H	−1.90	3.16	3.56
Meso *m*-H′	−1.28	2.12	3.15
Meso *p*-H	−1.45	2.40	−3.37

proton signals is due to slow rotation of the phenyl group and to the metal ion being above the porphyrin plane. The factorization performed through single crystal magnetic susceptibility data is essentially consistent with that performed from the analysis of the temperature dependence of the isotropic shifts. From the pattern of the contact shifts it appears that some π spin density is present on the phenyl rings as a result of spin density on the meso position, thus suggesting metal-to-ligand unpaired spin transfer. ^{13}C data (see Section 7.4.2) confirm the presence of π spin density, of positive sign, on the meso carbon. A concomitant σ spin density delocalization mechanism accounts for the large downfield shift of the signal due to pyrrole protons. A recent theoretical analysis of spin density has warned about neglecting ligand-centered dipolar shifts [27].

High spin iron(III) complexes of natural hemin derivatives show spectra consistent with the above scheme (Figure 7.4), although their low symmetry may complicate the analysis of the spectral data. The resonances of methyl protons have shifts of the same sign as those for pyrrole protons, confirming the dominance of σ delocalization in the contact shifts [28]. Meso protons are extremely sensitive to the coordination number (Figure 7.4), with shifts being around 50 or −50 ppm for six- or five-coordinated systems, respectively; the methyl/pyrrole proton shift ratio also reflects a change in coordination number, but to a lesser extent. Such differences are ascribed to larger π delocalization contributions in six-coordinated complexes due to coplanarity of the metal with the ligand [28], as confirmed by ^{13}C NMR measurements (see Section 7.4.2).

An understanding of such complexes has been invaluable in the investigation of heme-containing proteins. The ^1H NMR spectra of ferricytochrome *c′* from *Rhodospirillum rubrum* shown in Figure 7.5 are comparable to those of Figure 7.4, and therefore they are considered typical of high spin iron(III). This is true in the pH range 6.7 to 10.7, where two pH-dependent species are observed [29]. According to an early report [30], the low pH form is a quantum mechanical admixture of $S = \frac{3}{2}$ and $S = \frac{5}{2}$, but NMR data, together with Mössbauer [31] and near infrared magnetic circular dichroism [32] data are indicative of an essentially high spin species. The four high intensity signals, represented by c to f of ferricytochrome c′

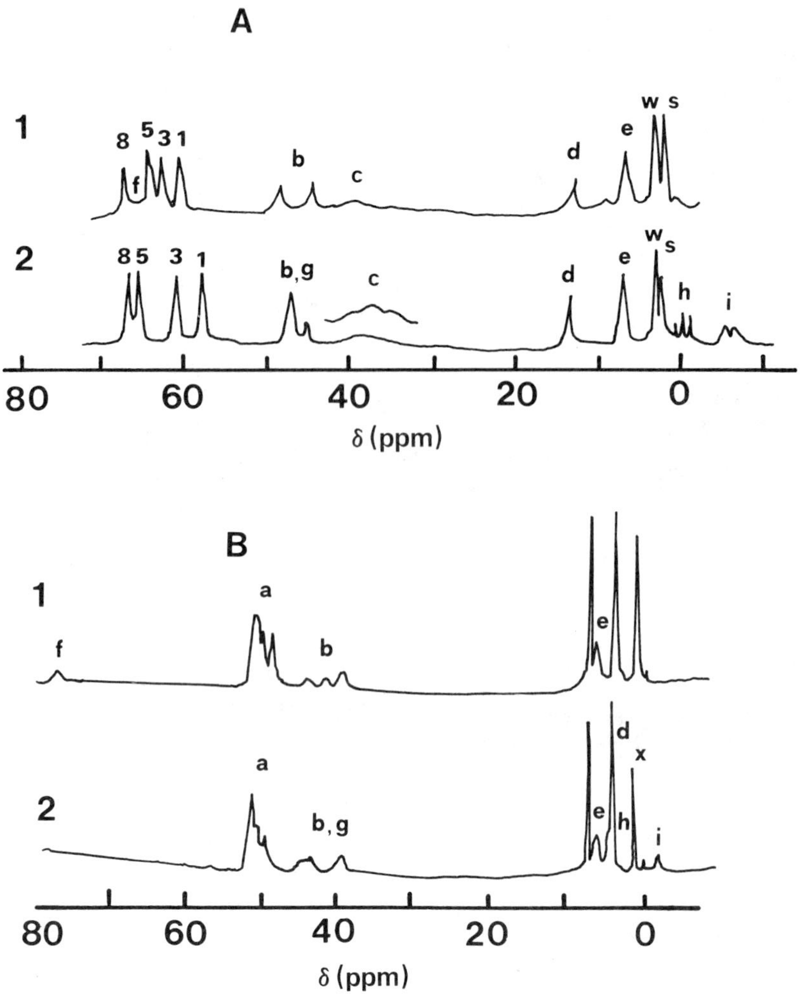

Figure 7.4 270 MHz ^{1}H NMR spectra. **A** Six-coordinated high spin iron(III) complexes of deuteroporphyrin IX, R = R′ = H (1), and protoporphyrin IX, R = R′ = —CH=CH$_2$ (2). **B** Five-coordinated high spin iron(III) complexes of the corresponding dimethylester derivatives—the meso signal at −40 ppm is not shown. The assignment is as follows: a: methyl protons; b: α-CH$_2$; c: meso protons; d: COOH (**A**) or COOCH$_3$ (**B**); e: β-CH$_2$; f: 2,4-H; g: 2,4-H$_\alpha$; h: 2,4-H$_\beta$(cis); i: 2,4-H$_\beta$ (trans); w: water; s: solvent (DMSO); x: impurity [28]. Chemical shifts are in ppm from TMS.

assigned to the methyl protons, while the eight signals, g to n—n is around −20 ppm, but is not shown—with an intensity of unity are collectively ascribed to the two vinyl α protons, the four β-methylene protons from the 6, 7 propionic acid chains, and the two methylene protons of the axial histidine. The histidine ring protons have not been assigned. Resonances a and b have been proposed to arise

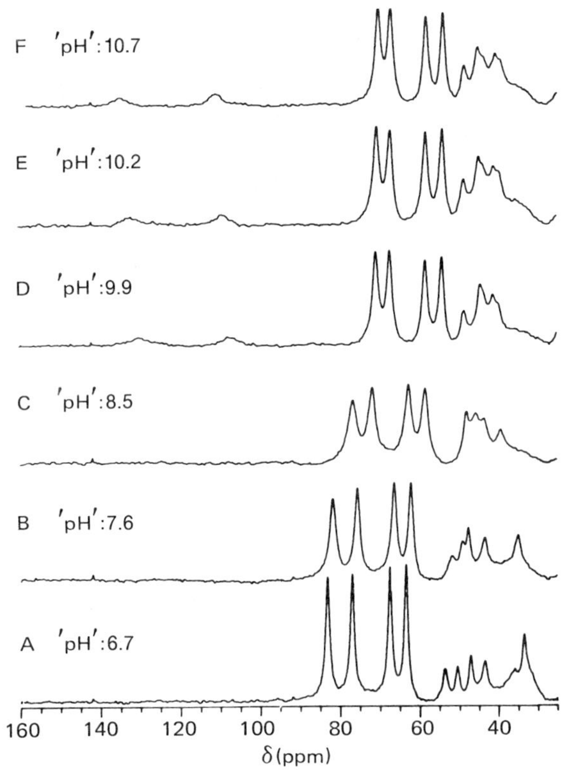

Figure 7.5 360 MHz ^1H NMR spectra of ferricytochrome *c′* in D_2O at different pH values. Chemical shifts are in ppm from DSS. Reprinted from [29] with permission; copyright © 1981 by American Chemical Society.

from a protein residue coordinating as a sixth ligand at high pH; however, subsequent studies on ferricytochrome *c′* from *R. palustris* [33] suggest that these resonances arise from the methylene protons of the only axial histidine of the five coordinated chromophore. The observed dramatic increase in downfield shift could then be explained by an increase in the imidazolate character of such a residue [33] (see Section 7.3.10). Unfortunately, the meso protons, which would have been helpful in discriminating between five and six coordination, have not been identified.

The electronic properties of this system are best investigated by analyzing the field dependence of the ring methyl proton linewidths [29]. The observed quadratic field dependence is a striking example of Curie spin relaxation (see Section 3.4) in high spin iron(III), and is analogous to the findings observed in high spin iron(II) hemoproteins (see Section 7.3.6). The field-independent part of the linewidths has been extracted and used to estimate the electronic relaxation time of the metal ion. Although the result may be affected by a large error owing to neglect

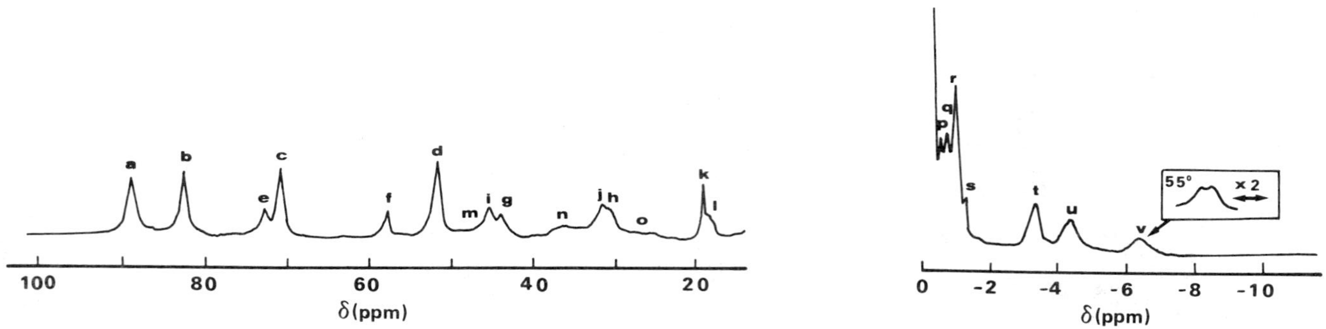

Figure 7.6 360 MHz ^{1}H NMR spectrum of met-myoglobin at 30 °C. Chemical shifts are in ppm from DSS. Reprinted from [35] with permission; copyright c 1980 by American Chemical Society.

of possible ligand-centered contributions (see Sections 3.12 and 7.4.2), a qualitative comparison of the T_2 values for the ring methyl groups of the various pH-dependent species suggests an increase in the electronic relaxation times by a factor of two on passing from the low to the high pH form. This finding may be consistent with the proposed increase in the coordination number from five to six [29, 34], or with an increase in the imidazolate character of the axial histidine [33]. Both are in fact expected to decrease the value of the parameter D and therefore to increase T_{1e} (Equation 3.27). This point will be further discussed in Section 7.7.

The ^{1}H NMR spectrum of met-myoglobin is reported in Figure 7.6. The signals are well resolved [35] as a result of large isotropic shifts. Using the technique of reconstitution with partially deuterated porphyrins, up to twelve signals can be unambiguously assigned [35]: a to d are assigned to the ring methyl groups in the order 8, 5, 3, and 1; m to o to three of the four meso protons; i and j to the two vinyl H_α; v to the two vinyl H_β: k to one H_β of the propionic chains [153]. The temperature dependence of the isotropic shifts only shows very small deviations from Curie behavior, indicating that the dipolar contribution to the isotropic shifts is probably smaller than that found in model complexes. The shifts of the meso protons are indicative of six coordination. Finally, it should be noted that, compared to low spin ferric systems (see Section 7.3.4), the contact shifts are less sensitive to the heme asymmetry because of the large σ contributions to the overall unpaired spin density on the heme ring [28].

7.3.4 Low Spin Iron(III)

In tetragonal iron(III) porphyrins the unpaired electron presumably resides in the d_{xz}, d_{yz} orbitals:

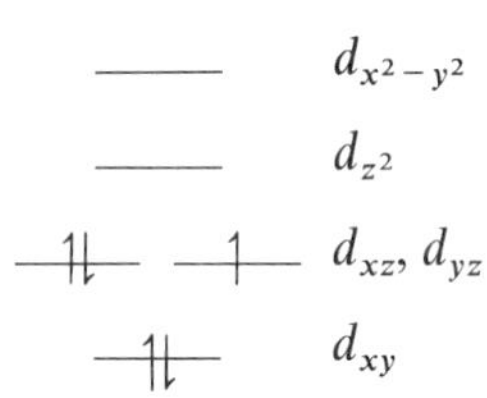

At first glance no σ spin delocalization is predicted, since the unpaired electron resides in a π orbital. Low temperature EPR spectra provide three g values which can be used to estimate the dipolar shifts. In some cases, it has been suggested that the two in-plane g values average to a single value at room temperature, probably due to the Jahn-Teller effect on the ^2E ground state [36] (dynamical Jahn-Teller effect). The error involved in using the g values, rather than the χ values, is estimated to range from 10 to 15% [37]. In several cases, it has been assumed and proved *a posteriori* that the isotropic shift in meso aryl groups of meso aryl substituted porphyrins is only due to dipolar shifts [20, 38]. Indeed, even in the high spin iron(III) derivatives, the phenyl rings of TPP experience relatively small contact shifts due to orthogonality of the phenyl and porphyrin planes. By making such an assumption, the authors have calculated the dipolar shifts for the pyrrole protons or substituents and then determined the contact

Table 7.3 Dipolar and Contact Contributions to the Isotropic Shifts for Low Spin $FeP(Im)_2$ Derivatives [20, 38]

Position	Isotropic Shift (ppm)	Dipolar Shift (ppm)[a]	Contact Shift (ppm)[a]
Meso *o*-H[b]	−3.09	−3.09	0
Meso *m*-H[b]	−1.49	−1.44	~0
Meso *p*-H[b]	−1.37	−1.27	~0
Meso *p*-CH_3[b]	−0.94	−0.94	0
Pyrrole H[b,c]	−25.4	−5.8	−19.5
Meso H[d]	−7.0	−9.3	2.3
Meso α-CH_2[c]	0.6	−4.5	5.1
Pyrrole α-CH_2[d]	2.0	−3.2	5.2
1-H(pyrrole-N)[e]	~2	−9.6	11.6
1-CH_3[e]	17.2	10.3	6.9
2-H[e]	−9.5	−28.0	18.5
2-CH_3[e]	~12	~12	~0
4-H[e]	9.7	−8.2	17.9
5-H[e]	4.0	−7.6	11.6
5-CH_3[e]	15.5	9.0	6.5

[a] Values in semiquantitative agreement would be found by using $2g_{zz}^2 - (g_{xx}^2 + g_{yy}^2)$ and Equation (2.19).

[b] For $FeTPP(Im)_2^+$ and $FeT\text{-}p\text{-}CH_3PP(Im)_2^+$.

[c] For $FeT\text{-}n\text{-}PrP(Im)_2^+$.

[d] For $FeOEP(Im)_2^+$.

[e] For $FeTPP(CH_3Im)_2^+$.

shifts (Table 7.3). In most cases the factorization of the isotropic shifts based on g anisotropy has provided results that are in satisfactory agreement with this procedure. Apparently, g anisotropy is strongly dependent on many factors, including the solvent, the ionic strength, and the temperature, and this results in some uncertainty in the factorization.

The spin density pattern, with small spin density at the meso positions and larger at the pyrrole positions, is consistent with a π delocalization in the $3e_\pi$ molecular orbitals, arising from ligand-to-metal π bonding. Binding of bases in axial positions in proteins or under restriction of rotation splits the $3e_\pi$ orbitals making the pyrrole rings nonequivalent pairwise [39] (Figure 7.7). Figure 7.7 is a general model for low spin iron(III) heme proteins, both in terms of shift pattern and asymmetry generated by restricted rotation of the axial base.

The proton signals of the axial ligands imidazole, pyridine, and their derivatives display sizeable dipolar and contact shifts [20] (Table 7.3). The Im shifts are upfield for ring hydrogens and downfield for CH_3 substituents, and are consistent with an unpaired electron being delocalized into π ligand orbitals. It is proposed

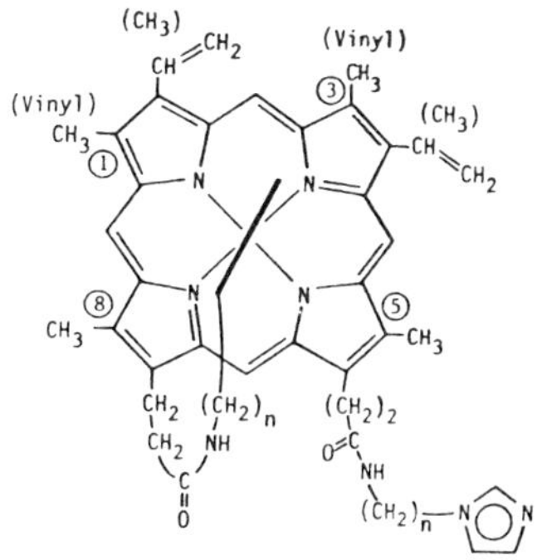

n = 3

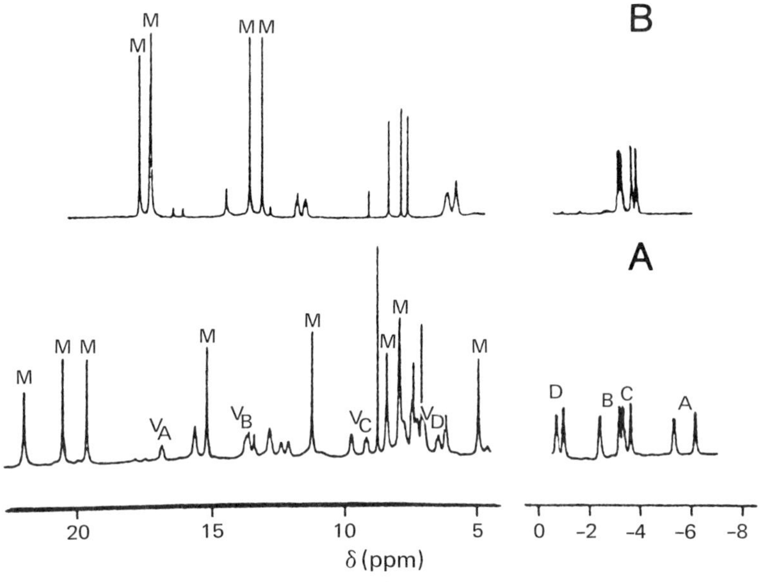

Figure 7.7 220 MHz ^{1}H NMR spectra of monocyanide iron(III) complexes. **A** Bischelated hemin (scheme). **B** 1-CH_3-imidazole complex. M: methyl protons; V_{A-D}: methine protons; A–D: methylene protons. Chemical shifts are in ppm from TMS. Reprinted from [39], with permission; copyright c 1980 by American Chemical Society.

that the unpaired electron is mainly delocalized into the lowest filled π ligand orbital [20].

The following discussion deals mainly with cytochromes *c* and cyanomet-myoglobin and hemoglobin; recently, the ^{1}H NMR spectra of the cyanide adduct of horseradish peroxidase have been reported [40].

The NMR spectra of the heme group of oxidized horse heart cytochrome *c* are shown in Figure 7.8 [41, 42]. The heme moiety, where $R = R' =$ —CH(CH_3)—S—(Cys) in Figure 7.1, is linked to the protein by thioether bridges

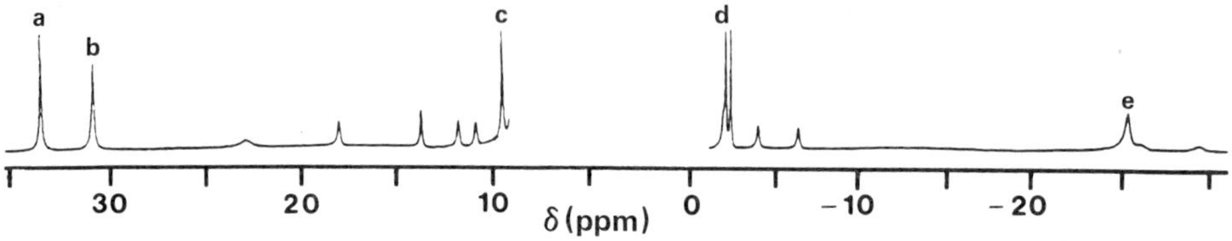

Figure 7.8 360 MHz ^1H NMR spectra of oxidized Horse Heart Cytochrome c. (L. S. Fe^{III}) [41, 42]. The labeled signals are assigned (see text) to: a = 8-CH_3, b = 3-CH_3, c = 5-CH_3, d = thioether bridge 2-CH_3, e = axial methionine S-CH_3; the resonances at 7.4 ppm (1-CH_3) and 3.1 ppm (thioether bridge 4-CH_3) are not shown. Chemical shifts are in ppm from DSS.

through Cys 14 and 17; the axial ligands are His 18 and Met 80 (Table 7.1). If this spectrum is compared to those of synthetic and natural porphyrins (Figure 7.7, for example) the features characteristic of a low spin Fe(III) system can be recognized; that is, relatively sharp signals and a narrow range of isotropic shifts. On the basis of their intensity, the four signals at about 34, 31, 10, and 7 ppm downfield (the latter not shown) are assigned to the ring methyl groups. The assignment is reported in Figure 7.8 and performed as explained in Section 7.3.9). A noticeable feature of this spectrum is the larger spreading of the ring methyl resonances with respect to isolated low spin Fe(III) porphyrins. Attempts to factorize dipolar and contact contributions to the isotropic shifts are available in the literature. An estimate of such contributions is reported in Table 7.4 according to a calculation by Horrocks and Greenberg [37] using Equation (2.18). The princi-

Table 7.4 Separation of Dipolar and Contact Shift Contributions for Some Heme and Axial Ligand Protons in Horse Heart Ferricytochrome *c* [37]

Proton[a]	$\left(\frac{\Delta\nu}{\nu_0}\right)_{ax}^{dip\,b}$	$\left(\frac{\Delta\nu}{\nu_0}\right)_{eq}^{dip\,c}$	$\left(\frac{\Delta\nu}{\nu_0}\right)^{con}$
8-CH_3	−3.9	2.9	32.9
3-CH_3	−3.9	2.3	29.1
5-CH_3	−3.9	−2.3	13.0
1-CH_3	−3.9	−2.3	10.0
Meso H(β, δ)	−10.1	1.4	
Meso H(α, γ)	−10.1	−1.4	−3.4, −5.6
Im 2-H	19.9	−1.5	
Im 5-H	18.9	−1.4	
Im N-H	12.8	−0.1	−8.1
Met S-CH_3	17.8	−1.2	−36.8

[a] The assignment is taken from reference [42].

[b] Contribution from the $3\cos^2\theta - 1$ term in Equation 2.18.

[c] Contribution from the $\sin^2\theta \cos 2\Omega$ term in Equation 2.18.

pal susceptibility values were calculated, the only check being the reproducibility of the experimental g values. Despite such a limitation it appears that the dipolar shifts are relatively small for all but the meso positions. The spreading of the ring methyl shifts is therefore due mainly to differences in the contact contributions, which may be ascribed to a protein-induced asymmetry of the unpaired electron spin density distribution in the heme group. This peculiarity was recognized in early studies of cytochrome c obtained from several different sources [43], and its possible relevance in biological redox functions hypothesized. On the basis of the complete assignment of the four ring methyl signals in the paramagnetic species (Table 7.5), it appears that methyl 3 and 8, which are attached to the pyrrole rings II and IV opposite to each other (Figure 7.1), experience larger isotropic shifts, probably because of larger contact contributions. The spin density on the ring carbon atoms 3 and 8 is estimated to be about four times larger than that on positions 1 and 5.

The origin of the asymmetric electron spin density in cytochrome c has been the subject of many investigations. To clarify the problem, the NMR spectra of cytochrome c derivatives in which the axial methionine is replaced by ligands like CN^-, N_3^-, and pyridine have been investigated; parallel studies have also been

Table 7.5 Resonance Assignments for Heme c in Horse Heart Cytochrome c [42][a]

	Ferrocytochrome c		Ferricytochrome c
Resonance Assignment	Chemical Shift (ppm from TSP[b] at 52 °C)	Irradiation Causes Negative NOE on the Following Heme c Resonances	Chemical Shift (ppm from TSP[b] at 35 °C)
Meso H(α)	9.32	3-CH_3, methine and methyl of thioether 2	
Meso H(β)	9.59	5-CH_3, methine and methyl of thioether 4	
Meso H(γ)	9.62		
Meso H(δ)	9.04	1,8-CH_3	
1-CH_3	3.52	Meso δ	7.4
3-CH_3	3.87	Meso α	31.2
5-CH_3	3.61	Meso β	10.5
8-CH_3	2.19	Meso δ	33.9
Thioether bridge 2:	5.24	Meso α	
methine-CH_3	1.53	1-CH_3, meso α	−2.1
Thioether bridge 4:	6.34	Meso β	
methine-CH_3	2.59	3-CH_3, meso β	3.1

[a] Nuclear Overhauser enhancements (NOE) observed upon homonuclear double irradiation of heme c resonance lines, on which the resonance assignments were based, are also indicated.

[b] TSP: sodium 3-trimethylsilyl-[2,2,3,3-$^2H_4$]propionate.

performed on the isolated octapeptide (Figure 7.9) adducts using the same anions or *N*-acetyl-*DL*-methionine [44] as the sixth ligand.

The pattern of the individual ring methyl signals for the octapeptide adducts, as well as for the cytochrome *c* derivatives, is invariably 5, 8, 1, and 3 going upfield, in contrast with the 8, 3, 5, 1 order (Table 7.5) in native cytochrome *c*. The isolated octapeptide shares a number of features with the native heme (Figure 7.9), including the orientation of the axial histidine ring. The only characteristic that cannot be reproduced, apart from the influence of the surrounding protein, is the fixed orientation of the axial methionine ligand. The latter has therefore been proposed as the source of the peculiar isotropic shift pattern [44].

Similar studies have been subsequently performed on cytochromes *c* [45], c_{551} [46–48], c_{552} [49, 50], c_{553} [51], c_2 [48, 52], c_{550} [53], c_{554} [54], and c_{555} [55]. Some behave like horse cytochrome *c*, but others show an opposite trend

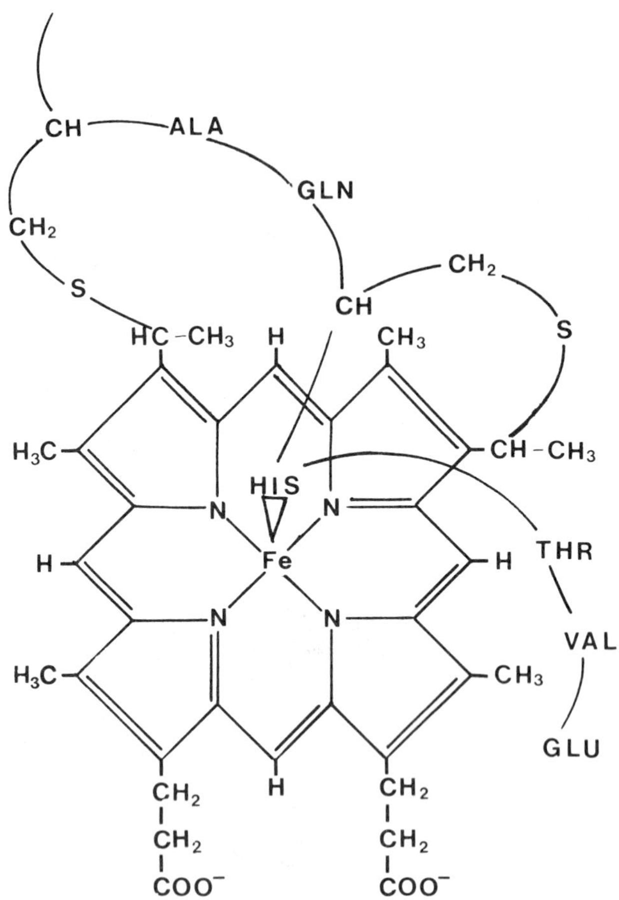

Figure 7.9 Schematic diagram of the cytochrome *c* heme octapeptide.

of the ring methyl shifts, corresponding to a 90° rotation of the x and y axes in the plane, which is due to opposite chirality of the methionine ligand. Some results are summarized in Table 7.6. Interestingly, the highest spin density found in all the systems investigated was in the pyrrole ring exposed to the solvent. The ring methyl shifts in *C. thiosulfatophilum* cytochrome c_{555} show patterns and magnitudes that deviate from those expected from the above considerations and have been tentatively attributed to an intermediate orientation of the sulfur lone pair in between two pyrrole nitrogens [55]. This methionine coordination geometry has been also observed in *D. vulgaris* cytochrome c_{553} [56].

The longitudinal relaxation of the ring methyl protons in horse heart cytochrome c has been found to deviate from a simple exponential behavior when measured in the presence of small amounts of cytochrome oxidase. This has been ascribed to specific binding of the molecule to the solvent-exposed edge of cytochrome c [57], which would thus be the electron transfer site.

In addition to the heme signals, other resonances from the protein can also be weakly isotropically shifted by the presence of the paramagnetic center, and their analysis can help detecting subtle changes in the protein conformation, not apparent from the X-ray structure. Such studies are intermediate between paramagnetic NMR and very high resolution NMR, in that the signals of interest may even fall within the diamagnetic spectral region, and relatively small chemical shift changes have to be detected, for instance, on passing from the oxidized to the reduced diamagnetic species. This was the case for horse heart cytochrome c for which careful studies [58–62] have led to the individual assignment of a large number of signals in both the reduced and oxidized protein. An analysis of the temperature dependence has revealed weak dipolar shift contributions for some of them; it has been proposed that a small conformational change occurs in a well-defined region of the protein upon change in oxidation state [154].

Somewhat related to this type of study is the characterization of a conformational heterogeneity in ferricytochrome c. The investigation was stimulated by the observation that the 3-CH_3 signal linewidth displayed an unusual temperature, pH, and magnetic field dependence [41]. The phenomenon was attributed to the presence of two possible conformations for the Phe 82 residue, which is in close contact with the methyl group. A split 3-CH_3 signal can indeed be observed at low temperature. Furthermore, the same behavior is found in cytochrome c obtained from other species which, like the horse protein, have a Phe residue in position 46; but no such effects are observed in those whose cytochrome c, like that from tuna, have a Tyr residue in position 46 [41].

Striking examples of heterogeneity have been revealed by NMR studies on native and heme-reconstituted cytochromes. For example, it has long been known that native cytochrome b_5 exists in two interconvertible forms: the minor component can be detected through NMR [63] but not by the X-ray structure. The heme group in cytochrome b_5 is not covalently attached to the protein as it is in cytochrome c. By reconstituting the protein with several partially deuterated porphyrins a number of signals can be assigned [64]. The results of such a study are

Table 7.6 Isotropic Shifts (ppm) of Heme Protons in Low Spin Ferricytochromes *c*

	Cyt *c*		Cyt c_2	Cyt c_{552}	Cyt c_{551}					Cyt c_{555}
Resonance Assignment	Horse Heart [42]	*S. cerevisiae* [45]	*R. rubrum* [48]	*E. gracilis* [49]	*P. aeruginosa* [49]	*P. mendocina* [48]	*P. stutzeri* [48]	*R. gelatinosa* [47]		*C. thiosulfatophilum* [55]
								iso1	iso2	
1-CH_3	3.9	4.4	6.8	9.5	21.1	18.2	18.4	22.4	26.3	24.1
3-CH_3	27.3	26.7	26.0	29.0	9.6	10.1	8.5	8.4	6.7	30.1
5-CH_3	6.9	7.6	11.8	13.8	28.8	28.0	28.7	28.9	32.3	36.0
8-CH_3	31.7	31.6	32.5	35.0	14.4	14.5	14.1	18.1	13.7	48.9
Thioether bridge: 2-CH_3	−3.6			−1.9	0.8					
Thioether bridge: 4-CH_3	0.5			0.7	−2.5					

shown in Figure 7.10 and Table 7.7. By comparing Figure 7.10A and 7.10B it appears that a larger share of the component Y is present immediately after reconstitution of the protein with protoporphyrin IX. Furthermore, the pattern of the isotropic shifts of the methyl protons indicates that the two components differ by a 180° rotation of the heme group about the α-γ meso axis. The different

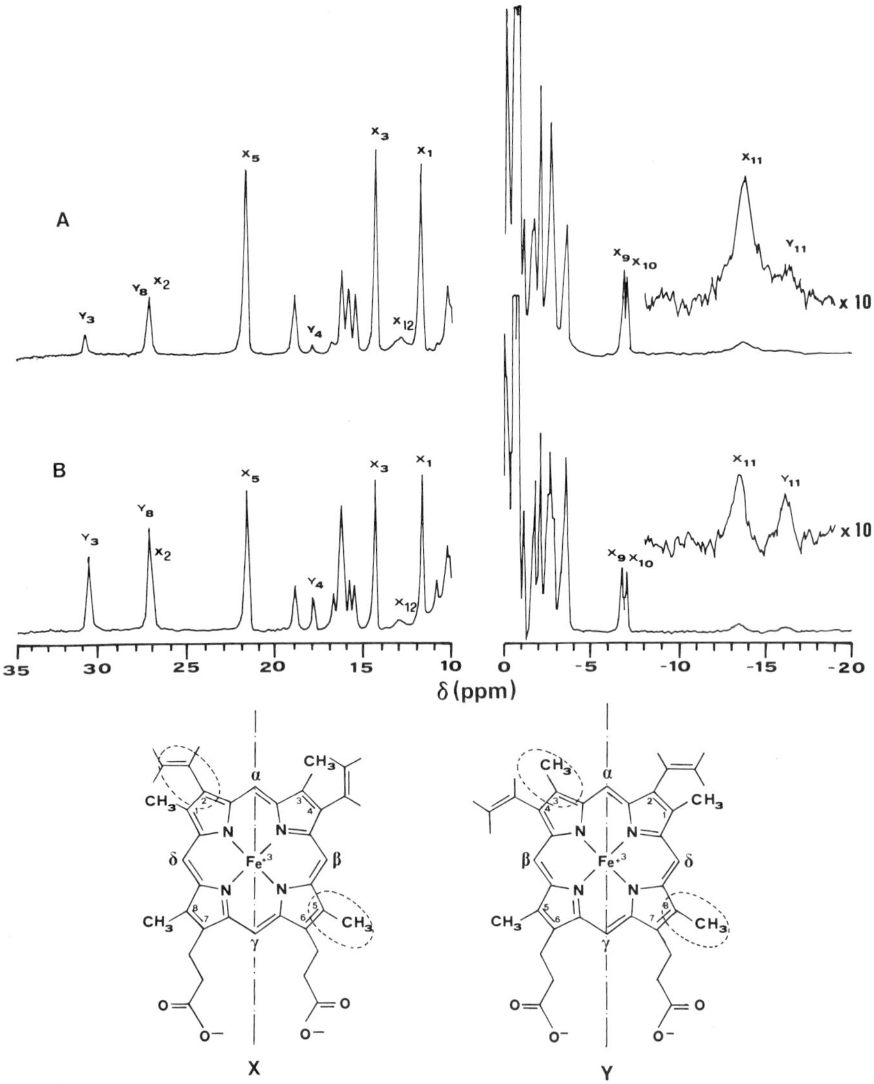

Figure 7.10 360 MHz ¹H NMR spectra of cytochrome b_5. **A**: native and **B**: freshly reconstituted with protoporphyrin IX. The proposed orientation of the heme in the major (X) and minor (Y) components is also shown [64]. Chemical shifts are in ppm from DSS.

Table 7.7 Chemical Shift and Assignment of Some Heme Signals from the Major and Minor Components of Cytochrome b_5[a]

Resonance Assignment	Chemical Shifts (ppm from DSS)	
	Major	Minor
1-CH_3	11.7 (X_1)	[b]
3-CH_3	14.4 (X_3)	30.8 (Y_3)
5-CH_3	21.8 (X_5)	[b]
8-CH_3	[b]	27.3 (Y_8)
2-$C^{\alpha}H$	27.1 (X_2)	[b]
2-$C^{\beta}H$ (cis)	−6.8 (X_9)	[b]
2-$C^{\beta}H$ (trans)	−7.0 (X_{10})	[b]
4-$C^{\alpha}H$	[b]	18.0 (Y_4)
4-$C^{\beta}H$	[b]	[b]
4-$C^{\beta}H$	[b]	[b]
2-H		
4-H		
Im 2-H[c]	13.6 (X_{11})	−16.1 (Y_{11})
Im 4-H[c]	12.9 (X_{12})	[b]

[a] Peak labelling as shown in Figure 7.10 [64].

[b] Not located.

[c] Tentative.

stability of the two conformers depends on the different interactions of the two vinyl groups and the 1- and 3-CH_3 groups in the two cases.

As a further example of low spin iron(III) the spectrum of cyanometmyoglobin is reported in Figure 7.11A. The spectrum is typical of this class of heme compounds, with well-resolved and relatively sharp resonances in the region $+26 \div -10$ ppm. The signals a, b, and e have been assigned to the 5-CH_3, 1-CH_3, and 8-CH_3 groups, respectively (see Section 7.3.9) [66]. The application of isotope labelling, NOE, and T_1 measurements permitted the assignment of several other signals [67, 68]: for example, signal *c* belongs to the 2-vinyl H_{α}, *l* to the 2-vinyl H_{β} (*cis*), *m* to the 2-vinyl H_{β} (*trans*), *n*, *o*, and *p* to the γ_2-CH_3, δ-CH_3, and γ_1-CH of the nearby Ile 99 residue, respectively. Among the exchangeable protons, signal *r* is assigned to the NH proton of the proximal histidine, and signal *q* to that of the distal histidine [65]. Heme heterogeneity has also been detected in this derivative upon reconstitution [69]. As in the case of cytochrome *c*, the dipolar contribution to the ring protons' shifts has been calculated [37] and found to be relatively small. The extent of the heme proton shift has been used to compare the strength of the axial ligand field in leg-hemoglobin cyanide and myoglobin cyanide. The former shows very much smaller hyperfine shifts consistent with a stronger axial field [70].

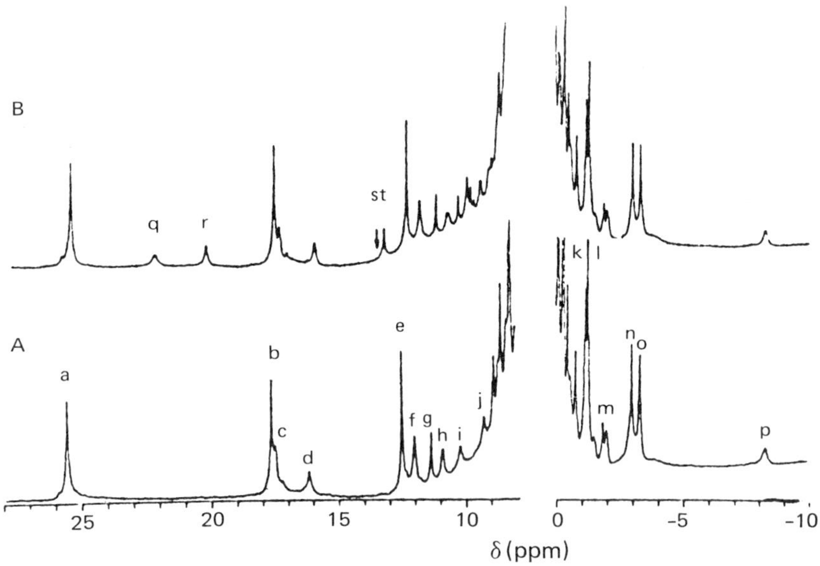

Figure 7.11 360 MHz ¹H NMR spectrum of sperm whale cyanometmyoglobin in **A** D_2O and **B** H_2O. Resonances labelled q to t arise from exchangeable protons. Chemical shifts are in ppm from DSS. Reprinted from [65], with permission; copyright c 1981 by American Chemical Society.

Usually, no field dependence of the linewidths is observed in low spin iron(III) heme proteins. This is explained by the fact that the Curie relaxation term is proportional to $S^2(S+1)^2$, which is 136 times smaller for an $S = \frac{1}{2}$ system compared to the $S = \frac{5}{2}$ systems discussed above.

7.3.5 Spin-Admixed Fe(III)-P and High Spin-Low Spin Equilibria

As already mentioned, high spin Fe(III)-PX compounds, where X is a very weak anion, have spin admixed ground states of $S = \frac{5}{2}$ and $S = \frac{3}{2}$ character. The pure $S = \frac{5}{2}$ and $S = \frac{3}{2}$ states are closely spaced [71], so spin-orbit coupling provides a mechanism for mixing them. The difference between the two spin states lies in the fact that the $d_{x^2-y^2}$ orbital is empty for $S = \frac{3}{2}$ species. Since unpaired spin in the $d_{x^2-y^2}$ orbital is associated with predominant σ spin delocalization and downfield pyrrole proton isotropic shifts, whereas unpaired spin in d_{xz} and d_{yz} orbitals results in upfield pyrrole proton isotropic shifts through π spin delocalization, proton NMR resonances are quite sensitive to the $S = \frac{5}{2}$ and $S = \frac{3}{2}$ contributions in a spin-admixed complex. In addition, temperature changes affect the mixing of the two states with dramatic effects on the pyrrole proton shifts [9]. A decrease in the magnitude of the isotropic shift with decreasing temperature is shown in Figure 7.12 for the pyrrole protons; this finding is contrary to that expected on

Figure 7.12 Temperature dependence of the isotropic shifts of pyrrole proton resonance of $FeTPPSO_3CF_3$ (●) [9] and β-pyridine proton resonance of $FeOEP(4\text{-ac-py})_2ClO_4$ (○) [6].

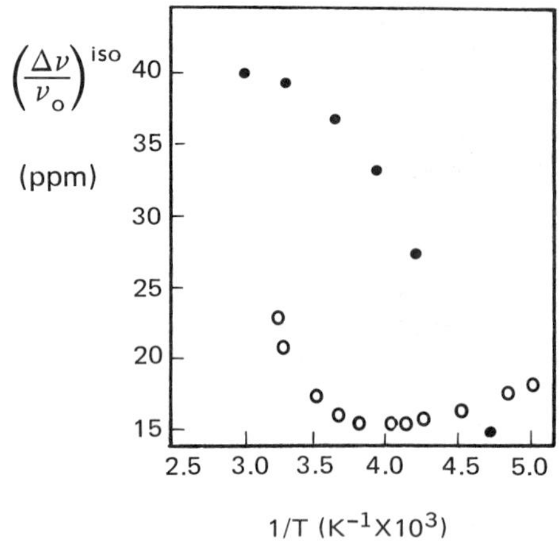

the basis of Equations (2.10) and (2.20). For a series of pyrrole-substituted octaethylporphyrins with formula $FeOEPX_2ClO_4$, where X is substituted pyridines, the magnetic and isotropic shifts patterns have been accounted for by considering spin admixing between $S = \frac{5}{2}$, $S = \frac{3}{2}$, and $S = \frac{1}{2}$ states [6].

The spectrum of horseradish peroxidase (HRP) is reported in Figure 7.13: assignment of the signals was performed through substitution with deuterated porphyrins [72]. The general pattern of the isotropic shifts is indicative of high spin iron(III); however, when the protein is substituted with deuterohemin, in which the two vinyl groups are substituted by pyrrole protons (R = R′ = H, Figure 7.1), the signals of the pyrrole protons are not observed in the usual region around 60 to 80 ppm downfield, but most probably lie within the diamagnetic re-

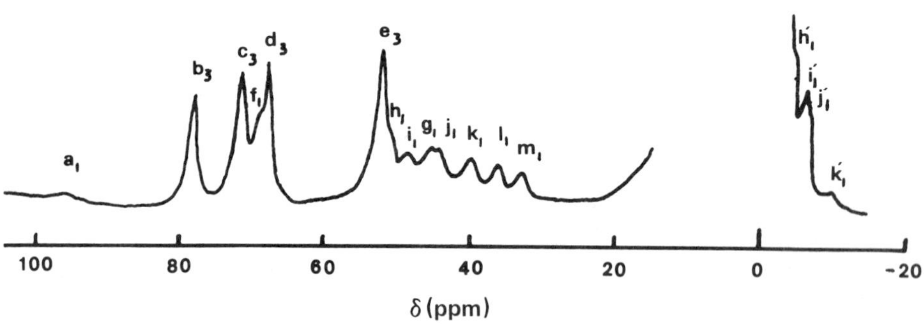

Figure 7.13 360 MHz ^{1}H NMR spectrum of horseradish peroxidase at 35 °C. Peak a is assigned to the axial histidine N-H while peaks b to e are assigned to the methyl groups in the order 5, 1, 8, and 3; peaks f and g are assigned to the vinyl H_α protons, peaks h to l to propionic acid H_α and axial histidyl H_β protons, and peaks i′ and j′ to the trans vinyl H_β protons [64]. The numbers indicate the integrated intensity. Chemical shifts are in ppm from DSS.

gion. This has been taken to be consistent with the proposed admixture of $S = \frac{5}{2}$ and $S = \frac{3}{2}$ ground states [72].

The temperature dependence of the isotropic shift is also not expected to follow Equations (2.11) and (2.22) when there is thermal equilibrium between two pure spin states. Such equilibria have been proposed to occur between $S = \frac{3}{2}$ and $S = \frac{1}{2}$ states [73], as well as between $S = \frac{1}{2}$ and $S = \frac{5}{2}$ states [74, 75]. It may often be hard to discriminate between spin equilibria and spin-admixed states unless independent measurements provide evidence for two species in equilibrium.

When imidazole is added to a high spin Fe(III) complex like TPPFeCl, six-coordinate *bis*-imidazole low spin species are obtained, as discussed before. The high spin-low spin equilibrium is slow on the NMR time scale ($\tau_M^{-1} = 10^2$ to $10^3\ s^{-1}$), and therefore the spectra of the two species can be detected. Their intensity depends on population partition (Figure 7.14) [76, 77].

Many hemoproteins are characterized by coordination of the heme iron to an imidazole group of a histidyl residue and to a water molecule (or hydroxy ion) or another amino acid side chain. Substitution of the sixth ligand by added ligands may be accompanied by changes in magnetic properties. Weak-field ligands, such as F^-, form high spin ferric hemoprotein complexes, whereas strong-field ligands,

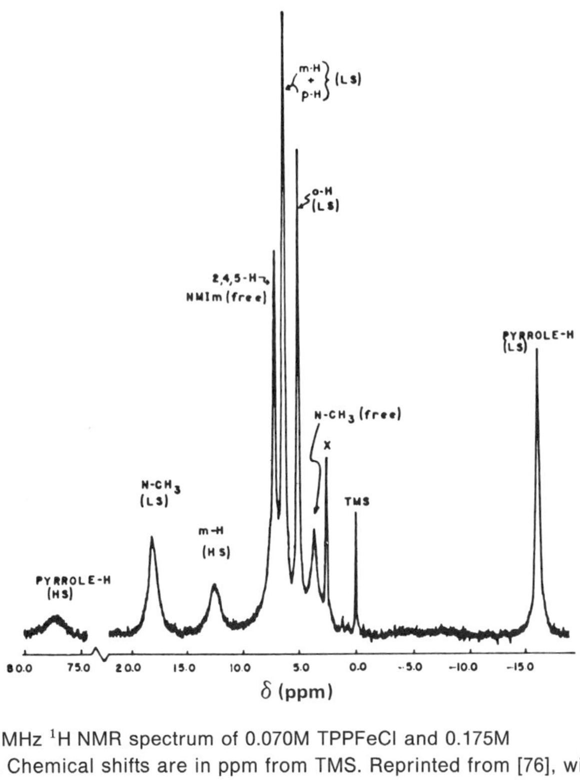

Figure 7.14 100 MHz ¹H NMR spectrum of 0.070M TPPFeCl and 0.175M N-CH_3-imidazole. Chemical shifts are in ppm from TMS. Reprinted from [76], with permission; copyright © 1972 by American Chemical Society.

such as CN^-, form low spin hexacoordinate ferric hemoprotein complexes. Ligands with intermediate field strength, such as azide, form complexes that exhibit temperature-dependent spin state equilibria [78]. Moreover, the protein structure may be important in determining the spin state equilibrium.

7.3.6 High Spin Fe(II)-P

High spin Fe(II)-P compounds are produced by creating a five-coordinate geometry for the metal ion—for example, by adding 2-methyl imidazole to Fe(II)-P prepared *in situ* [21]. Deoxyhemoglobin and deoxymyoglobin contain such species.

The NMR spectra of high spin iron(II) porphyrins, such as TPP derivatives (Figure 7.15D), show small phenyl proton shifts and large downfield pyrrole proton shifts. Natural porphyrins show small upfield meso proton shifts [79]. These features have been interpreted assuming that (1) dipolar shifts are negligible and (2) contact shifts arise from a mechanism similar to that occurring in Fe(III)-P, where the $d_{x^2-y^2}$ and d_{xz} and d_{yz} orbitals contain one unpaired electron each. The first orbital gives rise to a predominance of σ delocalization while the d_{xz}, d_{yz} orbitals account for moderate π delocalization at the meso positions through an M → L charge transfer mechanism. The electronic distribution and orbital pattern is prob-

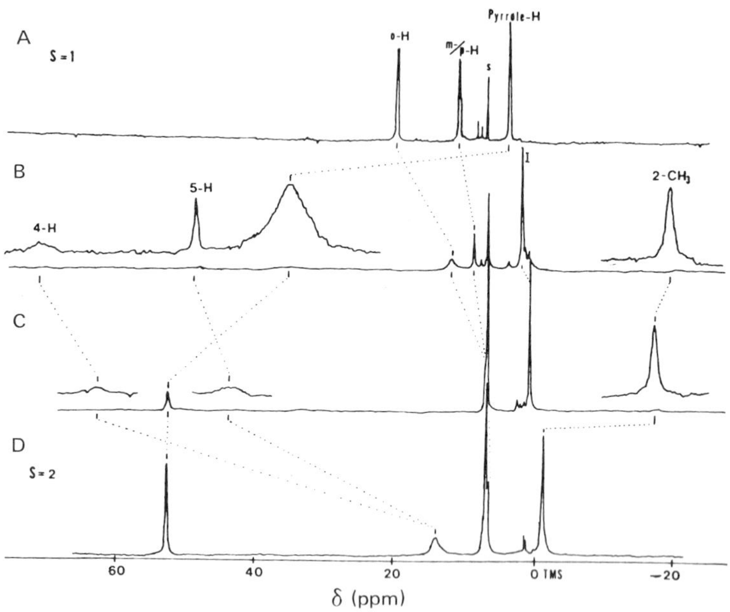

Figure 7.15 Titration of iron(II) porphyrins with 2-CH_3-imidazole. **A**: 3mM FeTPP; **B**: 8mM Fe-*p*-*i*-prop-TPP + 5mM 2-CH_3-Im; **C**: 8mM Fe-*p*-*i*-prop-TPP + 8mM 2-CH_3-Im; **D** 3mM FeTPP + 20mM 2-CH_3-Im; s: solvent. Chemical shifts are in ppm from TMS. The more soluble porphyrin in **B** and **C** is used to show the bound imidazole signals. Reprinted from [21], with permission; copyright © 1977 by American Chemical Society.

ably that shown in scheme A:

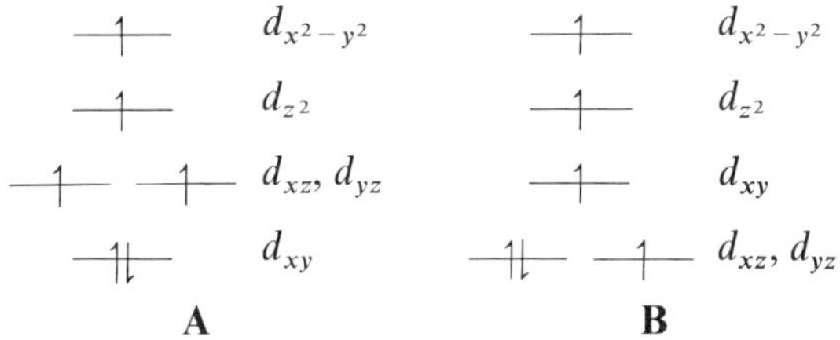

although that shown in B cannot be ruled out. ^{13}C data are also consistent with the former scheme (see Section 7.4). The 2-CH_3 signal of the axial imidazole shows an upfield shift of about 20 ppm, while the 4-H and 5-H signals are at ~70 and 50 ppm, respectively (Figure 7.15C). The proton downfield shifts for both 4-H and 5-H are indicative of dominant σ delocalization, but the smaller upfield 2-CH_3 shift may be due to spin polarization mechanisms.

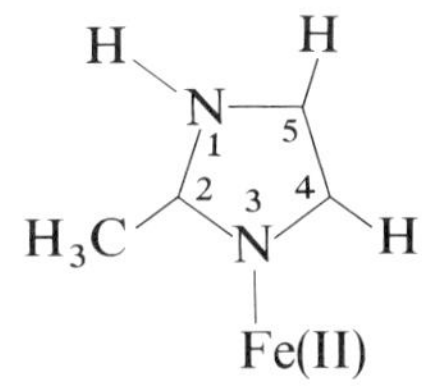

Several efforts have been made to synthetically produce a system that mimics the protein environment of five-coordinate high spin iron(II), both from the point of view of reactivity towards dioxygen and carbon monoxide [80], and in-plane asymmetry caused by the axial base inserted in a chain bridged over one of the porphyrin faces. Inequivalence of pyrrole proton shifts in such model complexes can be ascribed to either the removal of degeneracy of the d_{xz} and d_{yz} orbitals [81] or conformational equilibria between rotamers with respect to the Fe—N axial ligand bond [82], or both.

The spectrum of deoxymyoglobin high spin iron(II) complex with $S = 2$, is reported in Figure 7.16A [69, 83]. Several paramagnetically shifted resonances are observed [83] ranging from 16 ppm downfield to 8 ppm upfield. Of these, peaks X_1, X_3, and X_6 are assigned to ring methyl protons. A small upfield shift of a methyl group has also been observed in a model high spin iron(II) compound [21]; deoxyhemoglobin shows a similar spectrum [84] despite the known inequivalence of the α and β subunits (see Section 7.2.1). The peaks marked Y in Figure 7.16B are assigned to a minor component of deoxymyoglobin in which the heme is rotated by 180° about the α-γ meso axis. Indeed, reconstitution of the protein with heme yields about 40% of this component, which equilibrates slowly towards the equilibrium mixture ($\simeq$8%). Heme orientational disorder thus seems to be a general phenomenon in heme proteins.

Both deoxyhemoglobin and deoxymyoglobin have been studied in detail with the aim of investigating their relaxation properties [78]. By measuring the linewidth at different magnetic fields the Curie contribution to the overall line broadening can be factorized out. Using the data obtained for several protons both the

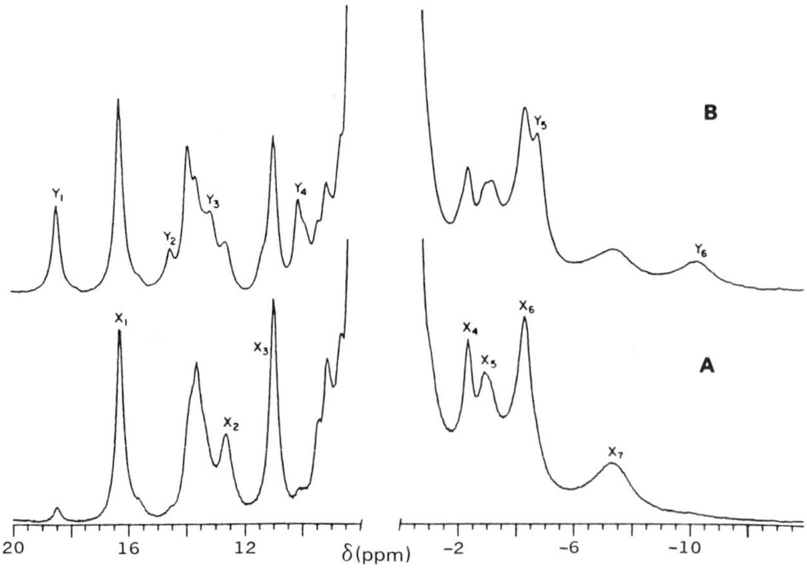

Figure 7.16 200 MHz ^{1}H NMR spectrum of deoxymyoglobin. **A**: At pH 7.0 in D_2O; **B**: The heme-reconstituted protein [69]. Chemical shifts are in ppm from DSS.

rotational correlation time, τ_r, and the metal-proton distance can be evaluated from Equation (3.20). The electronic relaxation time can then be obtained from the field-independent part of the relaxation, assuming negligible contact and ligand-centered dipolar contributions. The values of τ_r and τ_s obtained from the methyl protons of deoxyhemoglobin and deoxymyoglobin using this procedure are shown in Table 7.8. However, a warning is in order here concerning the independent measurement of the longitudinal relaxation times, T_1, for the same protons. This kind of measurement is more and more routinely accessible, also due to the development of suitable pulse sequences [84] (see Section 5.4.3). Unlike the linewidth, longitudinal relaxation times are only negligibly affected by Curie relaxation mechanism, and therefore an independent estimate of τ_s can be made,

Table 7.8 Rotational Correlation Times, τ_r, and Electronic Relaxation Times, τ_s, of Deoxyhemoglobin and Deoxymyoglobin Obtained from the Methyl Proton Linewidths[a]

Heme Protein	τ_r (s)	τ_s (s)
Deoxyhemoglobin	2.5×10^{-8}	6×10^{-12} $(7 \times 10^{-13})^b$
Deoxymyoglobin	$5\text{–}6 \times 10^{-9}$	6×10^{-12}

[a] Resonances X_1 and X_3 of Figure 7.16 have been analyzed [78].

[b] The value in parentheses is the electronic relaxation time estimated from T_1 measurements of the methyl protons in deoxyhemoglobin [84].

provided r is known and the Solomon Equation (3.8) holds. Such a value for deoxyhemoglobin is reported in Table 7.8: this value differs by about one order of magnitude, indicating that an estimate of the field-independent part of T_2^{-1} may be affected by a large error, being only a small fraction of the observed linewidth. It should be reemphasized that ligand-centered dipolar relaxation may also contribute to the overall relaxation. Thus, the electronic relaxation times reported in Table 7.8 are best regarded as upper limit values.

7.3.7 Intermediate Spin Fe(II)-P

In situ reduction of Fe(III)-PCl results in Fe(II)-P compounds with $S = 1$. At variance with high spin Fe(II)-P, sizeable dipolar shifts are present in the isotropic shifts of these compounds [22] (Figure 7.15A). It is assumed, as for low spin Fe(III), that the phenyl proton shifts of TPP are only dipolar in origin, thus permitting [22, 23] factorization of the shifts. The resultant contact shift pattern is of the type discussed for low spin Fe(III)-P; that is, a π spin density mechanism through P $\rightarrow$ Fe(II) charge transfer. In most $S = 1$ porphyrins, unpaired electrons are in the d_{xz}, d_{yz} orbitals, thus giving rise to a $^3A_{2g}$ ground state with the ^3E_g state close in energy [85] (see Section 7.1).

Iron(II) $S = 1$ is not found in natural systems, at least as a stable species, probably because it requires the absence of axial ligands and oxidizing agents.

7.3.8 Iron Porphyrins in High Oxidation States

High spin (formally $S = 2$) iron(IV) porphyrins can be obtained by two-electron oxidation of oxobridged iron(III) porphyrin dimers with subsequent cleavage by hydrochloric acid to give monomeric oxidized species of formula $FePCl^+$. For TPP, the NMR signal of the pyrrole protons is at 66 ppm downfield, the phenyl ortho signals are at 37.6 and 34.4 ppm downfield, and the meta and para signals are at -12.4 and -29.5 ppm upfield, respectively [14]. Such unusually large phenyl shifts are indicative of large spin density at the meso position, suggesting that the complex could be described as a high spin iron(III) π a_{2u} cation radical. Strong antiferromagnetic coupling between the $S = \frac{5}{2}$ iron(III) and the $S = \frac{1}{2}$ porphyrin radical occurs through a breakdown of the planarity of the porphyrin and an overlap of the $d_{x^2-y^2}$ and a_{2u} orbitals [86]. However, this coupling does not occur when there is strict D_{4h} symmetry around the metal ion [86, 87]. The strong coupling of the metal and radical spin systems also explains the observation of well-resolved NMR lines; apparently the overall relaxation time of the system is close to that of the corresponding high spin iron(III) compound [13] (see Section 3.8).

Addition of imidazole to CH_2Cl_2 solutions of high-spin porphyrin iron(III) π-cation radicals permits generation of an imidazole complex that has been described as low spin iron(III) π-cation radical. The far upfield signals of the pyrrole protons are indicative of low spin iron(III), whereas the alternating shifts of the aromatic moiety of TPP are reminiscent of a π-radical. Magnetic moments estimates indicate that metal and radical spins are not strongly coupled [88].

Reaction of peroxobridged iron(III) porphyrin dimers with nitrogen bases, such as *N*-methylimidazole, pyridine, and piperidine, yields a monomeric porphyrin in which the iron is coordinated to the base and an oxygen atom in the axial positions as a result of cleavage of the peroxo bridge. Such a compound is best described as a low spin ($S = 1$) iron(IV) porphyrin [11]:

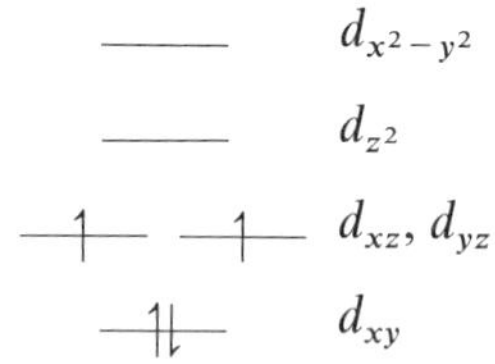

The NMR spectrum of a complex in which the porphyrin is *meso*-tetra-*m*-tolyl porphyrin has the pyrrole signal at 5 ppm downfield from TMS [11]. For natural porphyrins, the only resonance shifted well outside the diamagnetic region (~ 15 to 20 ppm downfield) is that of the meso protons [152].

Compound II of horseradish peroxidase is an unstable species obtained by one-electron reduction of compound I, which is obtained by two-electron oxidation of the iron(III)-containing resting system (see below). Its NMR spectrum (Figure 7.17B) is consistent with the presence of a low spin ($S = 1$) oxoiron(IV) species [89, 17] which only shows two isotropically shifted signals at 14.6 and 16.1 ppm downfield. These signals have been assigned to meso protons through reconstitution with deuterated hemins [17].

Low spin oxoiron(IV) species have also been proposed, on similar grounds, for ferryl myoglobin, which is obtained by oxidation of met-myoglobin with hydrogen peroxide [17].

Compound I of cytochrome *c* peroxidase (CCP) is obtained as the two-electron oxidation product of the native iron(III)-containing species, and is therefore formally iron(V). Its NMR spectrum [91] (Figure 7.17C) has a remarkable similarity with that of compound II of horseradish peroxidase; moreover, ESR and ENDOR experiments have shown the presence of a noninteracting radical. Therefore, CCP compound I is likely to contain an $S = 1$ oxoiron(IV) and a free radical on an amino acid side chain. No NMR spectrum has been reported for the labile CCP compound II [91].

Oxidation of chloro-*meso*-tetramesitylporphyrinato iron(III) with *m*-chloroperoxybenzoic acid produces an intermediate characterized as an iron(IV) porphyrin π cation radical [15]. The iron(IV) would be in the $S = 1$ state, but the overall system would have $S = \frac{3}{2}$. Its unique NMR spectrum shows an upfield shift of the β pyrrole protons and downfield shifts for the meta-protons and ortho and para methyl groups. The shifts on the mesityl moiety are consistent with π delocalization as a result of π spin density at the meso carbons. Such a situation is expected when an electron is lost from a porphyrin π orbital. An $a_{1u}\pi$ cation radical would account for the large upfield shift of the pyrrole protons, compared with the small downfield shifts in iron(IV) porphyrins.

Oxidation of horseradish peroxidase with hydrogen peroxide gives rise to compound I, which formally contains iron(V). This species has been assigned either as an iron(IV) porphyrin π cation [92] or a high spin iron(IV) species with a non-interacting free radical on an amino acid side chain [89]. By comparing its NMR spectrum (Figure 7.17A) with that of the deuterohemin-reconstituted species ($R = R' = H$, Figure 7.1) and pointing out the opposite sign of the isotropic shifts for the CH_3 signals ($55 \div 70$ ppm) and the 2,4 pyrrole proton signals (-25 and -35 ppm), La Mar *et al.* have argued against the high spin iron(IV)/amino acid radical formulation. It is suggested that the difference in linewidth of the methyl protons, which had been taken as evidence for this formulation [89], could be accounted for by the different contact contributions on the various positions [90]. In fact, the recent synthesis of a model compound and a theoretical study [93] supports the description of horseradish peroxidase compound I as an $S = 1$ iron(IV)

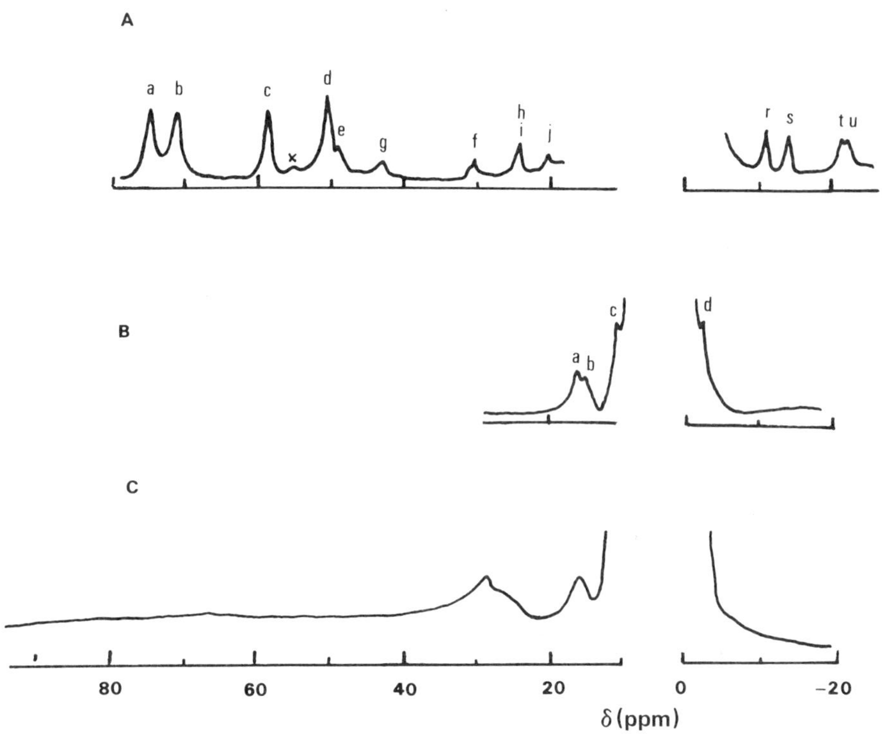

Figure 7.17 **A** 360 MHz ^1H NMR spectrum of horseradish peroxidase compound I at 25 °C. Peaks a to d are assigned to the methyl groups in the order 8, 3, 5, and 1; peaks e and f to the two vinyl H_α; peaks g to j to the α-CH_2 protons from the propionic chains; and peaks r to u to the four vinyl H_β [90]. **B** 360 MHz ^1H NMR spectrum of horseradish peroxidase compound II at 25 °C [17]. **C** 360 MHz ^1H NMR spectrum of cytochrome *c* peroxidase compound I at 21 °C [91]. Chemical shifts are in ppm from DSS. Spectrum **B** reproduced from [17], with permission; copyright c 1983 by American Chemical Society.

porphyrin/$S = \frac{1}{2}$ π cation radical, ferromagnetically coupled. Native compound I would have an a_{2u} π cation structure, while the deuterohemin-reconstituted species would be of a_{1u} π cation type [93].

7.3.9 Electron Transfer

Figure 7.18 shows the spectra of *Rhodospirillum rubrum* cytochrome c' in the reduced (A) and oxidized (C) forms, together with that of a 50% reduced sample (B). The latter spectrum shows that the intermolecular electron transfer is slow on the NMR time scale. By using the equations for chemical exchange reported in Chapter 4 and the observed difference in chemical shift of the methyl signals in the two forms, an upper limit of 10^4 s^{-1} for the exchange rate can be estimated [34]. As the exchange rate is much higher for other cases, small additions of one of the two forms to the other causes dramatic exchange broadening effects. The following equation has been developed [94] to obtain the electron exchange time, τ, as a function of the molar fraction of the two species, p_A and p_B, the intrinsic transverse relaxation rate of a signal in the two environments, T_{2A}^{-1} and T_{2B}^{-1}, and

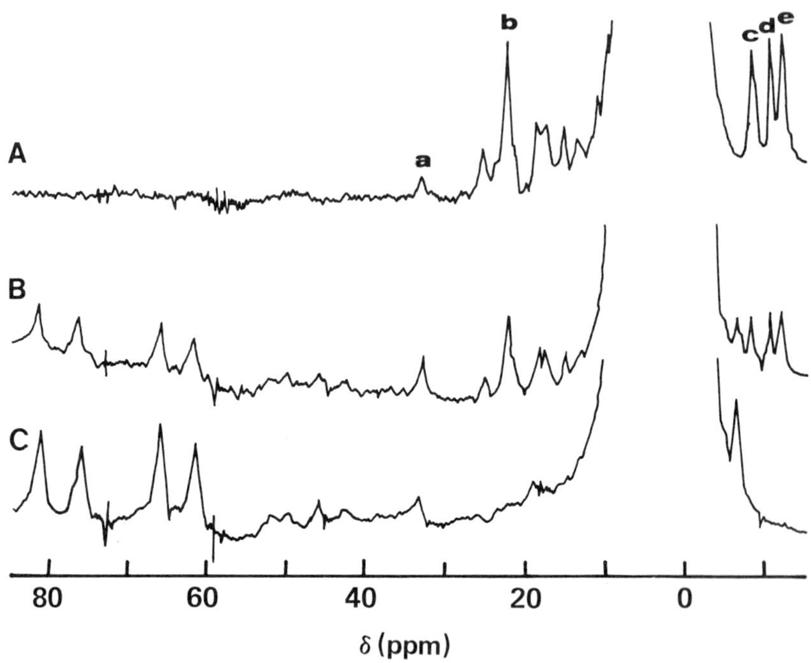

Figure 7.18 270 MHz ^{1}H NMR spectra of cytochrome c' at pH 7.0 in D_2O. **A**: Fully reduced sample; **B**: 50% reduced sample; **C** Fully oxidized sample (Figure 7.5). Peaks b to e are assigned to the ring methyl protons; peak a and an exchangeable proton peak observed at 93 ppm in H_2O (not shown) are assigned to the axial histidine (see Section 7.3.10). Chemical shifts are in ppm from DSS. Reproduced from [34], with permission; copyright c 1981 by American Chemical Society.

the chemical shift difference, $\omega_A - \omega_B$. T_2^{-1} is the observed relaxation rate and p_M is the molar fraction of the minor component:

$$T_2^{-1} = 2p_M(P_A T_{2A}^{-1} + P_B T_{2B}^{-1})(\omega_A - \omega_B)^2\tau$$

This equation holds in the limits $p_A \gg p_B = p_M$ or $p_B \gg p_A = p_M$.

Line-broadening analysis has recently provided a detailed description of the electron transfer between hexacyanoferrate(II) and ferricytochromes *c* from horse and *C. krusei*. The electron transfer rate constant is about $8 \times 10^2\ s^{-1}$ in both cases, although the association constants are significantly different [95].

In the slow-exchange limit, where no appreciable exchange broadening occurs, the saturation transfer method (Section 5.3.2) is applicable, provided that the electron exchange rate is of the order of T_1^{-1}. This method, which has been routinely applied since the successful determination of the electron exchange rate of cytochrome *c* in 1965 [96], enables the paramagnetic signals to be assigned by relating them to the signals in the reduced diamagnetic proteins [42]. The latter can in turn be assigned with the aid of NOE (Section 5.3.2), while spin decoupling allows signals belonging to the same spin multiplet to be related; NOE is used to identify spatial interactions between neighboring protons. The signal assignment in most low spin cytochromes has been performed through such techniques [42, 45, 47–49, 55].

Besides the time scale limits that are intrinsic in saturation transfer experiments, a practical limitation arises in systems undergoing several oxidation-reduction steps: that is, the need of performing many time-consuming experiments to identify all the pairs of signals connected by electron transfer processes at different solution redox potentials. This is exemplified by the recent efforts of elucidating the electron transfer mechanisms in *D. gigas* cytochrome c_3 [97], which contains four hemes and therefore exists in five different oxidation states. This system has thus stimulated the development of a two-dimensional saturation transfer experiment that seems very promising for applications of this kind [98]. According to the general scheme described in Section 5.3.2, the following pulse sequence is applied

$$(90^\circ - t_1 - 90^\circ - \tau_m - 90^\circ - t_2)_n$$

where t_1 is the evolution period, τ_m is a mixing time, and t_2 is the observation period. In this way a 2D exchange spectrum is obtained where the off-diagonal peaks connect all the pairs of signals involved in electron transfer. Only four 2D experiments at the appropriate solution redox potential would then be needed to completely resolve the problem of *D. gigas* cytochrome c_3. This example shows that 2D experiments can be performed on paramagnetic systems, provided that the T_1's of the signals are not too short (25–80 ms in this case) [98].

7.3.10 Axial Ligands

The protein ligands in the axial position(s) of heme groups are (1) histidine nitrogens, (2) methionine sulfurs, (3) cysteine sulfurs, and (4) tyrosine oxygens. Although all of these ligands are expected to give rise to hyperfine shifted proton resonances, only signals from histidines and methionines have been detected so far. Table 7.9 summarizes the isotropic shift ranges in which the proton signals of these axial ligands are found for the various oxidation and spin states of iron.

Observation of the histidine N-H signal depends not only on the general factors determining the detectability of paramagnetically shifted and broadened signals, but also on the proton exchange rate (Section 5.4.2). Even if this rate is much smaller than that required to average out the signal with the water peak, it may still be fast compared to T_1^{-1}. Under these conditions, the signal can be quenched, if saturation techniques are applied to suppress the water signal [29, 65].

The axial histidine residue in hemoglobin is of particular interest because of its possible involvement in the T → R conversion that takes place upon dioxygen binding (see Section 7.2.1). It is relatively easy to assign the N-H proton signal, as it is eliminated in D_2O and is in slow exchange with the bulk solution. This signal undergoes a sizeable downfield shift (60 ÷ 90 ppm) in deoxyhemoglobin, which arises mainly from spin transfer via Fe—N σ bonding [101]. The changes observed during the T → R transition are therefore sensitive indicators of the mode of coordination of the axial histidine. In high spin iron(III) met-myoglobin [34] and in cytochrome *c'* at low pH [33], the N-H signal has been observed at 102 ppm. In cytochrome *c'* the His β-CH_2 protons have also been located around 37 ppm downfield, and are proposed to undergo a dramatic downfield shift (~120 ÷ 140 ppm) during the transition between the low and intermediate pH forms [33]. This is probably due to an increase in contact contribution to the shift, which has been ascribed to an increase in the imidazolate character of the

Table 7.9 Chemical Shift Range (ppm) for Axial Ligand Protons in Hemoproteins

	His				Met
Iron Species	2-H	4-H	N-H	$-CH_2$	$-CH_3$
Iron(III) $S = \frac{1}{2}$	−13.6[a]	12.9[a]	15[b]	8 − 15[c]	−15[d], −30[e]
Iron(III) $S = \frac{5}{2}$			102[f]	30[g], 140[h]	
Iron(II) $S = 2$	30 ÷ 50[i,j]		60 ÷ 90[i,j,k,l]		

[a] In cytochrome b_5 [64].
[b] In cyanometmyoglobin [99].
[c] In cytochrome *c* [100].
[d] In cytochrome c_2 [52].
[e] In cytochrome *c* [43].
[f] In met-myoglobin [35] and cytochrome *c'* [33].
[g] In cytochrome *c'* (low pH) [33].
[h] In cytochrome *c'* (intermediate pH) [33].
[i] In deoxymyoglobin [79].
[j] In deoxyhemoglobin [101].
[k] In reduced horseradish peroxidase [102].
[l] In reduced cytochrome oxidase [103].

axial residue. No 2-H and 4-H protons shift data could confirm this interesting hypothesis.

In low spin cytochrome b_5 there are two axial histidines. The 2-H and 4-H signals from one of these histidines have been tentatively located at -13.6 and 12.9 ppm, respectively [64]. Again, these signals are sensitive to changes in the nearby environment, such as the conformational heterogeneity arising from a 180° rotation about the α-γ meso axis (see Section 7.3.4).

In the NMR spectrum of ferricytochrome c (low spin, $S = \frac{1}{2}$) (Figure 7.8), the methionine methyl signal has been assigned as the one at 25 ppm upfield [43] on the basis of its intensity and larger linewidth with respect to the heme methyl protons. This signal disappears upon addition of cyanide, indicating that of the two axial ligands it is the methionine and not the histidine that is replaced by the anion.

Cyanide. Cyanide is the most common reactant employed to modify the heme moiety. All the monocyano and dicyano derivatives of isolated porphyrins and hemoproteins are low spin and contain either diamagnetic iron(II) or paramagnetic ($S = \frac{1}{2}$) iron(III). ^{13}C and ^{15}N NMR spectra have been reported for the latter systems and, as shown in Table 7.10, dramatic paramagnetic effects have been measured for both nuclei: the ^{13}C isotropic shifts range from $-2{,}000$ to $-2{,}500$ ppm upfield whereas the ^{15}N shifts range from 600 to 1,200 ppm downfield. The shifts have been interpreted as being mainly contact in origin: possible point dipolar contributions estimated from the known metal-nucleus distance and magnetic anisotropy values are of the order of 100 to 300 ppm downfield [105]. The spin delocalization would be dominantly spin polarization, through either σ or π bonds, which also accounts for the opposite signs of the carbon and nitrogen shifts [104, 105]. It should be recalled, however, that theoretical studies [106] have estimated sizeable ligand-centered dipolar contributions to the shifts of nitrogen in $Fe(CN)_6^{3-}$ (Section 2.4).

Although the magnitude of the shifts is only slightly sensitive to variations in the nature of the heme and to its contacts with the surrounding protein, it is strongly dependent on the identity of the trans ligand. For example, the ^{15}N signal in protohemin cyanide shifts from 945 to 989 to 1,070 ppm downfield on passing from 4 acetylpyridine to pyridine to 3,5-lutidine as the trans ligand. Both ^{13}C and ^{15}N isotropic shifts are also very sensitive to changes in the solvent. Aprotic solvents usually give the largest absolute values, which decrease by as much as 200 ppm on passing from DMSO to methanol to water [105]. In general, the axial ligand field strength of cyanide, and hence the extent of unpaired spin delocalization, decreases with the increasing hydrogen bonding capabilities of the solvent.

$C^{15}N$ NMR data are available for cyanometmyoglobin isolated from horse and sperm whale, cytochrome c, and human adult cyanomethemoglobin (Table 7.10). The large observable differences may be related to the different coordination strength of the trans histidine and/or to the different hydrogen bonding of the coordinated cyanide with other protein residues—such as the distal histidine present in myoglobin and hemoglobin but not in cytochrome c. The difference in the ^{15}N chemical shift (~ 70 ppm) of the α and β subunits in hemoglobin (Table 7.10)

Table 7.10 ^{13}C [104] and ^{15}N [105] Shifts (ppm) for Coordinated Cyanide in Low Spin Iron(III) Porphyrins and Hemoproteins

Iron Species	Porphyrin 2,4-R Group	^{13}C Chemical Shift[a]
Iron(III) protoporphyrin IX	$-CH{=}CH_2$	−2,393
Iron(III) mesoporphyrin IX	$-CH_2CH_3$	−2,381
Iron(III) deuteroporphyrin IX	—H	−2,375
Iron(III) 2,4-dibromodeuteroporphyrin IX	—Br	−2,359
Hemin *c*	$-CH(CH_3)SCH_2-CH(NH_2)COO^-$	−2,300
Iron(III) 2,4-disulfonate-deuteroporphyrin IX	$-SO_3^-$	−2,167
Iron(III) *meso*-tetrakis (4-carboxyphenyl) porphyrin		−1,968
Myoglobin (C_2H_5NC) (diamagnetic)		171.3

Iron Species	Solvent	^{15}N Chemical Shift[b]
Protohemin $(C^{15}N^-)_2$	Me_2SO	732
	Py	696
	CH_3OD	506
	$CH_3OD + D_2O$ (1:1)	480
	H_2O (pH 9.4)	451
	H_2O (pH 9.2)	448
Deuterohemin $(C^{15}N^-)_2$	Me_2SO	738
	$Me_2SO + D_2O$ (20:1)	716
	CH_3OD	509
Mesohemin $(C^{15}N^-)_2$	Me_2SO	720
	CH_3OD	496
Hematohemin $(C^{15}N^-)_2$	CH_3OD	495
OEP $(C^{15}N^-)_2$	Me_2SO	718
	CH_3OD	474
Protohemin (4-AcPy) $(C^{15}N^-)$	4-AcPy + D_2O	945
Protohemin (Py) $(C^{15}N^-)$	Py + D_2O	989
Protohemin (3,5-Lu) $(C^{15}N^-)$	3,5-Lu + D_2O	1,070
Protohemin (N-MeIm) $(C^{15}N^-)$	N-MeIm + Me_2SO	926

[a] Shifts from TMS.

[b] Shifts from $^{15}NO_3^-$.

Table 7.10 (*continued*)

Iron Species	pH	^{15}N Chemical Shift[b]
Myoglobin ($C^{15}N^-$) (horse)	5.7	931
	6.0	939
	6.9	948
	8.0	936
	10.0	940
	11.7	940
Myoglobin ($C^{15}N^-$) (sperm whale)	8.8	945
Cytochrome *c* ($C^{15}N^-$)	5.2	841
	6.6	842
	7.8	847
	9.0	848
Hemoglobin ($C^{15}N^-$)	7.3	975 (α)
		1,047 (β)
	7.7	985 (α)
		1,055 (β)

illustrates the sensitivity of the cyanide nuclei to the protein environment around the axial positions. Clearly, explaining the origins of these changes would be of great help in the understanding of the system.

A final comment is in order here concerning the relaxation rates of the coordinated cyanide nuclei. Linewidths ranging from 100 to 600 Hz have been observed for both ^{13}C and ^{15}N. The paramagnetic broadening may arise from dipolar (metal and ligand centered) and contact contributions. From the magnitude of the observed broadening, especially for ^{15}N, simple metal-centered dipolar effects seem to play a negligible role. On the other hand, a rough estimate of the magnitude of the contact coupling constant, $A_c/\hbar$, can be made by using Equation (2.4) and the experimental value of the isotropic shift. The value so obtained is 2×10^7 rad s^{-1}, which gives a reasonable τ_s value of 3×10^{-12}s when substituted in the contact term of the transverse relaxation (Equation 3.16). Therefore, the relaxation can be ascribed to the contact term, although additional contributions from ligand-centered dipolar relaxation may be operative.

7.4 ^{13}C NMR SPECTRA OF Fe-P

^{13}C NMR spectra are available for iron(III) porphyrins as high spin, low spin, and spin-admixed species, as well as for high spin iron(II) porphyrins [107–112]. Three typical spectra are reported in Figure 7.19. The spectra of low spin iron(III)

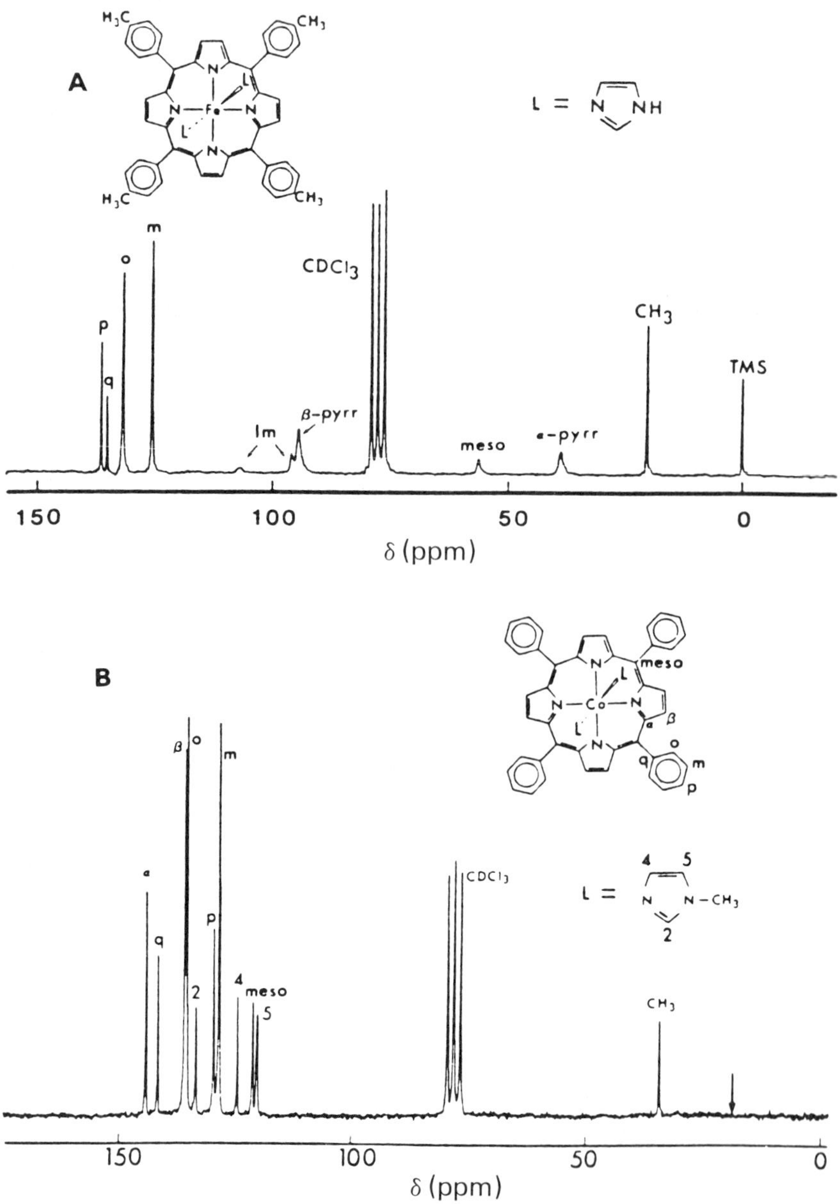

Figure 7.19 ^{13}C NMR spectra of **A** low spin [107] and **C** high spin [108] iron(III) porphyrins; **D** high spin [110] iron(II) porphyrins; **B** diamagnetic cobalt(III) derivative [107], where the arrow indicates the *p*-CH_3 signal position in the corresponding Co(III) T*p*-CH_3PP derivative. Chemical shifts are in ppm from TMS. Reproduced from [107] (**A**, **B**), with permission; copyright © 1981 by American Chemical Society. Reproduced from [108] (**C**), with permission; copyright 1981 by The Royal Society of Chemistry. Reproduced from [110] (**D**), with permission; copyright © 1983 by American Chemical Society.

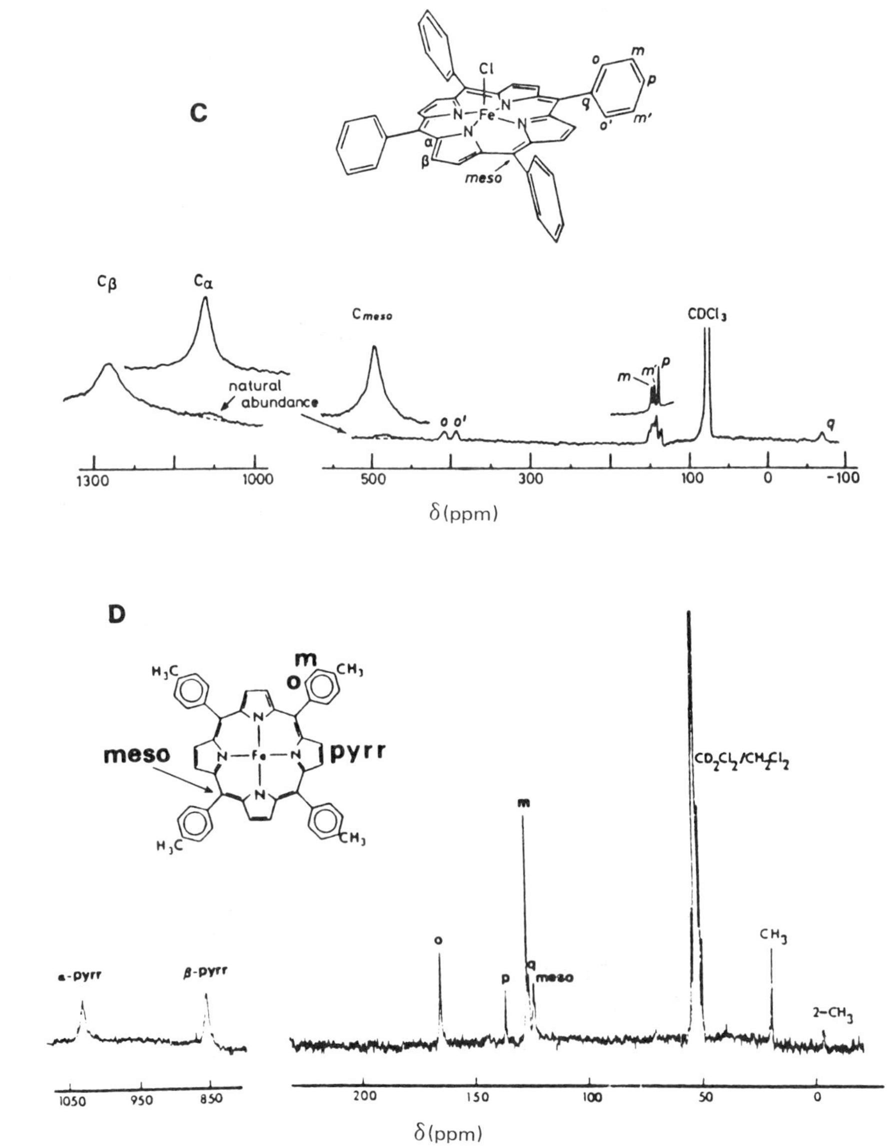

Figure 7.19 (*continued*)

and high spin iron(II) porphyrins are of good quality due to short electronic relaxation times. For these, the assignment of signals is performed through residual ^{13}C-^{1}H spin-spin coupling in off-resonance proton double irradiation experiments [107]. This is an easy task because the proton resonances are widely separated. In the case of high spin iron(III) species, unfavorable electronic relaxation times result in spectra of low quality [108, 109]. The spectrum shown in Figure 7.19C has been obtained on isotopically enriched samples. Final assignment in this case is based on selective enrichment [108]. A striking feature of this spectrum is the large isotropic shift of the resonances of the pyrrole carbon nuclei, which is even larger than that in high spin iron(II) [110]. The spin-admixed FeTPP ClO_4 is reported to display smaller linewidths and isotropic shifts than either the high or low spin species [112]. Particularly sharp are the signals in the sulphate-bridged Fe(III)TPP dimer [113], in which weak magnetic coupling causes a sizeable decrease in the electronic relaxation rates, although the shifts are typical of an $S = \frac{5}{2}$ species.

For all of the above cases, the origin of the isotropic shifts is both contact and dipolar. The shifts in low spin iron(III) are relatively small, since they are due to a single electron delocalized through a π mechanism. In high spin iron(III) and iron(II), on the other hand, the shifts are large owing to large σ delocalizations. As mentioned in Section 2.4, the dipolar contribution to the isotropic shift can be factorized into a metal-centered dipolar shift and a ligand-centered dipolar shift—the latter shift is due to the presence of spin density on the ligand and may be sizeable for nuclei other than hydrogen. Although the separation of contact, metal-centered, and ligand-centered dipolar shifts is not an easy task, some attempts at a semiquantitative analysis are available. Such an analysis is helpful in trying to understand the details of the electronic structure of these systems.

7.4.1 Factorization of the ^{13}C Isotropic Shifts in Low Spin Iron(III)

In order to understand the nature of the isotropic shifts, the low spin FeTPP $(1\text{-}CH_3\text{-}Im)_2$ complex has been analyzed in some detail [107]. It is assumed in this analysis that the shifts of the aromatic groups attached at the meso positions are entirely dipolar in origin—the same assumption was found to be correct in the analysis of the proton isotropic shifts. In fact, the delocalization in the $3e_\pi$ molecular orbitals provides small spin density at the meso positions. Indeed, with the exception of the aromatic carbon attached at the meso position, the other carbon shifts are proportional to the geometrical factors, which is as expected. This procedure enabled Goff [107] to estimate the metal-centered dipolar shifts. The values obtained by this method can be checked by making a comparison with those calculated using the g values according to Equation (2.19). The discrepancy between the two methods is presumably due to neglect of the second-order Zeeman effect in the latter case.

Since the unpaired spin density is located in the π porphyrin system, the ligand-centered dipolar shift for each ith carbon may be estimated by assuming that it is directly related to the π spin density at the same carbon:

$$\left(\frac{\Delta\nu_i}{\nu_0}\right)_i^{\mathrm{LC}} = d\rho_i^\pi \tag{7.1}$$

The contribution of π electron spin density on neighboring atoms is considered negligible. Furthermore, it is assumed that the proportionality constant, d, is the same for the various positions. The π spin density, ρ^π, has been calculated from the proton contact shift through Equation (2.14). By taking $Q = -63$ MHz, ρ^π is 0.012 for the β carbon and -0.0015 for the meso carbon.

The contact contribution to the shift of the ith carbon arises from unpaired spin density in the $1s$ and $2s$ orbitals of the nucleus. The presence of this spin density on the nucleus is due to several contributions (Section 2.3.1). The first contribution is the spin polarization by the unpaired electron in the ρ^π orbital of the same carbon atom, and is therefore proportional to its π spin density:

$$\left(\frac{\Delta\nu}{\nu_0}\right)_i^{\text{con}} = (S + 2Q_{\text{CC}'} + Q_{\text{CH}})\rho_i^\pi \cdot B; \qquad B = \frac{g_e\mu_B S(S+1)}{3\hbar\gamma_C kT}$$

where S reflects polarization of the $1s$ orbital, and $Q_{\text{CC}'}$ and Q_{CH} reflect polarization of the three sp^2 bonds. The polarization of the two carbon-carbon bonds ($2s$ orbital of the observed ith carbon) by π spin density on the neighboring hth and lth carbon atoms should also be taken into account:

$$\left(\frac{\Delta\nu}{\nu_0}\right)_i^{\text{con}} = [(S + 2Q_{\text{CC}'} + Q_{\text{CH}})\rho_i^\pi + Q_{\text{C}'\text{C}}(\rho_h^\pi + \rho_l^\pi)] \cdot B \tag{7.2}$$

where ρ_h^π and ρ_l^π are the π spin density on the two adjacent carbon atoms and $Q_{\text{C}'\text{C}}$ is the appropriate proportionality constant. Actual evaluation of $(\Delta\nu/\nu_0)_i^{\text{con}}$ is rather speculative. For π radical systems, the constants S, $Q_{\text{CC}'}$, Q_{CH}, and $Q_{\text{C}'\text{C}}$ have been assigned some numerical values (Section 2.3.1), which are reported in Table 7.11 for convenience.

Equations (7.1) and (7.2) can be combined to give:

$$\left(\frac{\Delta\nu}{\nu_0}\right)_{\beta\text{ position}}^{\text{LC}} + \left(\frac{\Delta\nu}{\nu_0}\right)_{\beta\text{ position}}^{\text{con}} = d\rho_\beta^\pi + [(S + 2Q_{\text{CC}'} + Q_{\text{CH}})\rho_\beta^\pi + Q_{\text{C}'\text{C}}(\rho_\beta^\pi + \rho_\alpha^\pi)] \cdot B \tag{7.3}$$

$$\left(\frac{\Delta\nu}{\nu_0}\right)_{\text{meso}}^{\text{LC}} + \left(\frac{\Delta\nu}{\nu_0}\right)_{\text{meso}}^{\text{con}} = d\rho_{\text{meso}}^\pi + [(S + 2Q_{\text{CC}'} + Q_{\text{CH}})\rho_{\text{meso}}^\pi + 2Q_{\text{C}'\text{C}}\rho_\alpha^\pi] \cdot B \tag{7.4}$$

In both of these equations, the left-hand terms represent experimental values minus the metal-centered dipolar shifts, which are easily determined from the geometrical factors of the various carbon atoms once the dipolar shifts for the protons are known. The unknowns are d and ρ_α, since the α position does not have any attached proton. Once d and ρ_α are estimated from the equations, ρ_N^π can also be determined from the isotropic shifts of the α carbon through expressions analogous to Equations (7.3) and (7.4). The value of ρ_N^π is obtained by taking $Q_{\text{CN}} = Q_{\text{CC}'}$

Table 7.11 Values for the Karplus and Fraenkel Constants (MHz) [114]

S^{C}	$Q^{\text{C}}_{\text{CC}'}$	Q^{C}_{CH}	$Q^{\text{C}}_{\text{C}'\text{C}}$
−35.5	+40.3	+54.6	−39

Table 7.12 Contributions to the Isotropic Shifts for Selected Carbon Atoms in Low Spin 1-CH_3-Im Complexes of Iron(III) Porphyrins [107][a]

Contribution	α Pyrrole ^{13}C	β Pyrrole ^{13}C	Meso ^{13}C	Pyrrole $^{13}CH_3$
$(\Delta\nu/\nu_0)^{iso}$	101.5	36.0	73.2	46.8
$(\Delta\nu/\nu_0)^{MC}_{dip}$	31	11	22	5.0
$(\Delta\nu/\nu_0)^{LC}_{dip}$	36	78	−10	
$(\Delta\nu/\nu_0)^{con}$	34	−53	61	47

[a] The values for α and β carbons refer to TPP; those for meso and pyrrole CH_3 carbons refer to etioporphyrin ($R = CH_3$, $R' = CH_2CH_3$ in Figure 7.1).

and $Q_{NC} = Q_{C'C}$. The estimated values are summarized in the following scheme:

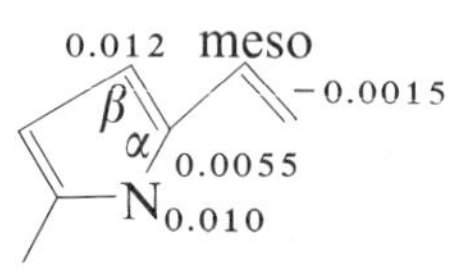

Such a procedure has enabled [107] a prediction of the ring methyl carbon contact shift. This carbon atom is not a part of the π system and to first order should have a contact shift term reflecting only polarization of its $2s$ electrons by π spin density at the β pyrrole carbon atom. Thus, the relation

$$\left(\frac{\Delta\nu}{\nu_0}\right)^{con}_{\text{ring } CH_3} = -Q_{C'C}\rho^{\pi}_{\beta} \cdot B \tag{7.5}$$

has been used to calculate a ring methyl carbon contact shift of −48 ppm, which is in good agreement with experimental findings (−42 ppm).

Despite some oversimplification, this treatment should give an idea of the relative contributions to the isotropic shifts of porphyrin carbon nuclei in low spin iron(III) systems (see Table 7.12).

Ring $^{13}CH_3$ of hemin c dicyanide ($R = R' = —CH(CH_3)—S—CH_2—CHNH_2—COO^-$, see Figure 7.1) shows shifts upfield [115], whereas the corresponding proton shifts are downfield, consistent with π spin delocalization in the pyrrole rings. On the meso positions, only small dipolar shifts are expected to be operative, and indeed both ^{13}C and ^1H shifts are slightly upfield [115].

7.4.2 Factorization of the ^{13}C Isotropic Shifts in High Spin Iron(III)

A different approach has been used for high spin Fe(III)-P compounds [108] owing to the simultaneous presence of both σ and π spin density. Metal-centered dipolar shifts are determined from the geometrical factors and the equations used for the factorizations of the proton isotropic shifts. Equation (2.22), with $g = 2$, can be written as

$$\left(\frac{\Delta\nu}{\nu_0}\right)^{MC}_{dip} = -\frac{\mu_0}{4\pi}\frac{112D\mu_B^2}{9(kT)^2}\frac{3\cos^2\theta - 1}{r^3} \tag{7.6}$$

where D is the zero field splitting parameter, equal to 5.9 cm^{-1}. The same value of D is used in the expression for the ligand-centered dipolar shift caused by spin density on the ρ^{π} carbon orbital:

$$\left(\frac{\Delta\nu}{\nu_0}\right)^{\text{LC}}_{\text{dip}} = -\frac{56\, D\mu_B}{9\gamma_C(kT)^2}(2b/\hbar)\rho^{\pi} \tag{7.7}$$

where $2b/\hbar$ is 1.35×10^9 rad s^{-1}. This equation is similar to that used for low spin Fe(III)-P. The contact contribution to the isotropic shift depends, as usual, on Equation (2.4), which is written as

$$\left(\frac{\Delta\nu}{\nu_0}\right)^{\text{con}} = \frac{35\mu_B}{6\gamma_C kT}(A_c/\hbar) \tag{7.8}$$

However, as the σ spin density does not permit $A_c/\hbar$ to be related to ρ^{π}, both of these parameters are unknown.

When the linewidths are also considered, further experimental data are available. Again, the contributions to T_{2M}^{-1} are

$$T_{2M}^{-1} = (T_{2M}^{-1})^{\text{con}} + (T_{2M}^{-1})^{\text{dip}}$$

where $(T_{2M}^{-1})^{\text{con}}$ is given by Equation (3.16) and $(T_{2M}^{-1})^{\text{dip}}$ is given by Equation (3.9). Equation (3.9) for this specific case takes the form

$$(T_{2M}^{-1})^{\text{dip}} = \frac{1}{15}\left(\frac{\mu_o}{4\pi}\right)^2 S(S+1)\gamma_C^2 g_e^2 \mu_B^2 (\rho_{\text{Fe}}^2 r_M^{-6} + \tfrac{4}{25}\langle r^{-3}\rangle_{2p}^2 \rho^2$$
$$- \tfrac{2}{5}\langle r^{-3}\rangle_{2p}\rho\rho_{\text{Fe}} r_M^{-3})\sum J(\omega, T_{1e}) \tag{7.9}$$

where the first term in the parentheses represents the contribution from the iron orbitals and the other two from the orbitals of the carbon nucleus of interest. The ρ_{Fe} value is reasonably settled to be around $3.5 \div 5$. Therefore, for each carbon, two equations are available—one for the shift and one for the linewidth—and the unknowns are ρ^{π}, A_c, and T_{1e}; the latter is the same for all the carbon nuclei. These are easily solved by recalling that the meso carbon has little direct σ spin density. If this is therefore assumed to be zero, Equation (7.2) holds for this carbon. The Fermi coupling constant can then be written as

$$\frac{A_c}{\hbar} = (S + 3Q_{\text{CC}'})\rho_{\text{m}}^{\pi} + 2Q_{\text{C}'\text{C}}\rho_{\alpha}^{\pi}$$

and the system can now be solved because the unknown A_c is expressed as a function of the unknowns ρ^{π} [108]. T_{1e} has been found to be 10^{-11} s; the values for the other parameters are reported in Table 7.13. It appears that the ligand-centered dipolar broadening is very effective; it is also effective for the meso carbon on which there is no direct σ spin delocalization but there is some π spin density. In this way the peculiar feature that the α carbon signal is narrower than that of the β carbon, can be accounted for on the basis of larger π spin density on the latter carbon atom [108].

Through a knowledge of ρ_{β}^{π} and the β proton T_2 data, the contact and metal- and ligand-centered contributions to the proton relaxation have been factorized out (Table 7.13).

Table 7.13 Contribution to the Isotropic Shift (ppm) and Linewidth (Hz) of the ^{13}C Resonances in FeTPPCl, and to the Linewidth of the Pyrrole Protons [108]

Contribution		α Pyrrole ^{13}C	β Pyrrole ^{13}C	Meso ^{13}C	β Pyrrole ^1H
$(\Delta\nu/\nu_0)^{iso}$		+1,010	+1,150	+374	
$(\Delta\nu/\nu_0)^{MC}_{dip}$		+24	+9	+17	
$(\Delta\nu/\nu_0)^{LC}_{dip}$		−38	−70	−67	
$(\Delta\nu/\nu_0)^{con}$	π	< -350	+430	+424	
	σ direct	$> +1{,}400$	+781	assumed 0	
	total	+1,024	+1,211	+424	
$\Delta\nu^{tot}$		450	900	450	251
$\Delta\nu^{MC}_{dip}$		136	18	66	84
$\Delta\nu^{LC}_{dip}$		16	467	333	151
$\Delta\nu^{con}$		298	415	51	16
ρ^a		5.0×10^{-3}	9.2×10^{-3}	8.8×10^{-3}	

[a] The ρ values are normalized to 1.

^{13}C NMR analysis has provided evidence for large π electron spin density on the β carbon. This observation has gone undetected in proton NMR because of the superimposition of the effect of σ spin delocalization.

7.5 IRON-SULFUR PROTEINS

Iron-sulfur proteins have attracted the attention of researchers since the late 1960s; that is, since the application of the isotropic shift technique to biological systems. These proteins represent fascinating, naturally occurring systems with antiferromagnetically coupled metal ions (Figure 7.2), the metal ions being iron(II), iron(III), or both. Despite their extensive analysis, however, the assignment of the signals is still not completely defined. Models have been elegantly synthesized by Holm *et al.* [116]. The only monometallic protein is rubredoxin, for which no isotropically shifted signals could be detected [117]. Other iron-sulfur proteins provide, in general, well-shaped ^1H NMR spectra with patterns that can be related to the type of cluster and overall oxidation state. Detection of ^1H NMR signals from residues bound to iron(III) is probably due to magnetic coupling of the latter ion with iron(II).

7.5.1 ^1H NMR Spectra of Cysteine Protons in 2Fe-2S Ferredoxins

Reduced 2Fe-2S ferredoxins containing one Fe(III) and one Fe(II) show seven signals, each with an intensity of unity, between 10 and 45 ppm downfield from 2,2-dimethyl-2-silapentane-5-sulphonate (DSS) [118–120]. Ferredoxins from *A.*

variabilis and spinach also exhibit a further signal slightly upfield from DSS [121, 122]. A most striking property of the signals is that four of them move away from the diamagnetic region with increasing temperature (Figure 7.20). Presumably, this arises from magnetic coupling of two nonequivalent iron sites. Protons associated with iron(III) experience contact shifts that decrease with increasing temperature, while those associated with iron(II) have contact shifts that increase

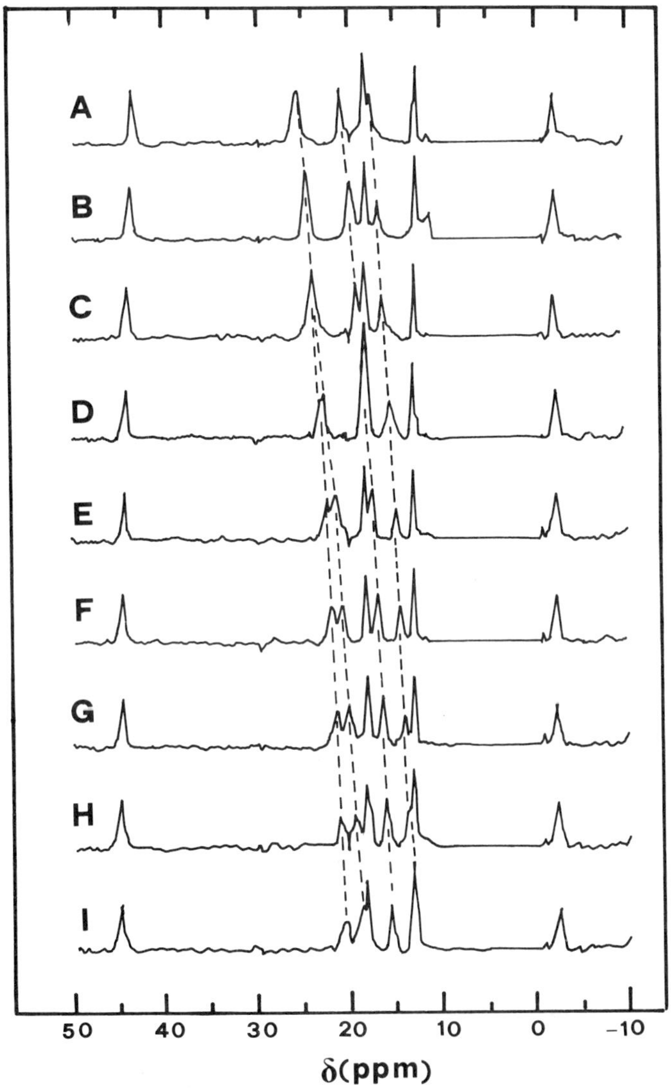

Figure 7.20 360 MHz ^{1}H NMR spectra of reduced *A. variabilis* (2Fe-2S) ferredoxin as a function of temperature: **A** 30; **B** 26; **C** 22; **D** 20; **E** 16; **F** 14; **G** 12; **H** 10; **I** 8 °C. Chemical shifts are in ppm from DSS. Reproduced from [121], with permission; copyright © 1983 by American Chemical Society.

with increasing temperature. In early studies, there was a tendency to assign all the signals to β-CH_2 protons; the inequivalence between the two protons of each CH_2 would arise from different orientation [123], according to Equation (2.16). Actually, the β-CH_2 protons close to iron(III) are expected to experience larger shifts (see Section 7.5.2); therefore the possibility that some of the signals associated with iron(III) are due to α-CH has been considered. In particular, the most downfield of the signals (around 40 ppm) has been assigned to the α-CH associated with iron(III) [121]. Recently, evidence has been found in spinach ferredoxin of four additional resonances above 100 ppm downfield, whose shifts decrease with increasing temperature [122] (Figure 7.21). These signals could arise from the β-CH_2 protons of the cysteines coordinated to iron(III) (see next section).

Oxidized 2Fe-2S proteins are diamagnetic at low temperatures and become paramagnetic at higher temperatures—>77 K for algal ferredoxins. At room temperature the NMR spectra of several oxidized ferredoxins show a broad signal between 30 and 40 ppm [116, 119, 120] and another signal at 15 ppm; if the latter has an intensity of unity, the broad signal has an intensity of six to eight [120]. The shifts increase with increasing temperature.

7.5.2 A Theoretical Model for 2Fe-2S Ferredoxins

A theoretical treatment is available for 2Fe-2S ferredoxins which enables the isotropic shift patterns and the temperature dependence of the shifts to be predicted [124]. This treatment assumes that the zero field splitting of each metal ion in either the oxidized or reduced state is small compared to the magnetic coupling, and that the resultant spin is quantized about the applied field. Only the contact term is considered; furthermore, it is assumed that iron 1 is responsible for the shifts of protons of cysteines bound to it and iron 2 for the shifts of protons of its own ligands.

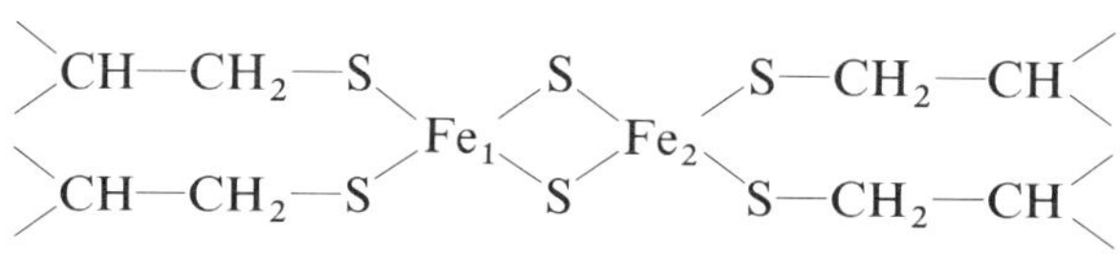

Magnetic coupling is taken into account by evaluating $\langle S_{iz}\rangle (i = 1, 2)$, as defined in Equation (2.3), over the magnetic coupled states characterized by the resultant spin quantum number S'. The contact shifts of protons, i, coupled with iron 1 are given by

$$\left(\frac{\Delta\nu}{\nu_0}\right)_i^{\text{con}} = a_i \langle S_{1z} \rangle I_{iz}$$

and those of protons, j, coupled with iron 2 by

$$\left(\frac{\Delta\nu}{\nu_0}\right)_j^{\text{con}} = a_j \langle S_{2z} \rangle I_{jz}$$

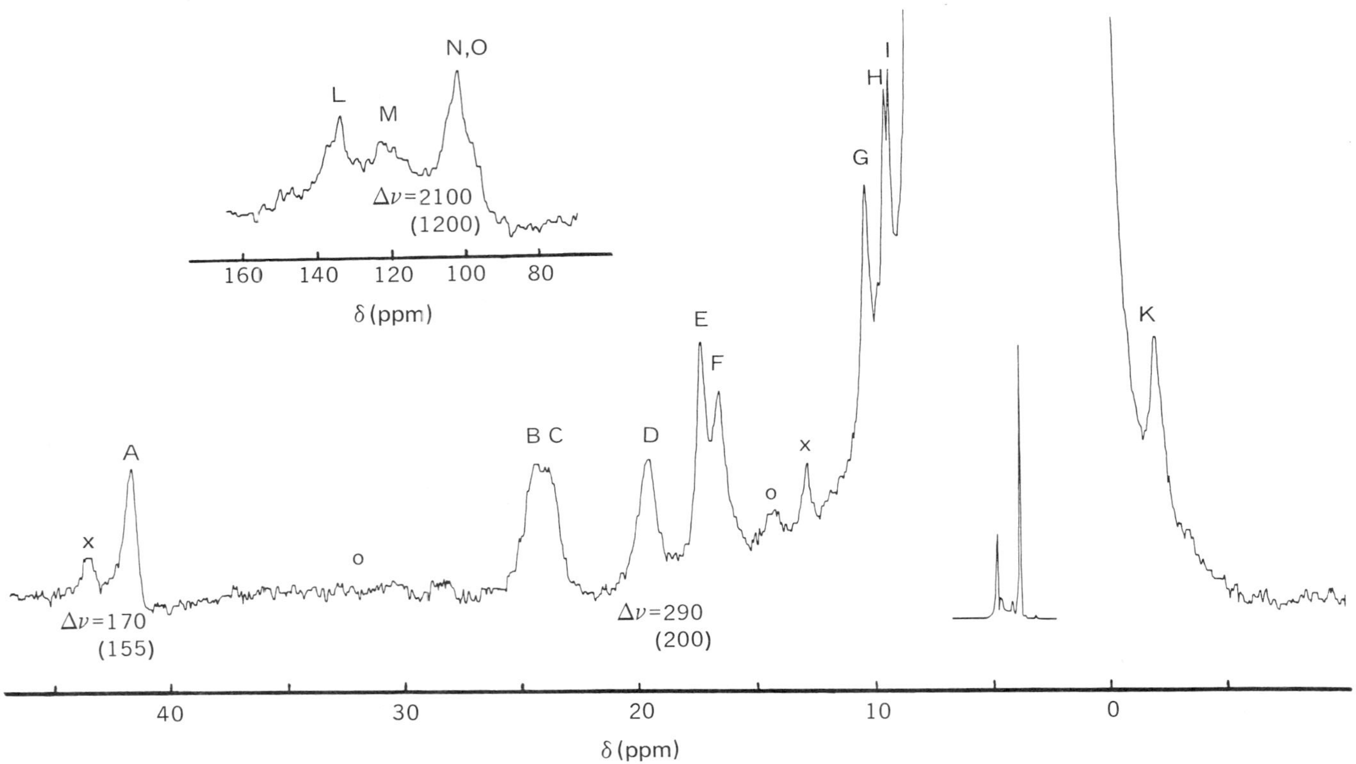

Figure 7.21 300 MHz ^{1}H NMR spectra of 2Fe-2S ferredoxin from spinach at 24°C [122]. Signals A and E are assigned to α-CH's of cysteines bound to iron(III). Signals B, C, D, F are assigned to β-CH's of cysteines bound to iron(II). For signals L through O, see text. Chemical shifts are in ppm from TMS. The linewidths, $\Delta\nu$ (Hz), at 300 MHz and 60 MHz (values in parentheses) for some signals are also shown.

$\langle S_{iz} \rangle$ is then expressed as

$$\langle S_{iz} \rangle = \frac{\sum_{S'M_{s'}} \exp\left(-\frac{E_{S'M_{s'}}}{kT}\right) \langle S'M_{s'}|S_{iz}|S'M_{s'} \rangle}{\sum_{S'M_{s'}} \exp\left(-\frac{E_{S'M_{s'}}}{kT}\right)}$$

where S' is the total spin quantum number, with $M_{s'}$ ranging from $-S'$ to $+S'$, and $E_{S'M_{s'}}$ is the energy of the $S'M_{s'}$ states—in the absence of an external magnetic field—which is a function of the exchange coupling constant, J.

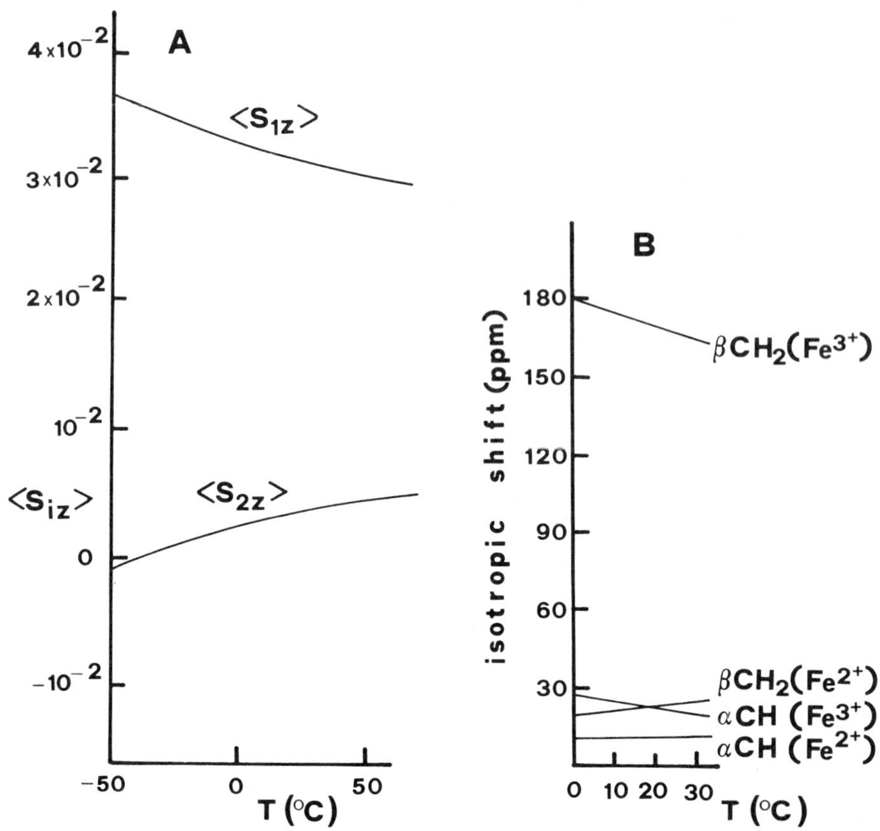

Figure 7.22 **A**: Temperature dependence of $\langle S_{iz} \rangle$ at the two irons in a high spin ferric (1) system antiferromagnetically coupled to high spin ferrous (2) system. **B**: Predicted isotropic shift range and temperature dependence for the cysteine protons in a reduced 2Fe-2S system. The calculated curves include coupling of the Fe(III) ion with the $^5A_1$ ground state of the Fe(II) ion and the excited $^5B_1$ level, which is about 400 cm^{-1} higher in energy. The two levels arise from the degenerate ^5E state in T_d symmetry. The J values for the two states have been taken to be 110 and 350 cm^{-1}, respectively. Reproduced from [124], with permission; copyright 1971 by Elsevier Biomedical Press B.V..

The matrix elements of the type $\langle S'M_{s'}|S_{1z}|S'M_{s'}\rangle$ are evaluated by applying the operator S_z over the states S_1M_{s1} of iron 1 present in the total $|S'M_{s'}\rangle$ function. The same holds for $\langle S_{2z}\rangle$. The average value of S_{iz} over the different S functions depends on the value of S', and may even change sign for different values of S'. Therefore, $\langle S_{iz}\rangle$ dramatically changes with temperature, since the different S states will have different weight in the summation.

In the case of reduced ferredoxins, the zero field splitting parameters of the isolated ions do not sensibly affect the $\langle S_{iz}\rangle$ values. The expected temperature dependences of $\langle S_{iz}\rangle$ and therefore of the contact shifts that are in accord with the experimental patterns are reported in Figure 7.22. By further assuming that $a_i = a_j$—that is, the proton hyperfine coupling constants are the same at both iron atoms despite the different oxidation states—the relative magnitude of the contact shifts for i and j protons can be estimated. In reduced 2Fe-2S systems, the β protons attached to iron(III) are expected to be shifted far downfield, whereas the α protons of iron(III) are expected to fall in the range of the β protons of iron(II). Recent data on spinach ferredoxin are quite consistent with this theoretical prediction [122]. T_1 measurements could also prove useful in discriminating between α- and β-CH protons.

7.5.3 Other Iron-Sulfur Systems

Reduced *Clostridium acidi urici* ferrodoxin, containing 2Fe(III)6Fe(II) in two 4Fe-4S clusters, shows sixteen well-resolved resonances in the range of 12 to 45 ppm downfield [125]. Eight of these move upfield while eight shift markedly downfield with increasing temperature. This finding is surprisingly similar to that observed for reduced 2Fe-2S ferredoxin. The spectra of oxidized *Clostridium acidi urici* ferredoxin, on the other hand, show much less shifted signals [126, 127] (Figure 7.23). By isotopic labeling, the peaks at 10.0 and 9.8 ppm from DSS are assigned to cysteine α protons resonances, while the others are assigned to some of the sixteen cysteine β protons [127]. Two doublets in the proton coupled ^{13}C NMR spectra of the same protein (obtained with ^{13}C enriched cysteines) located at 113 and 115 ppm from CS_2 collapse into singlets when the α proton resonances are selectively decoupled. The isotropic shifts increase with increasing temperature, presumably as a result of increasing magnetic susceptibility [127]. The temperature behavior of reduced High Potential Iron Proteins, HiPIPs, which also contain 2Fe(II)-2Fe(III) in a single 4Fe-4S cluster, is quite similar [128, 129], although the signal pattern is somewhat different.

Several oxidized HiPIPs containing three iron(III) and one iron(II) show six resonances in the region 12 to 50 ppm downfield from DSS, which are assigned to α-CH or β-CH_2 protons of cysteine residues [129]. In the case of *Chromatium* HiPIP, only five signals are observed: four of them are reported [128] to exhibit an almost Curie law dependence over the accessible temperature range, while the fifth, which has an intensity greater than unity, increases in shift with temperature [128]. Three other signals, one at about 100 ppm downfield and two at about 30 ppm upfield, are observed for *C. vinosum* and *R. gelatinosa* oxidized HiPIP

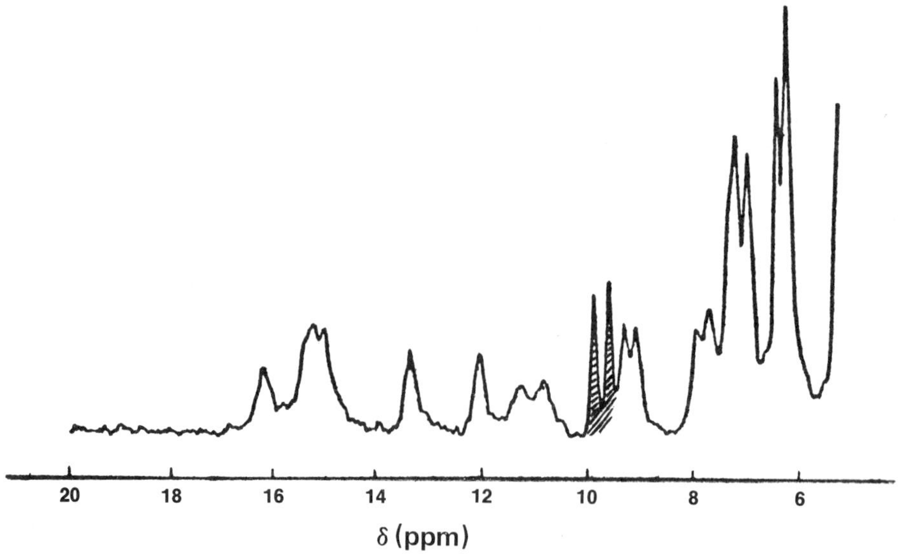

Figure 7.23 220 MHz ^1H NMR spectrum of oxidized *Clostridium acidi urici* $(4Fe–4S)_2$ ferrodoxin at 15 °C [127]. Dashed signals are assigned to α-CH protons. Chemical shifts are in ppm from DSS.

[129]. Whereas the first signal could be due to a β-CH, the assignment of the upfield signals is less straightforward. It has been proposed that a proton hydrogen bonded to inorganic sulfide is responsible for such shifts [129]. It is also possible that the latter signals are unique for the high potential oxidized 4Fe centers.

If both the oxidized and reduced forms of an iron-sulfur protein are present in solution, the electron exchange rate is generally slow on the NMR time scale, unless the proteins contain two 4Fe-4S clusters. In the latter case, an intramolecular process has been proposed to aid electron transfer [127]. Under slow-exchange conditions, saturation transfer techniques could be of use in relating the signals of the oxidized and reduced forms.

Anaerobic treatment of holoproteins with excess thiol removes *n*Fe-*n*S cores providing $Fe_2S_2(SR)_4^{2-}$ or $Fe_4S_4(SR)_4^{2-}$. If the thiol is p-$CF_3C_6H_4SH$, the ^{19}F contact shift varies depending on whether the extruded products are 2Fe-2S or 4Fe-4S clusters. Therefore, identification and quantification of Fe-S cores becomes possible. This procedure has been successfully applied to milk xanthine oxidase [130].

Studies have also been recently performed on proteins containing 3Fe clusters (either 3Fe-3S or 3Fe-4S). The peaks assigned to the 3Fe moiety for two different proteins [120, 131, 132] in their oxidized states—that is, with three iron(III) ions—are reported in Figure 7.24. These signals are assigned to α-CH protons of the cysteines. The two most shifted signals undergo decreasing shifts with increasing temperature; the reverse holds for the remaining signals. The reduced clusters do not show any isotropically shifted signal in ^1H NMR spectra [120].

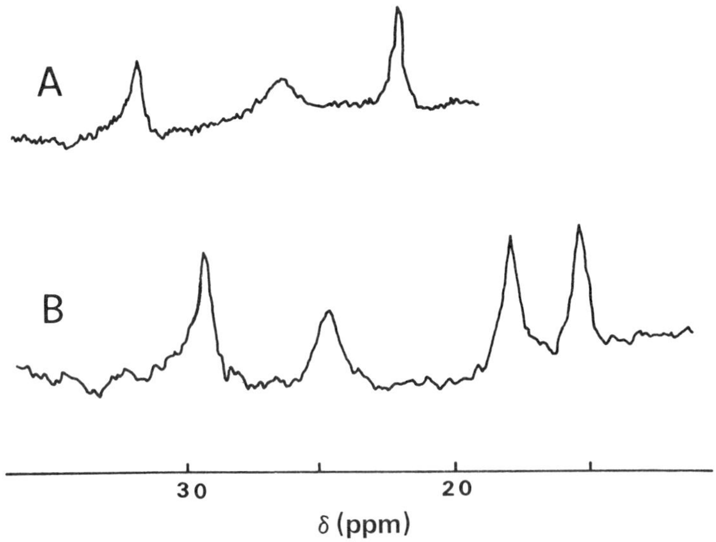

Figure 7.24 270 MHz ^{1}H NMR spectra of the oxidized forms of 3Fe ferredoxins from **A** *A. vinelandii* and **B** *D. gigas* [131, 132]. Chemical shifts are in ppm from DSS. Spectrum **B** reproduced from [132], with permission; copyright 1977 by Elsevier Biomedical Press B.V..

7.6 OTHER IRON(III) SYSTEMS

With the exception of heme and iron-sulfur proteins, NMR has seldom been used for the investigation of proteins and models containing paramagnetic iron. However, a few systems containing high spin iron(III) have been recently studied, and these are considered here. High spin iron(III) generally gives rise to broad NMR signals, but the linewidth is dependent on the particular system—for example, no isotropically shifted signal can be detected for iron(III)-transferrin-bicarbonate. The spectrum of FeSALEN-catecholate is shown in Figure 7.25 [133]. This spectrum belongs to a series of spectra of FeSALEN-phenolate and FeSALEN-catecholate complexes, which are models for dioxygenases. In each case, the spectra show an alternation of the shifts indicating π spin density on both salycylaldimine and phenolate moieties [133].

The binding mode of catecholate type of substrates to cathechol dioxygenases and protocatechuate 3,4-dioxygenase has been investigated through ^{1}H and ^{2}H NMR, showing that the substrate binds in a monodentate fashion to the former and in a bidentate fashion to the latter enzyme [134].

The ^{1}H NMR spectrum of reduced uteroferrin, which contains both Fe(II) and Fe(III) antiferromagnetically coupled, is reported in Figure 7.26 [135]. From the temperature dependence of the shifts, a value for J of about 20 cm^{-1} has been calculated through an equation of the type depicted by Equation (2.28) [135]. Furthermore, from a consideration of model compounds and the finding from

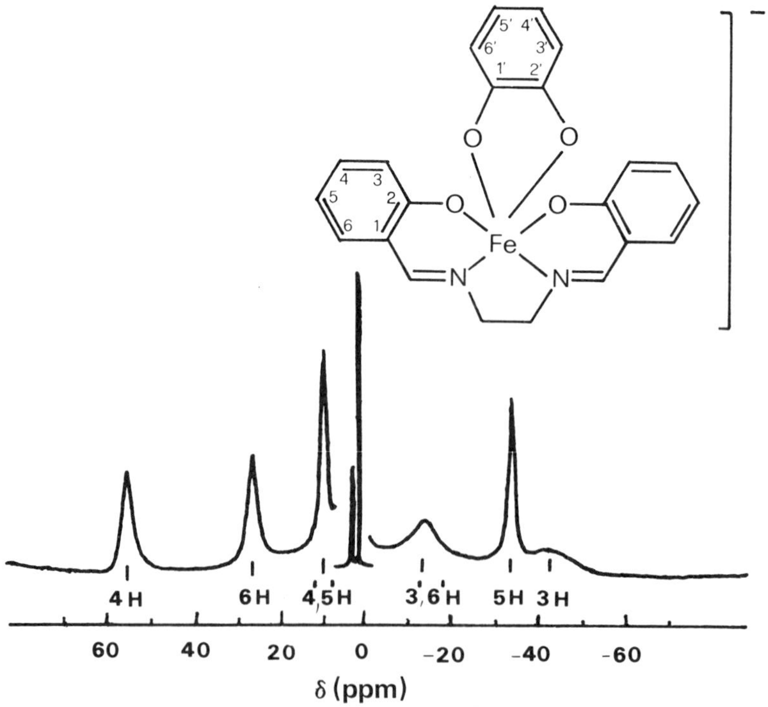

Figure 7.25 300 MHz ^{1}H NMR spectrum of [FeSALEN-catecholate]$^-$ in acetone-d_6. The signal assignment refers to the ring numbering shown in the scheme. Chemical shifts are in ppm from TMS. Reproduced from [133], with permission; copyright © 1982 by American Chemical Society.

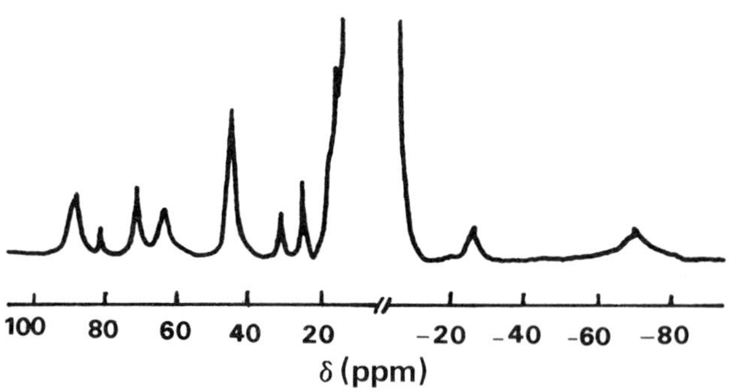

Figure 7.26 300 MHz ^{1}H NMR spectrum of reduced uteroferrin in H_2O (100 mM acetate buffer) at pH 4.9 and 30 °C [135]. Chemical shifts are in ppm from DSS.

resonance Raman spectra that there is a phenolate group in the coordination sphere, the signal at −70 ppm has been assigned to ortho protons of tyrosines and those at 63 and 70 ppm to meta protons. The tyrosinate residue would coordinate to the ferric site. Three exchangeable protons at 89, 44, and −25 ppm have also been recognized from a comparison between spectra in D_2O and H_2O, and the two downfield signals tentatively assigned to histidine protons.

The ^{1}H NMR spectra of some hemerythrins are now under investigation: a number of signals due to exchangeable protons is observed for iron(III)-iron(III) and iron(II)-iron(III) derivatives. Information on the structure of the metal site and on the magnetic coupling are obtained [136].

7.7 WATER PROTON RELAXATION

Water proton relaxation measurements of iron proteins are almost entirely concerned with high spin iron(III) systems because they are relatively effective relaxation reagents. High spin iron(III) has the same electronic configuration as manganese(II), whose relaxation properties are discussed in Chapter 8 in terms of mean squared fluctuations of zero field splitting. High spin iron(III) in porphyrins experiences much larger static zero field splitting, which is expected to cause shorter and possibly more spread out τ_s values.

Water proton relaxation data at different degrees of sophistication are available for catalase [137–139], methemoglobin [140], and their fluoride adducts, met-myoglobin [139], horseradish peroxidase [141], and cytochrome c' [142, 34] (Figure 7.27). The striking difference in relaxation enhancements between fluoromethemoglobin, catalase, fluorocatalase, and cytochrome c' at pH 2, on one hand, and methemoglobin, met-myoglobin, and cytochrome c' at pH 4, on the other, could at first glance be related to the presence or absence of water in the axial coordination position; however, present knowledge on such systems is strongly against this interpretation. For example, aquamethemoglobin shows much smaller water proton relaxation rates than fluoromethemoglobin, in which fluoride substitutes water in the axial position [140].

Another possible explanation for this difference could be that there are large variations in electronic relaxation times, τ_s, from one system to another. If the data are fitted to the Solomon equation, τ_s values ranging from 2×10^{-9} to 3×10^{-11} s are in fact obtained. However, the data cannot be quantitatively analyzed in terms of the Solomon equation because of the presence of large zero field splitting [143, 144]: as discussed in Section 3.5, for half-integer spins, when the zero field splitting energy is larger than $\hbar\tau_s^{-1}$, a large fraction of the low field relaxivity is quenched because some of the electron transitions occur at much larger frequency than the electron Larmor frequency, ω_s. The NMRD profiles are expected to again reach the limit predicted by Solomon only when $g_e\mu_B B_0 \gg 2D$ [143, 144]; however, for $2D$ values typical of high spin iron(III) porphyrins (as high as 40 cm^{-1}), such a limit is never reached in the range of magnetic fields

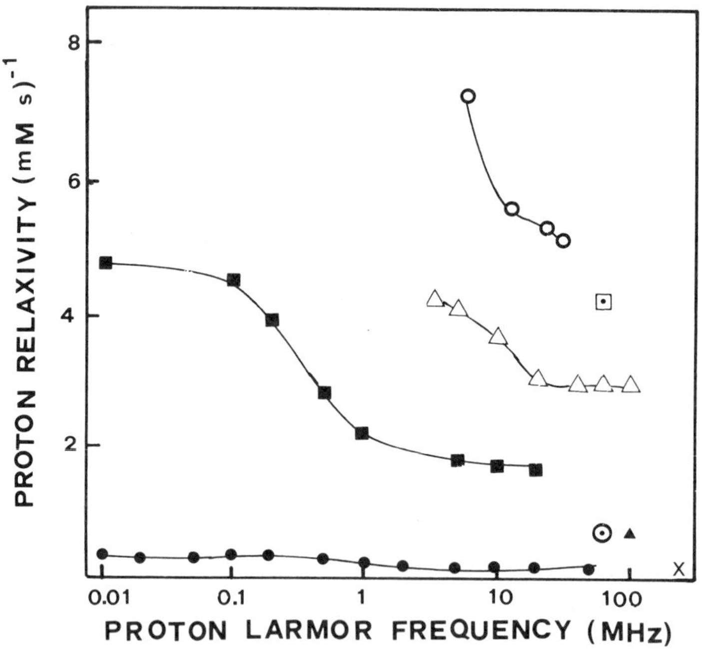

Figure 7.27 Water proton NMRD data for several high-spin heme proteins: ■ fluoromethemoglobin [140]; ● methemoglobin [140]; X metmyoglobin [139]; △ catalase [137–139]; ○ catalase + fluoride [138]; ⊙ cytochrome *c′* pH 4 [142]; ⊡ cytochrome *c′* pH 2 [142]; ▲ horseradish peroxidase [141].

covered by the data of Figure 7.27. Other possible sources of deviations from the Solomon behavior, like the presence of hyperfine coupling with the metal nucleus [143] (see Section 6.5.1), should be relatively less important, as long as the electronic relaxation time, τ_s, is not longer than 10^{-10} s. Moreover, g anisotropy is probably only a minor perturbation (Section 3.5) when large zero field splitting is present [143]. The main source of deviation from the Solomon behavior seems therefore to be the presence of zero field splitting.

However, further complications may arise if the water in the first coordination sphere of iron(III) is in a slow-exchange situation, so that most of the water proton relaxation enhancement comes from second sphere or outer-sphere water. From a comparative analysis of aquamethemoglobin and fluoromethemoglobin NMRD data, including outer-sphere effects, it has been proposed that a second sphere water, hydrogen bonded to the coordinated fluoride, is responsible for the relatively large ^{1}H T_1^{-1} values observed in fluoromethemoglobin [140].

For all of the foregoing reasons, ^{1}H NMR data on systems at a single frequency or in a limited range of frequencies have little meaning in the absence of an approximate, independent knowledge of τ_s. The uncertainty in defining a range of τ_s values for high spin iron(III) porphyrins could in principle be overcome by looking at the relaxation of protons other than those of water—that is, heme protons. In recent years attention has been devoted to this point, and the field dependence of the heme methyl linewidths is now available, for example, for met-myoglobin, and the acidic and basic forms of *R. rubrum* cytochrome c' [29, 35]. Curie spin relaxation (Sections 3.4 and 7.3.3) may also strongly contribute to the T_2 values of high molecular weight systems: this is indeed the case for ferricytochrome c' but not for met-myoglobin. From the field-independent contribution to the line broadening of the ferricytochrome c' (or the full linewidth of the met-myoglobin) and the known methyl proton-metal distance, rough estimates of τ_s of about 10^{-11} s have been made [29]. The expected relaxation rate of an axially coordinated, fast-exchanging water molecule could also be estimated, but such values may be affected if differences in relaxation efficiency along the z axis, which may arise from the presence of zero field splitting, are neglected. Furthermore, contact and ligand-centered dipolar contributions to the methyl proton relaxation might be operative, both of which are expected to depend on the amount of electron spin delocalization on the adjacent carbon, and hence to be reflected by the contact shifts. Although the methyl protons do in fact show different contact shifts, they nearly all have the same linewidth, thus suggesting that such contributions are not so important [29].

Now that T_1 measurements are routinely feasible on isotropically shifted signals in proteins (Section 5.4.3), τ_s may be estimated without the uncertainty introduced by factorizing out Curie spin contributions from the T_2 values. For example, an estimate of τ_s on horseradish peroxidase indicates a value of the order of 10^{-11} s [145]. Available room temperature EPR linewidths also place τ_s values in the range 10^{-10} to 10^{-11} s [139]. Therefore, the τ_s variability suggested from qualitative inspection of the NMRD data should actually be rather limited. Theoretical efforts should be directed towards an understanding of NMRD data in terms of parameters consistent with the indications available from independent experiments.

Besides heme proteins, water proton relaxation dispersion has only been measured on iron(III) transferrin, both in the mono- and the di-iron derivative [146, 147]. The data are shown in Figure 7.28 at two different temperatures. As a result of the complicated shape of the dispersion curves, it is impossible at the present time to do a quantitative analysis. However, the temperature dependence suggests that the important contribution is an underlying field-independent term arising from slowly exchanging coordinated water [146]. Such a contribution would account for as much as 70% of the overall relaxation at 25 °C and the low field part of the curve could then be treated as the $\omega_S^2\tau_S^2$ term, giving a τ_s of about 10^{-10} s. The high field rise could be due to the incoming $\omega_S^2\tau_v^2$ term, as observed in manganese(II) systems (see Section 8.2). Alternatively, the low field dispersion could be regarded as arising from outer-sphere effects. Whatever the case, the magnitude of the observed effect is consistent with the presence of exchangeable protons in the first coordination sphere. The two iron sites are almost indistinguishable from their relaxation behavior [147]. It is worthy noting here that the small EPR linewidth [148] and the failure to detect isotropically shifted signals in iron(III)$_2$ transferrin [145] point to a τ_s value somewhat longer than that of

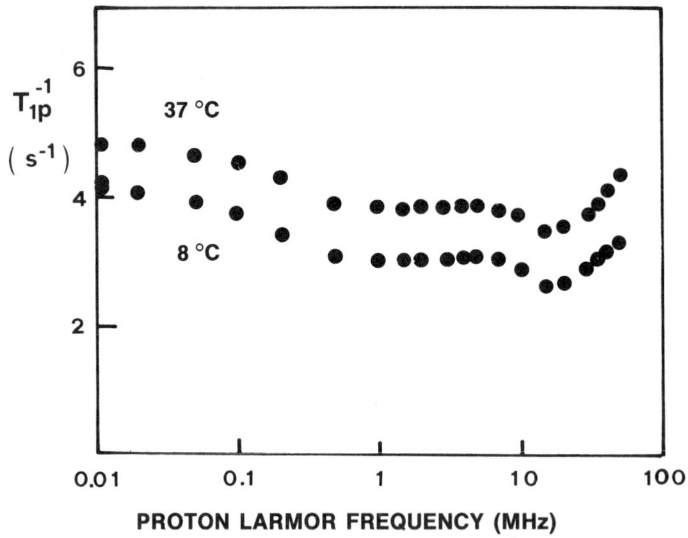

Figure 7.28 NMRD data for 1.4mM diferric transferrin at two different temperatures [147].

high spin iron(III) in heme proteins. This could arise from a relatively smaller zero field splitting in the nonheme system [149].

NMRD data are also available on low spin iron(III) systems. Oxidized cytochrome oxidase has been found to have a surprisingly high relaxation efficiency, that has been attributed to the heme a_3 moiety [150]. Frequency and temperature dependence of the data have yielded a very long τ_s value of 3×10^{-9} s, which is two to three orders of magnitude longer than that estimated for cyanohemoglobin and other low spin iron(III) proteins. However, τ_s values larger than 10^{-10} s have also been proposed for low spin cytochromes P-450 [151].

REFERENCES

1. Gibson, J. F., Hall, D. O., Thornley, J. H. M., and Whatley, F. R. *Proc. Natl. Acad. Sci. U.S.A.* **56** (1966) 987.
2. Gouterman, M. In *The Porphyrins*, D. Dolphin (ed.), Academic Press, New York, 1978, Volume 3, p. 1.
3. Phillippi, M. A., Baenzinger, N., and Goff, H. M. *Inorg. Chem.* **20** (1981) 3904.
4. Scheidt, W. R. In *The Porphyrins*, D. Dolphin (ed.), Academic Press, New York, 1978, Volume 3, p. 463.
5. Collman, J. P., Gagné, R. R., Reed, C. A., Robinson, W. T., and Rodley, G. A. *Proc. Natl. Acad. Sci. U.S.A.* **71** (1974) 1326.
6. Gregson, A. K. *Inorg. Chem.* **20** (1981) 81.
7. Scheidt, W. R., Cohen, I. A., and Kastner, M. E. *Biochemistry* **18** (1979) 3546.

8. Mashiko, T., Kastner, M. E., Spartalian, K., Scheidt, W. R., and Reed, C. A. *J. Am. Chem. Soc.* **100** (1978) 6354.
9. Boersma, A. D., and Goff, H. M. *Inorg. Chem.* **21** (1982) 581.
10. Wolberg, A., and Manassen, J. *J. Am. Chem. Soc.* **92** (1970) 2982.
11. Chin, D. H., Balch, A. L., and La Mar, G. N. *J. Am. Chem. Soc.* **102** (1980) 1446.
12. Chin, D. H., La Mar, G. N., and Balch, A. L. *J. Am. Chem. Soc.* **102** (1980) 5945.
13. Phillippi, M. A., and Goff, H. M. *J. Am. Chem. Soc.* **104** (1982) 6026.
14. Phillippi, M. A., and Goff, H. M. *J. Am. Chem. Soc.* **101** (1979) 7641.
15. Groves, J. T., Haushalter, R. C., Nakamura, M., Nemo, T. E., and Evans, B. J. *J. Am. Chem. Soc.* **103** (1981) 2884.
16. La Mar, G. N., and Walker, F. A. *J. Am. Chem. Soc.* **95** (1973) 6950.
17. La Mar, G. N., De Ropp, J. S., Latos-Grazynski, L., Balch, A. L., Johnson, R. B., Smith, K. M., Parish, D. W., and Cheng, R.-J. *J. Am. Chem. Soc.* **105** (1983) 782.
18. La Mar, G. N., and Walker, F. A. In *The Porphyrins*, D. Dolphin (ed.), Academic Press, New York, 1979, Volume 4, p. 61.
19. Walker, F. A., and La Mar, G. N. *Ann. N.Y. Acad. Sci.* **206** (1973) 328.
20. Satterlee, J. D., and La Mar, G. N. *J. Am. Chem. Soc.* **98** (1976) 2804.
21. Goff, H., and La Mar, G. N. *J. Am. Chem. Soc.* **99** (1977) 6599.
22. Goff, H., La Mar, G. N., and Reed, C. A. *J. Am. Chem. Soc.* **99** (1977) 3641.
23. Mispelter, J., Momenteau, M., and Lhoste, J. M. *Mol. Phys.* **33** (1977) 1715.
24. La Mar, G. N. *Pure Appl. Chem.* **40** (1974) 13.
25. La Mar, G. N., Eaton, G. R., Holm, R. H., and Walker, F. A. *J. Am. Chem. Soc.* **95** (1973) 63.
26. Behere, V., Birdy, R., and Mitra, S. *Inorg. Chem.* **21** (1982) 386.
27. Mun, S. K., Mallick, M. K., Mishra, S., Chang, J. C., and Das, T. P. *J. Am. Chem. Soc.* **103** (1981) 5024.
28. Budd, D. L., La Mar, G. N., Langry, K. C., Smith, K. M., and Nayyir-Mazhir, R. *J. Am. Chem. Soc.* **101** (1979) 6091.
29. La Mar, G. N., Jackson, J. T., and Bartsch, R. G. *J. Am. Chem. Soc.* **103** (1981) 4405.
30. Maltempo, M. M., Moss, T. H., and Cusanovich, M. A. *Biochim. Biophys. Acta* **351** (1974) 237.
31. Emptage, M. H., Zimmermann, R., Que, L. Jr., Münck, E., Hamilton, W. D., and Orme-Johnson, W. H. *Biochim. Biophys. Acta* **495** (1977) 12.
32. Rawlings, J., Stephens, P. J., Jafie, L. A., and Kamen, M. D. *Biochemistry* **16** (1977) 1725.
33. Jackson, J. T., La Mar, G. N., and Bartsch, R. G. *J. Biol. Chem.* **258** (1983) 1799.
34. Emptage, M. H., Xavier, A. V., Wood, J. M., Alsaadi, B. M., Moore, G. R., Pitt, R. C., Williams, R. J. P., Ambler, R. P., and Bartsch, R. G. *Biochemistry* **20** (1981) 58.
35. La Mar, G. N., Budd, D. L., Smith, K. M., and Langry, K. C. *J. Am. Chem. Soc.* **102** (1980) 1822.
36. Shulman, R. G., Glarum, S. H., and Karplus, M. *J. Mol. Biol.* **57** (1971) 93.
37. Horrocks, W. DeW. Jr., and Greenberg, E. S. *Biochim. Biophys. Acta* **322** (1973) 38.
38. La Mar, G. N., and Walker, F. A. *J. Am. Chem. Soc.* **95** (1973) 1782; *J. Am. Chem. Soc.* **95** (1973) 1790.
39. Traylor, T. G., and Berzinis, A. P. *J. Am. Chem. Soc.* **102** (1980) 2844.
40. De Ropp, J. S., La Mar, G. N., Smith, K. M., and Langry, K. C. *J. Am. Chem. Soc.* **106** (1984) 4438.

41. Burns, P. D., and La Mar, G. N. *J. Biol. Chem.* **256** (1981) 4934.
42. Keller, R. M., and Wüthrich, K. *Biochim. Biophys. Acta* **533** (1978) 195.
43. Wüthrich, K. *Struct. Bonding* **8** (1970) 53.
44. Smith, M., and McLendon, G. *J. Am. Chem. Soc.* **103** (1981) 4912.
45. Senn, H., Eugster, A., and Wüthrich, K. *Biochim. Biophys. Acta* **743** (1983) 58.
46. Keller, R. M., and Wüthrich, K. *Biochem. Biophys. Res. Commun.* **83** (1978) 1132.
47. Senn, H., and Wüthrich, K. *Biochim. Biophys. Acta* **743** (1983) 69.
48. Senn, H., and Wüthrich, K. *Biochim. Biophys. Acta* **746** (1983) 48.
49. Keller, R. M., Schejter, A., and Wüthrich, K. *Biochim. Biophys. Acta* **626** (1980) 15.
50. Keller, R. M., and Wüthrich, K. *Biochim. Biophys. Acta* **668** (1981) 307.
51. Ulrich, E. L., Krogmann, D. W., and Markley, J.L. *J. Biol. Chem.* **257** (1982) 9356.
52. Smith, G. M. *Biochemistry* **18** (1979) 1628.
53. Timkovich, R., Cork, M. S., and Taylor, P. V. *Biochemistry* **23** (1984) 3526.
54. Timkovich, R., and Cork, M. S. *Biochemistry* **23** (1984) 851.
55. Senn, H., Cusanovich, M. A., and Wüthrich, K. *Biochim. Biophys. Acta* **785** (1984) 46.
56. Senn, H., Guerlesquin, F., Bruschi M., and Wüthrich, K. *Biochim. Biophys. Acta* **748** (1983) 194.
57. Falk, K.-E., and Angström, J. *Biochim. Biophys. Acta* **722** (1983) 291.
58. Moore, R., and Williams, R. J. P. *Eur. J. Biochem.* **103** (1980) 493.
59. Moore, R., and Williams, R. J. P. *Eur. J. Biochem.* **103** (1980) 503.
60. Moore, R., and Williams, R. J. P. *Eur. J. Biochem.* **103** (1980) 513.
61. Moore, R., and Williams, R. J. P. *Eur. J. Biochem.* **103** (1980) 523.
62. Boswell, A. P., Moore, G. R., Williams, R. J. P., Chien, J. C. W., and Dickinson, L. C. *J. Inorg. Biochem.* **13** (1980) 347.
63. Keller, R. M., and Wüthrich, K. *Biochim. Biophys. Acta* **285** (1972) 326.
64. La Mar, G. N., Burns, P. D., Jackson, J. T., Smith, K. M., Langry, K. C., and Strittmatter, P. *J. Biol. Chem.* **256** (1981) 6075.
65. Cutnell, J. D., La Mar, G. N., and Kong, S. B. *J. Am. Chem. Soc.* **103** (1981) 3567.
66. Mayer, A., Ogawa, S., Shulman, R. G., Yamane, T., Cavaleiro, J. A. S., Rocha Gonsalves, A. M. D'A., Kenner, G. W., and Smith, K. M. *J. Mol. Biol.* **86** (1974) 749.
67. Ramaprasad, S., Johnson, R. D., and La Mar, G. N. *J. Am. Chem. Soc.* **106** (1984) 3632.
68. Ramaprasad, S., Johnson, R. D., and La Mar, G. N. *J. Am. Chem. Soc.* **106** (1984) 5330.
69. La Mar, G. N., Davis, N. L., Parish, D. W., and Smith, K. M. *J. Mol. Biol.* **168** (1983) 887.
70. Trehwella, J., and Wright, P. E. *Biochim. Biophys. Acta* **625** (1980) 202.
71. Bertrand, P., Theodule, F. X., Gayda, J. P., Mispelter, J., and Momenteau, M. *Chem. Phys. Lett.* **102** (1983) 442.
72. La Mar, G. N., De Ropp, J. S., Smith, K. M., and Langry, K. C. *J. Biol. Chem.* **255** (1980) 6646.
73. Hill, H. A. O., and Morallee, K. G. *J. Am. Chem. Soc.* **94** (1972) 731.
74. Degani, H. A., and Fiat, D. *J. Am. Chem. Soc.* **93** (1971) 4281.
75. Von Goldammer, F., and Zorn, H. *Biophys. Chem.* **3** (1975) 249.
76. La Mar, G. N., and Walker, F. A. *J. Am. Chem. Soc.* **94** (1972) 8607.
77. Satterlee, J. D., La Mar, G. N., and Frye, J. S. *J. Am. Chem. Soc.* **98** (1976) 7275.

78. Johnson, M. E., Fung, L. W.-M., and Ho, C. *J. Am. Chem. Soc.* **99** (1977) 1245.
79. La Mar, G. N., Budd, D. L., and Goff, H. *Biochem. Biophys. Res. Commun.* **77** (1977) 104.
80. Collman, J. P., Brauman, J. I., Collins, T. J., Iverson, B. L., Lange, G., Pettman, R. B., Sessler, J. L., and Walters, M. A. *J. Am. Chem. Soc.* **105** (1983) 3038.
81. Mispelter, J. T., Momenteau, M., and Lhoste, J.-M. *Biochimie* **63** (1981) 911.
82. Latos-Grazynski, L. *Biochimie* **65** (1983) 143.
83. La Mar, G. N., Budd, D. L., Sick, H., and Gersonde, K. *Biochim. Biophys. Acta* **537** (1978) 270.
84. Hochmann, J., and Kellerhals, H. *J. Magn. Reson.* **38** (1980) 23.
85. Mispelter, J., Momenteau, M., and Lhoste, J.-M. *J. Chem. Phys.* **72** (1980) 1003.
86. Buisson, G., Deronzier, A., Duée, E., Gans, P., Marchon, J.-C., and Regnard, J.-R. *J. Am. Chem. Soc.* **104** (1982) 6793.
87. Scholz, W. F., Reed, C. A., Lee, Y. J., Scheidt, W. R., and Lang, G. *J. Am. Chem. Soc.* **104** (1982) 6791.
88. Goff, H. M., and Phillippi, M. A. *J. Am. Chem. Soc.* **105** (1983) 7567.
89. Morishima, I., and Ogawa, S. *Biochemistry* **17** (1978) 4384.
90. La Mar, G. N., De Ropp, J. S., Smith, K. M., and Langry, K. C. *J. Biol. Chem.* **256** (1981) 237.
91. Satterlee, J. D., and Erman, J. E. *J. Biol. Chem.* **256** (1981) 1091.
92. Dolphin, D., Forman, A., Borg, D. C., Fajer, J., and Felton, R. H. *Proc. Natl. Acad. Sci. U.S.A.* **68** (1971) 614.
93. Loew, G. H., and Herman, Z. S. *J. Am. Chem. Soc.* **102** (1980) 6174.
94. Kimura, K., Peterson, J., Wilson, M., Cookson, D. J., and Williams, R. J. P. *J. Inorg. Biochem.* **15** (1981) 11.
95. Eley, C. G. S., Ragg, E., and Moore, G. R. *J. Inorg. Biochem.* **21** (1984) 295.
96. Kovalski, A. *Biochemistry* **4** (1965) 2382.
97. Santos, H., Moura, J. J. G., Moura, I., LeGall, J., and Xavier, A. V. *Eur. J. Biochem.* **141** (1984) 283.
98. Santos, H., Turner, D. L., Xavier, A. V., and LeGall, J. *J. Magn. Reson.* **59** (1984) 177.
99. Sheard, B., Yamane, T., and Shulman, R. G. *J. Mol. Biol.* **53** (1970) 35.
100. Moore, G. R., and Williams, G. *Biochim. Biophys. Acta* **788** (1984) 147.
101. Nagai, K., La Mar, G. N., Jue, T., and Bunn, H. F. *Biochemistry* **21** (1982) 842.
102. La Mar, G. N., and De Ropp, J. S. *J. Am. Chem. Soc.* **104** (1982) 5203.
103. Inubushi, T., and Yonetani, T. *FEBS Lett.* **160** (1983) 287.
104. Goff, H. *J. Am. Chem. Soc.* **99** (1977) 7723.
105. Morishima, I., and Inubushi, T. *J. Am. Chem. Soc.* **100** (1978) 3568.
106. Golding, R. M., Pascual, R. O., and Vrbancich, J. *Mol. Phys.* **31** (1976) 731.
107. Goff, H. M. *J. Am. Chem. Soc.* **103** (1981) 3714.
108. Mispelter, J., Momenteau, M., and Lhoste, J.-M. *J. Chem. Soc., Dalton Trans.* (1981) 1729.
109. Goff, H. M., Shimomura, E. T., and Phillippi, M. A. *Inorg. Chem.* **22** (1983) 66.
110. Shirazi, A., Leum, E., and Goff, H. M. *Inorg. Chem.* **22** (1983) 360.
111. Wüthrich, K., and Baumann, R. *Helv. Chim. Acta* **57** (1974) 336.
112. Goff, H. M., and Shimomura, E. *J. Am. Chem. Soc.* **102** (1980) 31.
113. Phillippi, M. A., Baenziger, N., and Goff, H. M. *Inorg. Chem.* **20** (1981) 3904.
114. Karplus, M., and Fraenkel, G. K. *J. Chem. Phys.* **35** (1961) 1312.

115. Goff, H. M., and Morgan, L. O. *Bioinorg. Chem.* **9** (1978) 61.
116. Holm, R. H., and Ibers, J. A. In *Iron-Sulfur Proteins*, W. Lovenberg (ed.), Academic Press, New York, 1977, Volume 3, Chapter 7.
117. Phillips, W. D., Poe, M., Weiher, J. F., McDonald, C. C., and Lovenberg, W. *Nature* (*London*) **227** (1970) 574.
118. Anderson, R. E., Dunham, W. R., Sands, R. H., Bearden, A. J., and Crespi, H. L. *Biochim. Biophys. Acta* **408** (1975) 306.
119. Poe, M., Phillips, W. D., Glickson, J. D., McDonald, C. C., and San Pietro, A. *Proc. Natl. Acad. Sci. U.S.A.* **68** (1971) 68. Saalmen, I., and Palmer, G. *Arch. Biochem. Biophys.* **150** (1972) 767.
120. Nagayama, K., Ozaki, Y., Kyogoku, Y., Hase, T., and Matsubara, H. *J. Biochem.* **94** (1983) 893.
121. Chan, T.-M., and Markley, J. L. *Biochemistry* **22** (1983) 6008.
122. Bertini, I., Lanini, G., and Luchinat, C. *Inorg. Chem.* **23** (1984) 2729.
123. Poe, M., Phillips, W. D., McDonald, C. C., and Lovenberg, W. *Nature* (*London*) **227** (1970) 574.
124. Dunham, W. R., Palmer, G., Sands, R. H., and Bearden, A. J. *Biochim. Biophys. Acta* **253** (1971) 373.
125. Poe, M., Phillips, W. D., McDonald, C. C., and Orme-Johnson, W. H. *Biochem. Biophys. Res. Commun.* **42** (1971) 705.
126. Phillips, W. D., and Poe, M. In *Iron-Sulfur Proteins*, W. Lovenberg (ed.), Academic Press, New York, 1973, Volume 2, p. 255.
127. Packer, E. L., Sweeney, W. V., Rabinowitz, J. C., Sternlicht, H., and Shaw, E. N. *J. Biol. Chem.* **252** (1977) 2245.
128. Phillips, W. D., Poe, M., McDonald, C. C., and Bartsch, R. G. *Proc. Natl. Acad. Sci. U.S.A.* **67** (1970) 682.
129. Nettesheim, D. G., Meyer, T. E., Feinberg, B. A., and Otvos, J. D. *J. Biol. Chem.* **258** (1983) 8215.
130. Wong, G. B., Kurtz, D. M. Jr., Holm, R. H., Mortenson, L. E., and Upchurch, R. G. *J. Am. Chem. Soc.* **101** (1979) 3078.
131. Sweeney, W. V. *J. Biol. Chem.* **256** (1981) 12222.
132. Moura, J. J. G., Xavier, A. V., Bruschi, M., and Le Gall, J. *Biochim. Biophys. Acta* **459** (1977) 278.
133. Heistand, R. H. II., Lauffer, R. B., Fikrig, E., and Que, L. Jr. *J. Am. Chem. Soc.* **104** (1982) 2789.
134. Lauffer, R. B., and Que, L. Jr., *J. Am. Chem. Soc.* **104** (1982) 7324.
135. Lauffer, R. B., Antanaitis, B. C., Aisen, P., and Que, L. Jr. *J. Biol. Chem.* **258** (1983) 14212.
136. Maroney, M. J., Lauffer, R. B., and Que, L. Jr. *J. Am. Chem. Soc.* **106** (1984) 6445.
137. Lanir, A., and Schejter, A. *Biochemistry* **15** (1976) 2590.
138. Vuk-Pavlović, S., and Williams-Smith, D. L. *Biochemistry* **16** (1977) 5465.
139. Hershberg, R. D., and Chance, B. *Biochemistry* **14** (1975) 3885.
140. Koenig, S. H., Brown, R. D., and Lindstrom, T. R. *Biophys. J.* **34** (1981) 397.
141. Lanir, A., and Schejter, A., *Biochem. Biophys. Res. Commun.* **62** (1975) 199.
142. Lanir, A., and Aviram, I. *Arch. Biochem. Biophys.* **166** (1975) 439.
143. Bertini, I., Luchinat, C., Mancini, M., and Spina, G. In *Magneto-Structural Correlations in Exchange Coupled Systems*, R. D. Willett, D. Gatteschi, and O. Kahn (eds.), Reidel, Dordrecht, 1985, p. 421.
144. Bertini, I., Luchinat, C., Mancini, M., and Spina, G. *J. Magn. Reson.* **59** (1984) 213.

145. Bertini, I., and Luchinat, C., unpublished results.
146. Koenig, S. H., and Shillinger, W. E. *J. Biol. Chem.* **244** (1969) 6520.
147. Unpublished results obtained at the IBM T. J. Watson Research Center, Yorktown Heights, New York.
148. Aisen, P., Aasa, R., Malmström, B. G., and Vänngård, T. *J. Biol. Chem.* **242** (1967) 2484.
149. Aasa, R. *J. Chem. Phys.* **52** (1970) 3919.
150. Blatt, Y., Benko, B., Pecht, I., and Vuk-Pavlović, S. *J. Biol. Chem.* **256** (1981) 2297.
151. Philson, S. B., Debrunner, P. G., Schmidt, P., and Gunsalus, I. C. *J. Biol. Chem.* **254** (1979) 10173.
152. Note added in proof. An $S = 1$ Fe(IV)P$(OCH_3)_2$ has been described and its ^1H NMR spectrum reported. The β-pyrrole protons were found at very high field, -37.5 ppm from TMS at -78 °C in CD_2Cl_2. All other resonances were observed near the expected diamagnetic positions [Groves, J. T., Quinn, R., McMurry, T. J., Nakamura, M., Lang, G., and Boso, B. *J. Am. Chem. Soc.* **107** (1985) 354].
153. Note added in proof. NOE measurements have been successfully used for peak assignment in high spin Fe(III)-containing met-myoglobin [Unger, S. W., LeComte, J. T. J., and La Mar, G. N. *J. Magn. Reson.* **64** (1985) 521].
154. Note added in proof. A detailed assignment of numerous signals in some ferricytochromes *c* has been elegantly performed which has allowed the authors to investigate the structure in solution or to compare the solution and crystal structures [Williams, G., Moore, G. R., Porteous, R., Robinson, M. N., Soffe, N., and Williams, R. J. P. *J. Mol. Biol.* **183** (1985) 409; Williams, G., Clayden, N. J., Moore, G. R., and Williams, R. J. P. *J. Mol. Biol.* **183** (1985) 447].

CHAPTER 8

Manganese

8.1 MANGANESE SYSTEMS

Manganese compounds exist in all the possible oxidation states up to +7, the most stable being +2. In higher oxidation states the element progressively loses the characteristics of a metal ion and exhibits those of the well-known oxoderivatives of manganese(VI) and manganese(VII).

Manganese(II) and manganese(III), which are the ions of biological relevance, are d^5 and d^4 ions, respectively, and generally give rise to high spin $S = \frac{5}{2}$ and $S = 2$ compounds. A few low spin complexes can be obtained with strong-field ligands, such as $Mn(CN)_6^{4-}$ $(S = \frac{1}{2})$ and $Mn(CN)_6^{3-}$ $(S = 1)$, but these will not be considered further. High spin manganese(II) and manganese(III) complexes show strikingly different nuclear relaxation properties: manganese(II) can be classified as a relaxation reagent, whereas manganese(III) is essentially a shift reagent.

8.2 THE ELECTRONIC PROPERTIES OF MANGANESE(II)

Manganese(II) has a $3d^5$ electronic configuration, giving rise to an orbitally nondegenerate ^6S ground term, as discussed in Section 1.2. The sixfold degeneracy of the ground state in O_h symmetry ($^6A_{1g}$) is only removed through spin-orbit coupling with excited levels that arise from other free ion terms. This perturbation splits the spin degeneracy into three Kramers' doublets in a symmetry lower than cubic (Figure 8.1). Manganese has only one isotope (^{55}Mn) with spin $I = \frac{5}{2}$; coupling of the nuclear magnetic moment with the unpaired electrons splits each ESR line into a sextet.

The ESR transitions of manganese(II) can be described by the following spin Hamiltonian:

$$\mathscr{H} = \mu_B \mathbf{B}_0 \cdot \mathbf{g} \cdot \hat{\mathbf{S}} + \hat{\mathbf{I}} \cdot \mathbf{A} \cdot \hat{\mathbf{S}} + D\left[\hat{S}_z^2 - \frac{1}{3}S(S+1)\right] + E(\hat{S}_x^2 - \hat{S}_y^2) \quad (8.1)$$

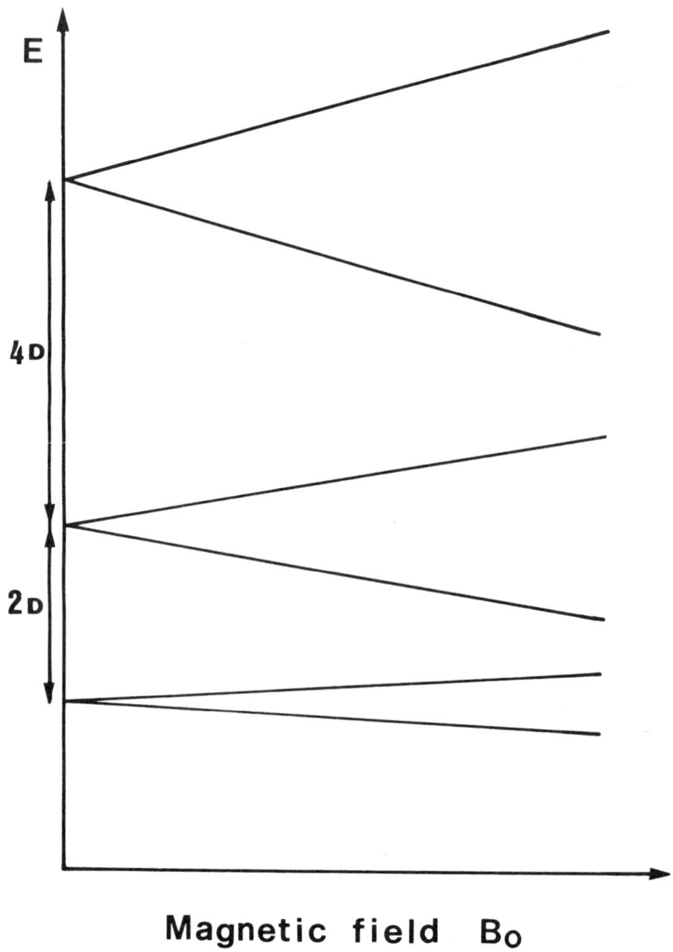

Figure 8.1 Splitting of the $^6A_1$ ground state of high spin manganese(II) ions in tetragonal symmetry. The magnetic field is applied along the z direction of the **D** tensor.

where the first term represents the Zeeman interaction, the second the hyperfine interaction with the metal nucleus, and the last two the zero field splitting characterized by the axial (D) and rhombic (E) components. The zero field splitting in manganese(II) is in general much smaller than that in high spin iron(III) systems (Chapter 7); typical D values for manganese(II) are in the range 0 to 1 cm^{-1} [1].

The Orbach and Raman mechanisms of electron relaxation (Section 3.7) are expected to be particularly inefficient for an ion with an S ground term and no other excited state with the same spin multiplicity. The only available relaxation mechanisms for such an ion are the time-dependent modulations of the interactions described by the spin Hamiltonian in Equation (8.1). It has long been established that of these, the dominant modulation for manganese(II) is that of the zero field splitting interaction [2], which may arise from rotation of the complex or, more probably, from distortions of the coordination sphere as a result of collisions with solvent molecules. Rubinstein *et al.* [3] have calculated the field dependence of T_{1e} and T_{2e} for solutions of manganese(II) ions: it is predicted that in general there are three longitudinal and three transverse relaxation times, the relative weights of which depend on the correlation time for electron relaxation, τ_v, and the electron Larmor frequency, ω_s. As long as $\omega_s\tau_v \ll 1$, there is only one longitudinal and one transverse relaxation time, and $T_{1e} = T_{2e}$. When $\omega_s\tau_v \geq 1$, the other contributions appear, but their relative weights for T_{1e} are still relatively small. The field dependence for T_{1e}^{-1} when $T_{1e} \neq T_{2e}$ is very similar to that given by Equation (3.27) [2]:

$$T_{1e}^{-1} = C\left(\frac{\tau_v}{1 + \omega_s^2\tau_v^2} + \frac{4\tau_v}{1 + 4\omega_s^2\tau_v^2}\right)$$

where C is proportional to the mean-squared fluctuation of the zero field splitting. From the wealth of EPR data on manganese complexes, typical T_{2e} values are in the range 10^{-8} to 10^{-9} s at X-band fields. These long values, coupled with the large values of the product $S(S+1)$ in the equations for nuclear relaxation, place manganese(II) among the most effective relaxation reagents. This very same property prevents the resolution of isotropically shifted signals in manganese(II)-containing proteins.

8.3 MANGANESE(II) IN SYSTEMS OF BIOLOGICAL RELEVANCE

Manganese(II) is a relatively abundant trace element. However, only a few proteins and enzymes can be classified as manganese(II) systems. From a chemical point of view, this is due to the lower stability constants of manganese(II) complexes with nitrogen and oxygen donors compared to, for instance, iron(III), cobalt(II), nickel(II), copper(II), and zinc(II). The biochemical consequence of this is that manganese(II) is often employed by living organisms as an activator of metabolic processes by forming labile complexes with proteins, enzymes, substrates, and simpler molecules. In a number of these cases, there is no specific requirement for manganese(II) since other metal ions, like magnesium(II) which is much more abundant, can activate the same process. Manganese(II) *in vivo* activates most of the kinases, over half of the ligases (synthetases), and some oxidoreductases. It is also found in a group of proteins, called lectins, which are able to bind carboyhydrates and subsequently agglutinate to the surface of various types of cells.

8.3.1 Kinases

The kinases constitute a class of metal-activated enzymes that transfer the terminal phosphoryl group of ATP to various substrates producing ADP and a phosphorylated molecule. For example, pyruvate kinase catalyzes the reaction

$$ATP^{4-} + CH_3{-}CO{-}COO^- \rightleftharpoons ADP^{3-} + CH_2{=}\underset{\displaystyle OPO_3^{2-}}{\underset{|}{C}}{-}COO^- + H^+$$

The enzyme from rabbit muscle is a tetramer of molecular weight 237,000. Each subunit has a strong binding site for either magnesium(II) or manganese(II) and a weaker binding site for potassium(I). The enzyme has an absolute requirement for both dipositive and monopositive cations.

8.3.2 Ligases

Pyruvate carboxylases are biotin-containing metalloenzymes that catalyze the reaction of pyruvate with HCO_3^- to give oxaloacetate. All of these metalloenzymes contain one mole of divalent metal ion per mole of biotin. The metal ion can be manganese(II), magnesium(II), or zinc(II). For example, the enzyme isolated from *S. cerevisiae* only contains zinc(II), the one from calf liver both magnesium(II) and manganese(II) in variable amounts, and the one from chicken liver only manganese(II).

Glutamine synthetase, which catalyzes the synthesis of glutamine from glutamate and NH_4^+ in a reaction involving ATP $\rightleftharpoons$ ADP transformation, has a molecular weight of 600,000 and contains twelve identical subunits. Each subunit requires manganese(II) for catalytic action. Other ligases, such as formyltetrahydrofolate synthetase and phosphoribosylpyrophosphate synthetase, require a metal-bound ATP molecule rather than, or in addition to, an enzyme-bound metal ion.

8.3.3 Other Manganese(II)-Containing Systems

Isocitrate dehydrogenase is a manganese(II) oxidoreductase that catalyzes the oxidative decarboxylation of isocitrate to α-ketoglutarate. This enzyme is a $NADP^+$-dependent monomer of molecular weight 60,000. Several manganese(II) hydrolases are also known, including enolase, arginase, and guanidinoacetate-amidino hydrolase.

Of the manganese(II)-specific metalloproteins, concanavalin A is probably the best characterized. Together with manganin and avimanganin, concanavalin A belongs to the group of proteins known as lectins. It is an oligomer with subunits of molecular weight 27,000 that probably exists as a dimer below pH 6 and tetramer above pH 7. Both of its properties—the carbohydrate binding and agglutination—require one manganese(II) and one calcium(II) ion per monomer.

Finally, many manganese-activated enzymes require ATP as a cofactor. Moreover, in some cases, manganese interacts with ATP instead of, or in addition to, the enzyme. For this reason, there is an abundance of NMR data characterizing both the Mn(II)-ATP system and the Mn(II)-ATP-enzyme system (ternary system).

The high relaxing efficiency of manganese(II) has also stimulated studies in which manganese(II) has been substituted for other divalent metal ions in metalloproteins, such as zinc enzymes or proteins containing metals like magnesium(II) or calcium(II).

8.4 ^{1}H RELAXATION DATA FOR THE $Mn(OH_2)_6^{2+}$ ION

^{1}H relaxation data for the simple hexaaqua complex will be analyzed in some detail and then more complicated biological systems will be considered. It should be recalled, however, that the dominant correlation time for the dipolar interaction in hexaaqua complexes is the rotational relaxation time of the molecule and that any conclusions drawn are not applicable as such to macromolecular complexes.

If manganese(II) ions are added to a water solution, even in millimolar concentrations, the water proton relaxation rates are dramatically increased. Figure 8.2 shows such effects for both T_2^{-1} and T_1^{-1} at two different temperatures [4]. For T_1^{-1}, the low frequency dispersion leads to an estimate of τ_c of $\sim 3 \times 10^{-9}$ s, which can only reflect the electronic relaxation time, since the rotational time is expected to be much shorter. Therefore, this dispersion arises from the contact term—see Equation (3.15)—with $\tau_e \simeq T_{2e}$, whereas the dipolar term is governed by τ_r. At high field, a second dispersion is observed in the range 8×10^6 to 1.5×10^7 Hz which is strongly dependent on temperature. If this is regarded as an $\omega_s \tau_c = 1$ dispersion, then a τ_c of 1.5 to 3×10^{-11} s is obtained, which is identified with the rotational time of the complex. As expected, its temperature dependence is rather pronounced, being proportional to η/T (see Equation 3.7). Both the dispersions observed are thus of the type $\omega_s \tau_c$. A further dispersion at $\omega_I \tau_c$ is of course expected, but this is at too high a field ($\sim$5,000 MHz) to be seen experimentally.

At low field, the transverse relaxation rate shows a $T_{2e}\omega_s = 1$ dispersion analogous to that of T_1^{-1}; however, the contact contribution does not disappear in this case but decreases to half of its original value due to the presence of the nondispersive term T_{1e} in Equation (3.16). Therefore, in the region around 1 MHz, the relaxation rate still contains the entire dipolar term and half the contact term operative in the fast-motion limit. At higher field, the $\omega_s \tau_r$ dispersion is barely detectable at 25 °C and is completely masked at 65 °C by a dramatic increase in the paramagnetic effect. The only possible origin for such an *increase* in the relaxation rate at high field is a field dependence of τ_s of the type described by Equation (3.27) in the Bloembergen and Morgan approximation [2]. In particular, since the $T_{2e}\omega_s = 1$ dispersion has already occurred, the field dependence must be due to the presence of the nondispersive T_{1e} contribution in the contact term. The correlation time, τ_v, for the field dependence of the electronic relaxation time can be estimated from the data to be about 10^{-11} s, which is in agreement with the idea of collisions with solvent molecules. If it were not for the finite residence time, τ_M, of the water molecules in the coordination sphere, the increase in T_2^{-1} could continue as long as T_{1e} increases. In practice, however, a plateau is reached when $T_{2M}^{-1} \gtrsim \tau_M^{-1}$ or T_{1e}^{-1} becomes smaller than τ_M^{-1}. At higher temperatures, the $\omega_s \tau_v$

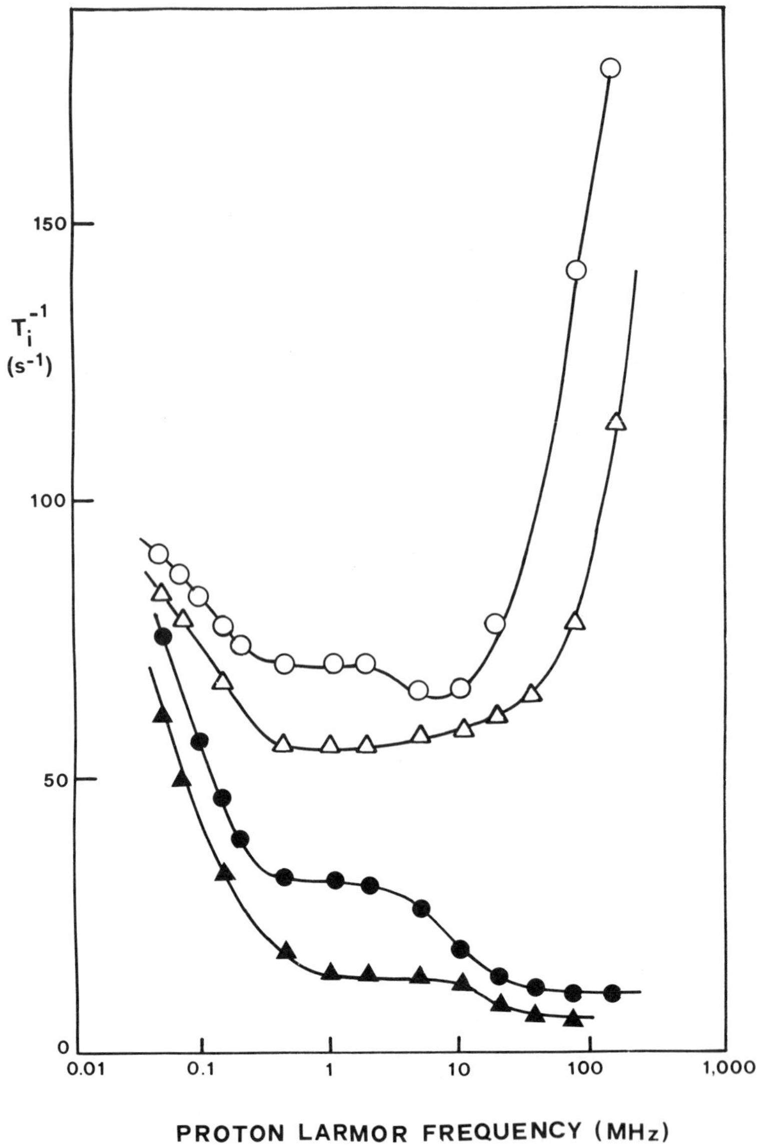

Figure 8.2 T_1^{-1} (● = 25 °C, ▲ = 65 °C) and T_2^{-1} (○ = 25 °C, Δ = 65 °C) values of water protons in 1.53 × 10^{-3}M $MnCl_2$ solutions as a function of the proton Larmor frequency [4].

dispersion occurs at higher field because the contribution of τ_M^{-1} to the electronic correlation rate

$$\tau_e^{-1} = \tau_M^{-1} + \tau_s^{-1}$$

is larger.

The increase in nuclear relaxation rates at high frequency is a characteristic of many of the manganese(II) systems that will be encountered in this chapter. Although it is well understood in terms of the available theory, it makes the analysis of NMRD data more difficult. Another characteristic of manganese(II) systems not reflected in the relaxation data of hexaaqua copper(II) is the presence of a sizeable contact contribution. The latter, however, is not observed in macromolecular systems because the dipolar contribution has a much larger weight, being also largely determined by the electronic relaxation time.

8.5 NMRD STUDIES ON MANGANESE(II) PROTEIN SYSTEMS

As NMRD data over a wide range of fields and experimental conditions are available for manganese(II) concanavalin A and manganese(II)-substituted carboxypeptidase A, these proteins will be considered in detail before reviewing the bulk of less complete data for other manganese proteins. Such an approach also serves to underline the subtleties of the electron-water proton interactions characteristic of such systems. It should also be borne in mind that some of the conclusions drawn on the less-defined systems are susceptible to changes should more extensive data become available in the future.

8.5.1 Concanavalin A

Figure 8.3 shows the NMRD data for concanavalin A at pH 6 [5]. A noticeable feature of this data is its striking difference to the T_1^{-1} data for $Mn(OH_2)_6^{2+}$

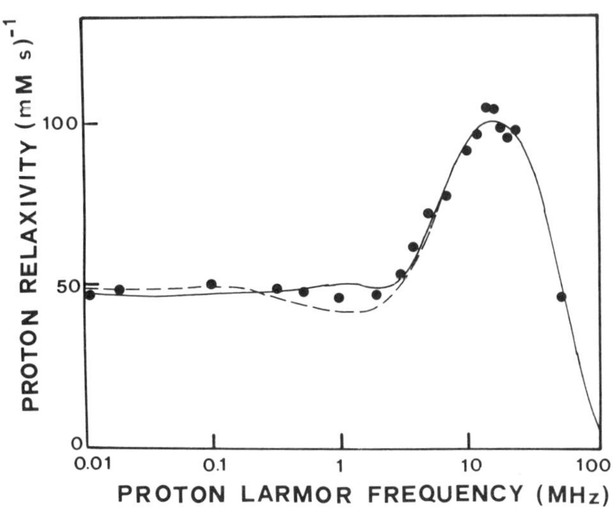

Figure 8.3 NMRD data for the binary complex of concanavalin A with manganese(II) at pH 6.0 and 25 °C. The dashed and solid lines represent five-parameter fitting with and without inclusion of zero field splitting effects, respectively. Reproduced from [5], with permission; copyright © 1980 by D. Reidel Publishing Company, Dordrecht, Holland.

reported in Figure 8.2. This observation is not surprising however because, as was anticipated, the rotational correlation time for the protein is much longer than that for the aquaion and is probably even longer than the electronic relaxation time. Since the electronic relaxation time is the relevant correlation time for the dipolar interaction, the relative importance of the dipolar term is therefore enhanced and the contact term does not need to be considered further. Nevertheless, the data in Figure 8.3 are far from the behavior expected from the Solomon equation. From a qualitative point of view, the high frequency increase in T_1^{-1} can be explained by the onset of a field-dependent increase in the electronic relaxation time, which is analogous to the rise observed for T_2^{-1} in the hexaaqua complex. At even higher fields, a sharp drop, at variance with the hexaaqua complex, is also observed. All the contributions to T_1^{-1} in the dipolar interaction are dispersive in nature, and therefore by increasing the field beyond the $\omega_I\tau_c = 1$ value a field region is eventually reached where T_{1M}^{-1} is proportional to $1/\omega_I^2\tau_c$. Since τ_c $(= T_{1e})$ is now proportional to $\omega_S^2\tau_v$—see Equation (3.27) for $\omega_S^2\tau_v^2 \gg 1$—T_{1M}^{-1} will be inversely proportional to $\omega_I^2\omega_S^2\tau_v$—the fourth power of the magnetic field—thus explaining the sudden drop.

In principle, analysis of the NMRD data in the peak region should yield values for the parameters G, as defined in Equation (6.1), and τ_c, which is expressed as a function of the constants C and τ_v in Equation (3.27). In practice, however, the analysis is complicated by two more unknowns; namely, the proton residence time, τ_M, and the rotational time, τ_r. When fast-exchange conditions are not fulfilled, the proton residence time is predicted to quench all the field-dependent variations in T_1^{-1} according to Equation (4.3):

$$T_{1p}^{-1} = \frac{f_M}{T_{1M} + \tau_M}$$

The rotational time has a similar effect, as far as the height of the peak is concerned, since the overall correlation time is given by

$$\tau_c^{-1} = \tau_r^{-1} + \tau_s^{-1}$$

Therefore the increase in τ_c with field due to τ_s will be quenched when $\tau_s^{-1} \lesssim \tau_r^{-1}$.

The results of a five-parameter fitting of the T_{1p}^{-1} dispersion data of manganese(II) concanavalin A solutions (Figure 8.3) are given in Table 8.1 [5]. The parameters τ_M and τ_r show a large covariance, as expected, and their values are thus purely indicative. However, an unreasonably large G value is obtained (corresponding to about eight water molecules!), together with a low field limit for τ_s of about 10^{-10} s. A further attempt has been made to rationalize the data [5] by qualitatively including the effect of the zero field splitting on the low field part of the data. At very low field, the Zeeman energy is probably smaller than the zero field splitting and so only the $\frac{1}{2} \leftrightarrow -\frac{1}{2}$ transition will be able to contribute to nuclear relaxation (see Section 3.5), assuming an axially symmetric system. Thus, the ω_s term is reduced to 9/35 of its original value [5], according to the fractional transition probability of the $\frac{1}{2} \leftrightarrow -\frac{1}{2}$ transition. At high fields, where zero field splitting $\ll$ Zeeman energy, the traditional equations should be again valid. A five-

Table 8.1 Values for the Five Parameters Obtained through a Least-Squares Fit of the T_{1p}^{-1} Dispersion Data of Manganese(II) Concanavalin A Solutions (Figure 8.3) With and Without Inclusion of Zero Field Splitting Effects [5][a]

	G (pm^{-6})[b]	τ_M (s)	τ_r (s)	$C(s^{-2})$	τ_v (s)
With zero field splitting	2.0×10^{-14}	1.6×10^{-7}	4.5×10^{-8}	1.51×10^{19}	3.9×10^{-11}
Without zero field splitting	3.2×10^{-14}	1.7×10^{-7}	1.73×10^{-7}	2.63×10^{19}	6.9×10^{-11}

[a] See note [54].

[b] A single regularly coordinated water molecule corresponds to $G = 4 \times 10^{-15}$ pm^{-6}.

parameter fitting has also been attempted (Figure 8.3 and Table 8.1) neglecting the data in the intermediate region, where zero field splitting $\approx$ Zeeman energy. Indeed, smaller G and longer τ_s values are obtained, but the results are still far from being satisfactory from a chemical point of view. Some improvement is obtained when ^{2}H data are considered [6].

8.5.2 Carboxypeptidase A

The NMRD data for manganese(II)-substituted carboxypeptidase A at pH 6.7 are shown in Figure 8.4 [7]. Under these conditions, the binding constant of manganese(II) for the apoenzyme is not high enough to ensure complete formation

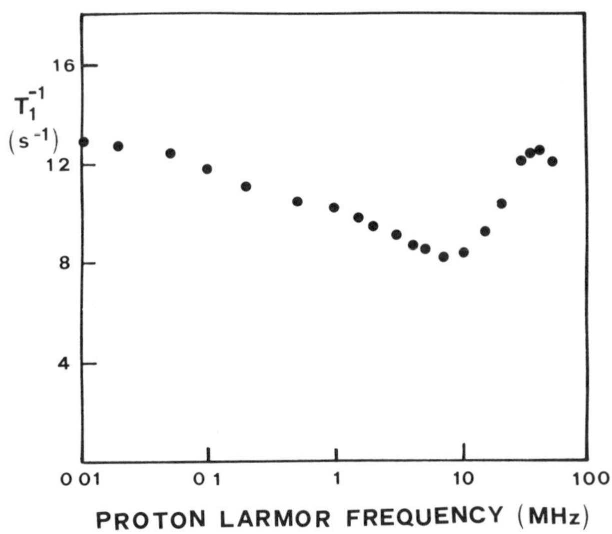

Figure 8.4 Water proton relaxation rates in solutions of manganese(II) substituted carboxypeptidase A at pH 6.7 and 25°C. The sample is 0.5 mM protein and 0.30 mM total Mn^{2+}. Reproduced from [7], with permission; copyright © 1983 by D. Reidel Publishing Company, Dordrecht, Holland.

of the complex [8]—the amount of free manganese(II) in solution can be estimated from EPR data. With this in mind, the first dispersion centered at about 0.1 MHz has been ascribed to the contribution of free manganese, being similar to that discussed for the aquaion (Figure 8.2). Both the decrease in T_1^{-1} around 2 MHz and the peak at about 30 MHz are features similar to those displayed by concanavalin A, although the peak is much less pronounced in this case. Attempts to fit the data of the noninhibited enzyme to the Solomon-Bloembergen-Morgan theories again yield too high values for G [7], and the results are not significantly improved by reducing the contribution of the ω_s term, as discussed for concanavalin A.

8.5.3 Critical Considerations

Although all the features of the NMRD data are qualitatively well understood, there is still a lack of quantitative agreement with reasonable chemical expectations. There are three possible reasons for this. (1) Zero field splitting, as discussed in Section 3.5. (2) An unpaired electron-metal nucleus hyperfine coupling larger than τ_c^{-1} and therefore larger than the Zeeman energy at low field. Such an interaction may have an effect similar to that of the zero field splitting in that it causes some transitions to be ineffective with respect to nuclear relaxation [9, 10]. (3) The fitting in the high field region is critically dependent on the form of the Bloembergen and Morgan equation for the field dependence of T_{1e}. As discussed in Section 3.7, this equation is valid for τ_v shorter than τ_s (Redfield limit), and although this seems to be true (see Table 8.1), τ_v in a macromolecule is not directly related to collision of solvent molecules with the chromophore; thus its physical meaning is rather obscure. In deriving this equation, as well as the more sophisticated numerical results of Rubinstein *et al.* discussed in Section 8.2, it is assumed that the fluctuations described by τ_v are isotropic, which is reasonable for the aquaion but may not be for a protein complex. If the fluctuations are anisotropic, Equations (3.26) and (3.27) are no longer valid. Attempts to better understand the experimental curves inevitably involve so many parameters that their determination is impossible.

8.5.4 Other Manganese(II) Proteins

NMRD data are also available for several other protein systems [11–21]. Some representative profiles are reported in Figure 8.5. The maximum field range used to study these systems is from about 3 to 100 MHz. In this region, all the systems display a more or less pronounced peak in T_1^{-1} values, thus indicating that this is a general characteristic of manganese-containing macromolecules. On the basis of the foregoing discussion, very little meaning should be attached to the G values (or to the hydration numbers q) derived from fitting the data to the Solomon-Bloembergen-Morgan equations, or to the observed changes in these parameters upon binding of coenzymes, substrates, or inhibitors. Nevertheless, some useful information can be obtained from the data if they are interpreted in the chemical

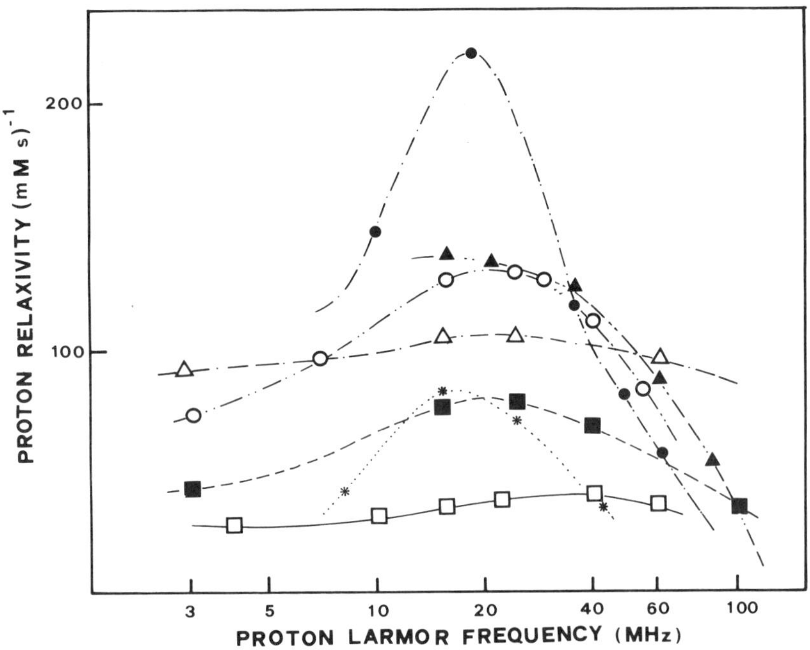

Figure 8.5 Room temperature NMRD data for some manganese(II) systems: (□) bovine carbonic anhydrase II [11]; (△) yeast enolase [12]; (∗) formyltetrahydrofolate synthetase-ADP [14]; (■) $(Na^+ + K^+)$-ATPase-P_i from sheep kidneys [15]; (▲) *E. coli* rRNA [16]; (○) *E. coli* tRNA [17]; (●) ATP-phosphofructokinase from rabbit muscle [18].

sense, rather than relying on the outputs of computer programs. For example, it may not be possible, or even wise, to discriminate between one or two coordinated water molecules, but it is almost always feasible to distinguish between some or none. Indeed, the simple observation of a peak in the NMRD data indicates that the relevant correlation time is the electronic relaxation time. This fact alone rules out the possibility that the relaxation is entirely of outer-sphere type. If this was the case (Section 4.5), nuclear relaxation would depend on the diffusional correlation time, τ_D, rather than on τ_s, since τ_s is often larger than τ_D for manganese(II). τ_D is of course field independent.

The binding of metals to macromolecules can also be studied by monitoring proton relaxation rates. Information on the number of binding sites and their affinity for metals can be gained by titrating the metal into enzyme solutions, and vice versa (Section 4.4), and making use of the enhancement factor ε, which is defined independently of the physical origin of the interaction. This technique has been particularly exploited with manganese(II) because of its favorable nuclear relaxation properties, and many excellent reviews are dedicated to the subject [22–28]. Detailed information has been obtained on many systems, including most of those reported in Figure 8.5.

8.5.5 Another Approach to the Analysis of the Nuclear Relaxation Rates of Water Nuclei

Since τ_c for manganese(II) in proteins can be safely set around 10^{-8} to 10^{-9} s at relatively high fields, it is possible, with the available routine NMR instruments, to perform measurements at magnetic fields high enough to fall beyond the $\omega_I\tau_c = 1$ dispersion. When $\omega_I\tau_c > 1$, the dipolar contributions to T_1 and to T_2 diverge—Equations (3.3) and (3.4). Furthermore, the relatively high magnetic field makes zero field splitting and hyperfine coupling with the metal nucleus small with respect to the electronic Zeeman energy in such a way that the Solomon equations are expected to hold. The high nuclear relaxing capability still warrants the observation of sizeable nuclear relaxation enhancements for both protons and deuterons in H_2O/D_2O mixtures.

Under such circumstances, four independent equations (T_1 and T_2 for ^1H and ^2H) are available. Assuming that contact contributions, particularly to T_{2M}^{-1}, are negligible—as it can be shown using the $A_c/\hbar$ value of the hexaaqua complex [29]—appropriate combinations of Equations (3.3) and (3.4) with Equation (4.3), where the hydration number q has been made explicit from the molar fraction f, can be written [30] in terms of three unknowns—q, τ_M, and τ_c. In principle, the numerical values of these three parameters can be obtained although, in general, the physical meaning of τ_c does not appear. The data should of course be supported by control experiments at different frequencies, where q and τ_M should remain constant, and different temperatures, where the expected variations of τ_M should be observed. An undesirable consequence of this approach is that outer-sphere relaxation effects become relatively significant in the suitable magnetic field region, since the contribution to T_{1M}^{-1} from the coordinated water is already dispersed down whereas that from outer-sphere nuclei has not, owing to its shorter correlation time. This procedure has been applied to manganese(II)-substituted carboxypeptidase A and bovine carbonic anhydrase [30]. In both cases, second sphere hydration had to be invoked to rationalize the data. The overall picture for both systems is that of a single regularly coordinated water molecule plus a number (3 to 4) of water molecules loosely bound in a second hydration shell.

This method has clearly some potential in clarifying the nature of the water-manganese interactions in proteins. A similar approach will be discussed for gadolinium(III) systems (Section 10.8.2). Its application to other metals is limited by the observability of deuteron relaxation enhancements beyond the $\omega_I\tau_c = 1$ dispersion, when attainable with suitable magnetic fields.

8.6 WATER ^{17}O NMR STUDIES

The nuclear spin of ^{17}O of water nuclei in $Mn(OH_2)_6^{2+}$ is relaxed mainly by a strong contact interaction with the unpaired electrons. This interaction is easily detected in the normal range of magnetic fields employed in ^{17}O NMR by measuring the transverse relaxation time, T_2, because of the presence of the nondispersive term in the relative Equation (3.16), whereas the T_1^{-1} values are dispersed down. The measured T_{2p}^{-1} values are affected by the chemical exchange—see Equation (4.4)—as shown by their temperature dependence (Figure 8.6) [31, 32]. Although the linewidth under these conditions is not directly related to T_{2M}, the latter has to be sizeable even for a metal concentration of 1.8×10^{-4}M. Manganese(II) is in fact one of the few ions suitable for detecting metal-water interactions through ^{17}O NMR.

Figure 8.6 also shows the temperature dependence of the ^{17}O chemical shift—Equation (4.2)—which although small is measurable for millimolar manganese solutions. At high temperature, fast exchange allows determination of the full isotropic shift of the bound water [32]. If this shift is assumed to be fully contact in origin, the coupling constant A_c/h can then be calculated. Its value, 5.3 MHz, is in agreement with that obtained from T_2^{-1} values, which are in turn calculated from τ_c values obtained from ^1H NMR data. An interesting feature of the Mn(II)-^{17}O interaction is the seeming insensitivity of A_c/h to the chemical environment of the manganese ion: values in the range 5 to 6 MHz are always found [32] for complexes like $MnEDTA(H_2O)^{2-}$, $Mn(NTA)(H_2O)_2^-$, Mn(Dacoda)(H_2O), and $MnPhDTA(H_2O)^{2-}$, where a variety of structures are known to be involved. These

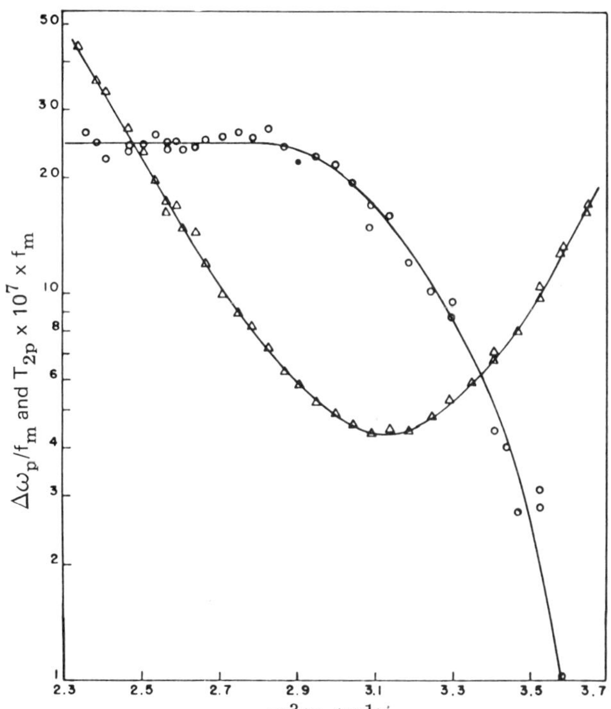

Figure 8.6 11.5 MHz water ^{17}O T_{2p} (△) and isotropic shift (○) data (log scale) for $Mn(OH_2)_6^{2+}$ solutions as a function of reciprocal temperature. Reproduced from [32], with permission; copyright © 1978 by American Chemical Society.

values are of course obtained by taking into account the different hydration numbers in each case. Measurements of A_c/h from the isotropic shift, if observable and provided that fast exchange conditions hold, would then give an estimate of the hydration number independently of a knowledge of τ_c.

Despite the increase in τ_c for macromolecules, as compared to small complexes, dipolar contributions to the ^{17}O transverse relaxation rates are still negligible. For typical ^{17}O Larmor frequencies—from 8.1 MHz, corresponding to 60 MHz for ^{1}H, to 68 MHz, corresponding to 500 MHz for ^{1}H—the $\omega_s\tau_c$ term is probably dispersed down for τ_c values around 10^{-9} s, while the $\omega_I\tau_c$ term is not. Therefore, the Solomon Equation (3.4) can be written as

$$T_{2M}^{-1} = \frac{7}{15}\left(\frac{\mu_0}{4\pi}\right)^2 \frac{\gamma_N^2 g_e^2 \mu_B^2 S(S+1)}{r^6}\,\tau_c$$

and the Bloembergen Equation (3.16) as

$$T_{2M}^{-1} = \frac{1}{3}\,S(S+1)\left(\frac{A_c}{\hbar}\right)^2 T_{1e}$$

Assuming $\tau_c = T_{1e}$, the dipolar and contact contributions to T_{2M}^{-1} are respectively proportional to

$$\frac{7}{5}\left(\frac{\mu_0}{4\pi}\right)^2 \frac{\gamma_N^2 g_e^2 \mu_B^2}{r^6} \quad \text{and} \quad \left(\frac{A_c}{\hbar}\right)^2$$

By taking $r = 210$ pm and $A_c/\hbar = 3.33 \times 10^7$ rad s^{-1}, as for $Mn(OH_2)_6^{2+}$, the contributions are proportional to

$$7.38 \times 10^{13}\ \text{s}^{-2} \quad \text{and} \quad 1.1 \times 10^{15}\ \text{s}^{-2}$$

Therefore, the contact term is still one order of magnitude larger. A further consequence is that T_1^{-1} will be negligibly affected by the interaction with the paramagnetic center, since it should depend entirely on the dipolar term, which in turn is $\frac{6}{7}$ of the dipolar contribution to T_2^{-1}. Indeed, no relaxation enhancement has ever been detected on ^{17}O T_1^{-1} in solutions of manganese proteins. The measured T_2^{-1} enhancements are again contact in origin and are complicated by chemical exchange [33].

8.7 OTHER NUCLEI

8.7.1 ^{19}F

The available information on the interaction of fluoride ions with manganese(II) in aqueous solutions is much more limited than that on $^{17}OH_2$. ^{19}F T_{1p}^{-1} and T_{2p}^{-1} for $Mn(H_2O)_5F^+$ at 94.1 MHz have been estimated to be 4.5×10^5 s^{-1} and 4.6×10^6 s^{-1}, respectively [34]. As the value for T_{2p}^{-1} is affected by the exchange rate, it therefore only gives a lower limit for T_{2M}^{-1}. Since T_{1p}^{-1} is one order of magnitude smaller than T_{2p}^{-1}, τ_M does not contribute to it; this is probably because $T_{2M}^{-1} > T_{1M}^{-1}$. The difference between T_{2M}^{-1} and T_{1M}^{-1} can

Table 8.2 ^{19}F Relaxation Parameters in Manganese(II) Proteins

Macromolecular Complex	T_{1p}^{-1}	T_{2p}^{-1}	Frequencies (MHz)	Reference
Mn(II) enolase F^-	3.5×10^5	5.5×10^5	94	[34]
Mn(II) enolase-P_i-F^-	4.5×10^4	5.2×10^4	94	[34]
Mn(II) carboxypeptidase F^-	8×10^6	8.3×10^6	94,56	[36]

therefore only be ascribed to contact contributions. From this difference, the value of the contact contribution to T_{1M}^{-1} can be estimated to be relatively small, and therefore T_{1M}^{-1} is essentially dipolar in origin. Moreover, its order of magnitude is consistent with a fluoride ion in the first coordination sphere [34, 35].

In macromolecular complexes the values of T_{1p}^{-1} and T_{2p}^{-1} are very similar. Table 8.2 shows the values of T_{1p}^{-1} and T_{2p}^{-1} measured for the interactions of $^{19}F^-$ with manganese(II) carboxypeptidase [36] and manganese(II) enolase [34] in both its binary complex and inactive ternary complex with inorganic phosphate. Both T_{2p}^{-1} and T_{1p}^{-1} are presumably in the exchange-limited region, as shown by their temperature dependence and insensitivity to frequency in the range 56 to 94 MHz. However, their similar values suggest T_{1M}^{-1} is close to T_{2M}^{-1}. It is possible, therefore, that dipolar relaxation contributions to T_2^{-1} are more important in large rather than in small complexes (for which $T_{2p}^{-1} > T_{1p}^{-1}$). Such a finding would be consistent with a larger τ_c in macromolecules than in smaller complexes. Using the dipolar equation for T_{1p}^{-1} and an upper limit for τ_c given by the rotational time of the molecules, the upper limits for the Mn(II)-^{19}F distances have been calculated and are found to be consistent with direct binding in the binary complex. However, no conclusions can be drawn for the ternary complex of enolase because of the much slower fluoride exchange rate.

As ligand-centered dipolar relaxation might be operative in Mn(II)-F systems, the distances estimated assuming that T_{1M}^{-1} is only metal-centered dipolar in origin might be shorter than the real ones. For example, from T_{1p}^{-1} for the $Mn(H_2O)_5F^+$ ion distances of 190 [34] and 210 [35] pm have been calculated. The latter value is consistent with X-ray data, while the former would indicate ligand-centered contributions to T_{1M}^{-1}. This discrepancy might be due to the different estimates for the dissociation constant of the $Mn(OH_2)_5F^+$ complex. Nevertheless, the possibility of unpaired spin delocalization on the ^{19}F orbitals is by itself evidence of direct metal coordination, which is the kind of meaningful information sought.

8.7.2 ^{31}P and ^{13}C

These nuclei are not usually directly bound to the paramagnetic center. In the case of phosphate or carboxylate groups coordinated to the metal, contact or ligand-centered contributions may be operative in addition to a dipolar coupling contribution, and should always be borne in mind when analyzing relaxation data to obtain structural information. Studies of this type are particularly numerous on manganese-containing systems: many investigations combining the use of ^{1}H, ^{13}C, ^{15}N, ^{31}P, and ^{17}O NMR have led to the complete mapping of substrate and coenzyme molecules in the active sites of binary and ternary systems. The general procedure has been outlined in Section 4.6; further details can be found in reviews and articles by major contributors in the field [22–28, 37, 38].

To illustrate how complex the analysis can be, the studies on the interaction of nucleotides with free manganese will be briefly considered. ATP and its analogs are universal cell components that exist *in vivo* as a mixture of variously ionized species, metal complexes, and different conformers. Furthermore, they are involved as substrates in many reactions catalyzed by manganese-containing or manganese-activated enzymes. In the latter systems, a manganese-nucleotide complex is sometimes formed even before the protein interacts with the metal and the specific substrate.

In 1957 it was proposed that ATP interacts with metal ions through a back-bound conformation [39]. In this 1:1 complex the metal would be bound to the β and γ phosphates and to the adenine ring at N7 and the amino NH_2 (Figure 8.7). After nearly thirty years of experimental work, mainly carried out through NMR, the situation is still not completely clear. Early ^{31}P line broadening and T_1 data were ascribed to dipolar relaxation and interpreted assuming equal distance of the manganese from each of the three phosphates [40]. It was later shown that the ^{31}P line broadening was contact in origin and probably exchange limited [41]. T_1 studies subsequently indicated that the α phosphate was somewhat less strongly interacting than the β and γ groups. Furthermore, T_{1p} is dominated by τ_M at low temperatures [41].

The metal binding to the ring is even more controversial. The use of ^{1}H NMR is limited by the presence of only two nonexchanging protons on the adenine ring, whereas ^{13}C line broadening data are hampered by sizeable contact contributions which prevent their use for distance measurements. However, ^{13}C T_1 data have suggested binding through N7 [42]. Moreover, it has been found that at the usually high nucleotide concentrations used in NMR studies the predominant species is a 1:2 complex [32]. Surprisingly, ^{1}H and ^{17}O relaxation measurements of the

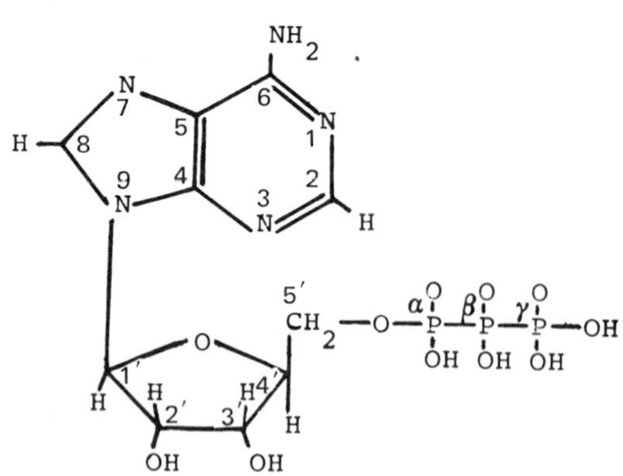

Figure 8.7 Structure and labelling of adenosinetriphosphate (ATP).

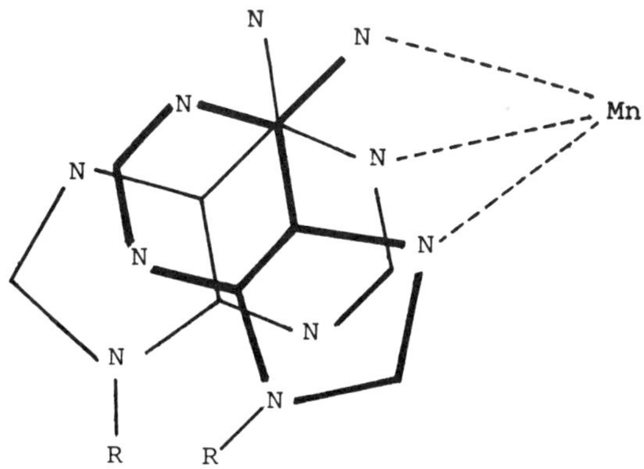

Figure 8.8 Proposed structure for the Mn(II)-AMP$_2$ stacked complex. Reproduced from [43], with permission; copyright © 1980 by American Chemical Society.

water nuclei still suggest a high hydration number for such species; ^{17}O data also indicate isotropic shifts consistent with 3 to 4 molecules in the first coordination sphere [32]. The 2:1 complex with AMP has been further investigated through ^{15}N NMR and a direct binding proposed for N1, N7, and amino NH_2 [43]. A stacking interaction between the adenine rings of two AMP molecules (Figure 8.8) could be consistent with all the experimental data [43]. Table 8.3 shows a summary [44, 45] of recent NMR data together with some considerations of the relative importance of contact and dipolar terms and exchange rates.

Table 8.3 Summary of NMR Relaxation Data of the Mn(II)-ATP Complex in Aqueous Solution [44, 45]

Nucleus	Exchange Conditions at rt (pH 7)	Paramagnetic Contribution	Further Remarks
^1H (ligand)	$T_{1M} \gg \tau_M$ $T_{2M} \gg \tau_M$	$T_{1p}^{-1} = fT_{1M}^{-1}$ $T_{2p}^{-1} = fT_{2M}^{-1}$	$(fT_{1p})^{-1} \simeq (fT_{2p})^{-1}$ Dipolar contribution only or an equal small contact contribution
^1H (water)	$T_{1M} \gg \tau_M$ $T_{2M} \gg \tau_M$	$T_{1p}^{-1} = fT_{1M}^{-1}$ $T_{2p}^{-1} = fT_{2M}^{-1}$	$(fT_{1p})^{-1} \simeq (fT_{2p})^{-1}$ T_{1p}^{-1} dipolar only T_{2p}^{-1} contact only at pH 4 $(fT_{1p})^{-1} \simeq (fT_{2p})^{-1}$
^{31}P	$\tau_M \gg T_{1M}$ $\tau_M \gg T_{2M}$	$(fT_{1p})^{-1} = (fT_{2p})^{-1} = \tau_M$	Exchange-controlled relaxation rates; kinetic information is possible
^{13}C	$T_{1M} \gg \tau_M$ $T_{2M} \gg \tau_M$	$T_{1p}^{-1} = fT_{1M}^{-1}$ $T_{2p}^{-1} = fT_{2M}^{-1}$	$(fT_{1p})^{-1} \ll (fT_{2p})^{-1}$ T_{1p}^{-1} dipolar only T_{2p}^{-1} contact only

Table 8.4 Isotropic Shift Data (ppm) for Manganese(III) Porphyrins in $CDCl_3$ Solutions [51–53]

Complex	Temperature (°C)	Pyrrole-H	Phenyl Protons				Propyl Protons			Ethyl Protons		Methyl Protons	*meso*-H
			o-H[a]	*m*-H	*p*-H	CH_3	α-CH_2	β-CH_2	γ-CH_3	α-CH_2	β-CH_3		
TPPMnCl	35	−30.3	—	+0.4	−0.4	—	—	—	—	—	—	—	—
o-CH_3TPPMnCl	35	−29.0	—	+0.5	−0.7	−0.4	—	—	—	—	—	—	—
m-CH_3TPPMnCl	35	−30.0	—	+0.5	−0.6	−0.22	—	—	—	—	—	—	—
p-CH_3TPPMnCl	35	−30.2	—	+0.6	—	+0.29	—	—	—	—	—	—	—
T-n-PrPMnCl	35	−29.3	—	—	—	—	+4.7	−0.5	~0	—	—	—	—
OEPMnCl	35	—	—	—	—	—	—	—	—	+18.2	+0.7	—	+41.4[b]
EPMnCl	21	—	—	—	—	—	—	—	—	+18.5	+0.7	+31.7	−20.6
MPDMEMnCl	21	—	—	—	—	—	—	—	—	+17.0	+0.9	+31.9, +36.0	−20.5, −23.5

[a] Unresolved signal.

[b] Assignment controversial [53].

Abbreviations: TPP: tetraphenylporphyrin; T-*n*-PrP: tetra-*n*-propylporphyrin;
OEP: octaethylporphyrin; EP: etioporphyrin; MPDME: mesoporphyrin IX dimethylester.

Finally, it should be mentioned that EPR data have shown [45] that the electronic relaxation time used in the analysis of NMR results is best regarded as a distribution of values due to a partial solid state behavior of the complex in solution, the rotational averaging not being complete.

8.8 MANGANESE(III)

Manganese(III) has been found in acid phosphatase from sweet potato and superoxide dismutase from *Escherichia coli.* It has also been suggested that manganese(III) is present as a porphyrin derivative in erythrocytes [46]. In all these systems, manganese(III) is high spin with $S = 2$. Furthermore, low symmetry or Jahn-Teller effects provide an orbitally nondegenerate ground state with the zero field splitting of the resulting quintuplet term being in general small ($\leq 1\ cm^{-1}$), except for porphyrin derivatives for which the value has been proposed to be a few wave numbers [47]. The electronic relaxation times in Mn(III) porphyrins can be set around 10^{-10} s [48]; in hexaaqua complexes they have been estimated more than one order of magnitude shorter [49].

^{31}P NMR data on the phosphate interaction with acid phosphatase have shown that the substrate closely interacts with the paramagnetic center [50]. The ^{31}P linewidth observed in the phosphate-enzyme 1:1 adduct and a comparison with the Mn(III) pyrophosphate complex are consistent with direct metal-oxygen bonding.

Manganese(III) porphyrin derivatives have been investigated with both natural [51] and synthetic porphyrins [48, 52, 53] in five- and six-coordinated adducts. The ^{1}H NMR spectrum in $CDCl_3$ of the five-coordinated tetra-*p*-tolyl porphynato manganese(III) chloride (Table 8.4) shows alternation of sign of the isotropic shifts of the meta and para protons and methyl substituents of the aryl moiety. These shifts, although very small, rule out the possibility of sizeable dipolar contributions. The large upfield shifts for the pyrrole protons, which compare with downfield shifts for α-CH_2 substituents (Table 8.4), indicate π spin density on pyrrole rings. Predominance of σ spin density is proposed for the meso protons. The π spin density distribution is consistent with porphyrin-to-metal π charge transfer as the delocalization mechanism. Furthermore, the π spin transfer mechanism is consistent with the decrease in the extent of spin transfer with increasing π donor properties of the axial halogen. Such a mechanism is analogous to that occurring in low spin Fe(III) porphyrins and in contrast to that of high spin Fe(III) porphyrins (see Chapter 7). In agreement with the zero field splitting values, ^{1}H NMR linewidths in manganese(III) porphyrins have been found to be two to three times larger than those in the corresponding iron(III) derivatives [48].

REFERENCES

1. Bencini, A., and Gatteschi, D. *Trans. Met. Chem.* **8** (1982) 1.
2. Bloembergen, N., and Morgan, L. O. *J. Chem. Phys.* **34** (1961) 842.

3. Rubinstein, M., Baram, A., and Luz, Z. *Mol. Phys.* **20** (1971) 67.
4. Hausser, R., and Noack, F. *Z. Phys.* **182** (1964) 93.
5. Koenig, S. H., Brown, R. D., and Brewer, C. F. *Proc. Natl. Acad. Sci. U.S.A.* **70** (1973) 475; Koenig, S. H., and Brown, R. D. III. In *ESR and NMR of Paramagnetic Species in Biological and Related Systems*, I. Bertini and R. S. Drago (eds.), Reidel, Dordrecht, 1980, p. 89.
6. Koenig, S. H., and Brown, R. D. III *J. Magn. Reson.* **61** (1985) 426.
7. Koenig, S. H., and Brown, R. D., III. In *The Coordination Chemistry of Metalloenzymes*, I. Bertini, R. S. Drago, and C. Luchinat (eds.), Reidel, Dordrecht, 1983, p. 19.
8. Brown, R. D. III., Brewer, C. F., and Koenig, S. H. *Biochemistry* **16** (1977) 3883.
9. Bertini, I., Luchinat, C., Mancini, M., and Spina, G. In *Magneto-Structural Correlations in Exchange Coupled Systems*, R. D. Willett, D. Gatteschi, and O. Kahn (eds.), Reidel, Dordrecht, 1985, p. 421.
10. Koenig, S. H., and Shillinger, W. E. *J. Biol. Chem.* **244** (1969) 6520.
11. Lanir, A., Gradstajn, S., and Navon, G. *Biochemistry* **14** (1975) 242. Bertini, I., Canti, G., and Luchinat, C, *Inorg. Chim. Acta* **56** (1981) 99.
12. Nowak, T., Mildvan, A. S., and Kenyon, G. L. *Biochemistry* **12** (1973) 1690.
13. Wydrzynski, T. J., Marks, S. B., Schmidt, P. G., Govindjee, G., and Gutowsky, H. S. *Biochemistry* **17** (1978) 2155.
14. Buttlaire, D. H., Reed, G. H., and Himes, R. *J. Biol. Chem.* **250** (1975) 261.
15. Grisham, C. M., and Mildvan, A. S. *J. Biol. Chem.* **249** (1974) 3187.
16. Peacocke, A. R., Richards, R. E., and Sheard, B. *Mol. Phys.* **16** (1969) 177.
17. Danchin, A., and Gueron, M. *J. Chem. Phys.* **53** (1970) 3599.
18. Jones, R., Dwek, R. A., and Walker, I. O. *Eur. J. Biochem.* **47** (1974) 285.
19. Chapman, B. E., Sullivan, W. J., Scopes, R. K., and Reed, G. H. *Biochemistry* **16** (1977) 1005.
20. Gupta, R. K., Fung, C. H., and Mildvan, A. S. *J. Biol. Chem.* **251** (1976) 2421.
21. Pillai, R. P., Marshall, M., and Villafranca, J. J. *Arch. Biochem. Biophys.* **199** (1980) 21.
22. Mildvan, A. S., and Cohn, M. *Adv. Enzymol.* **33** (1970) 1.
23. Mildvan, A. S., and Engle, J. L. *Methods Enzymol.* **C26** (1972) 654.
24. Mildvan, A. S., and Gupta, R. K. *Methods Enzymol.* **F49** (1978) 322.
25. Burton D. R., Forsén, S., Karlström, G., and Dwek, R. A. *Prog. NMR Spectrosc.* **13** (1979) 1.
26. Dwek, R. A., Williams, R. J. P., and Xavier, A. V. In *Metal Ions in Biological Systems*, H. Sigel (ed.), Marcel Dekker, New York, 1974, Volume 4, p. 61.
27. Mildvan, A. S. *Acc. Chem. Res.* **10** (1977) 246.
28. Mildvan, A. S., Granot, J., Smith, G. M., and Liebman, M. N. In *Advances in Inorganic Biochemistry*, D. W. Darnall and R. G. Wilkins (eds.), Elsevier North Holland, New York, 1980, Volume 2, p. 211.
29. Navon, G., *Chem. Phys. Lett.* **7** (1970) 390.
30. Kushnir, T., and Navon, G. *J. Magn. Reson.* **56** (1984) 373.
31. Ducommun, Y., Newman, K. E., and Merbach, A. E., *Inorg. Chem.* **19** (1980) 3696.
32. Zetter, M. S., Lo, G. Y.-S., Dodgen, H. W., and Hunt, J. P. *J. Am. Chem. Soc.* **100** (1978) 4430.
33. Meirovitch, E., and Kalb, A. J. *Biochim. Biophys. Acta* **303** (1973) 258.
34. Nowak, T., and Maurer, P. J. *Biochemistry* **20** (1981) 6901.
35. Mildvan, A. S., Leigh, J. S., and Cohn, M, *Biochemistry* **6** (1967) 1805.

36. Navon, G., Shulman, R. G., Wyluda, B. J., and Yamane, T. *J. Mol. Biol.* **51** (1970) 15.
37. Lanir, A., and Navon, G. *Biochemistry* **11** (1972) 3536. Zizapel, N., Navon, G., and Sokolovsky, M. *Eur. J. Biochem.* **52** (1975) 487.
38. Henry, B., Boubel, J.-C., and Delpuech, J.-J. Polyhedron **1** (1982) 113.
39. Szent-Györgyi, A. In *Bioenergetics*, Academic Press, New York, 1957.
40. Sternlicht, H., Shulman, R. G., and Anderson, E. W. *J. Chem. Phys.* **43** (1965) 3123.
41. Brown, F. F., Campbell, I. D., Henson, R., Hirst, C. W., and Richards, R. E. *Eur. J. Biochem.* **38** (1973) 54.
42. Lam, Y. F., Kuntz, G. P. P., and Kotowycz, G. *J. Am. Chem. Soc.* **96** (1974) 1834.
43. Levy, G. C., and Dechter, J. J. *J. Am. Chem. Soc.* **102** (1980) 6191.
44. Niccolai, N., Tiezzi, E. and Valensin, G. *Chem. Rev.* **82** (1982) 359.
45. Basosi, R., Niccolai, N., Tiezzi, E., and Valensin, G. *J. Am. Chem. Soc.* **100** (1978) 8047.
46. Hancock, R. G. V., and Fritze, K. *Bioinorg. Chem.* **3** (1973) 77.
47. Brackett, G. C., Richards, P. L., and Caughey, W. S. *J. Chem. Phys.* **54** (1971) 4383.
48. La Mar, G. N., and Walker, F. A. *J. Am. Chem. Soc.* **95** (1973) 6950.
49. Luz, Z., Silver, L., and Fiat, D. *J. Chem. Phys.* **46** (1967) 469.
50. Kawabe, H., Sugiura, Y., and Tanaka, H. *Biochem. Biophys. Res. Commun.* **103** (1981) 32.
51. Janson, T. R., Boucher, L. J., and Katz, J. J. *Inorg. Chem.* **12** (1973) 940.
52. La Mar, G. N., and Walker, F. A. *J. Am. Chem. Soc.* **97** (1975) 5103.
53. La Mar, G. N., and Walker (Jensen), F. A. In *The Porphyrins*, D. Dolphin (ed.), Academic Press, New York, 1979, Volume 4, p. 61.
54. Note added in proof. A complete treatment of the zero field splitting effect is becoming available in the literature (Banci, L., Bertini, I., Briganti, F., and Luchinat, C. *J. Magn. Reson.*, in press). The best fitting parameters for $D = 0.04\ \text{cm}^{-1}$ are: $G = 1.3 \times 10^{-14}\ \text{pm}^{-6}$, $\tau_M = 4.7 \times 10^{-7}$ s, $\tau_r = 2.0 \times 10^{-8}$ s, $C = 7.1 \times 10^{19}\ \text{s}^{-2}$, $\tau_v = 4.4 \times 10^{-11}$ s.

CHAPTER 9

Other Transition Metal Ions

9.1 INTRODUCTION

3*d* metal ions not treated in the preceding chapters include scandium, titanium, cobalt, nickel, vanadium, and chromium. Of these, scandium(III) and titanium(IV) are diamagnetic, and titanium(III) is a reducing agent of scarce, if any, biochemical relevance.

Paramagnetic metal ions among the heavier transition metals are less common. Although molybdenum is a biologically important metal, especially in nitrogenases, the paramagnetic molybdenum(V) species is mostly studied by EPR since complexes with large molecular weights do not favor NMR investigations. Nevertheless, NMR studies of nuclei of ligands in rapid exchange can provide useful information. Ruthenium(III) has become popular in the study of electron transfer with several metalloproteins. The metal ion is paramagnetic and has electronic relaxation times suitable for NMR investigations. As far as the other paramagnetic metal ions are concerned, it should be apparent by now that every paramagnetic metal ion, under the proper conditions, can be used as a shift reagent, relaxing agent, or both.

9.2 COBALT

Cobalt is a $3d^7 4s^2$ transition metal with oxidation numbers +1, +2, and +3, of which the latter two are by far the more common. Cobalt(III) is almost exclusively low spin and diamagnetic, except when bound to very weak ligands, such as fluoride. Cobalt(I) is generally a reducing agent and is often low spin diamagnetic—if it is paramagnetic it exhibits the properties of high spin nickel(II) (d^8 configuration). Cobalt(II), which is the ion of interest here, may be low spin with one unpaired electron or high spin with three unpaired electrons. The low spin metal ion is mostly encountered in planar complexes or in association with more than one strong ligand, such as CN^-. It occurs, for example, in cobalt(II) porphyrin systems or in vitamin B_{12} coenzyme. High spin cobalt(II) is commonly encountered in four- (tetrahedral), five- (either square pyramidal or trigonal bipyramidal), and six-coordinated (octahedral) chromophores. High spin cobalt(II) is also found in geometries of this type when the ion is associated with proteins.

9.2.1 High Spin Cobalt(II)

High spin cobalt(II) is a $S = \frac{3}{2}$ ion. The splitting of the ^4F free ion ground term (see Figure 9.1) indicates that there are relatively close excited states in every geometry. Furthermore, as spin-orbit coupling is relatively strong, the electronic relaxation times will be rather short—these times have in fact been related to the coordination number of the ion, as will be shown in Section 9.2.2. Such properties of high spin cobalt(II) make it suitable for investigation through NMR, mostly from the shift point of view. As the metal ion can also be fruitfully analyzed by electronic spectroscopy, it is therefore a really first rank paramagnetic probe in the study of metalloproteins.

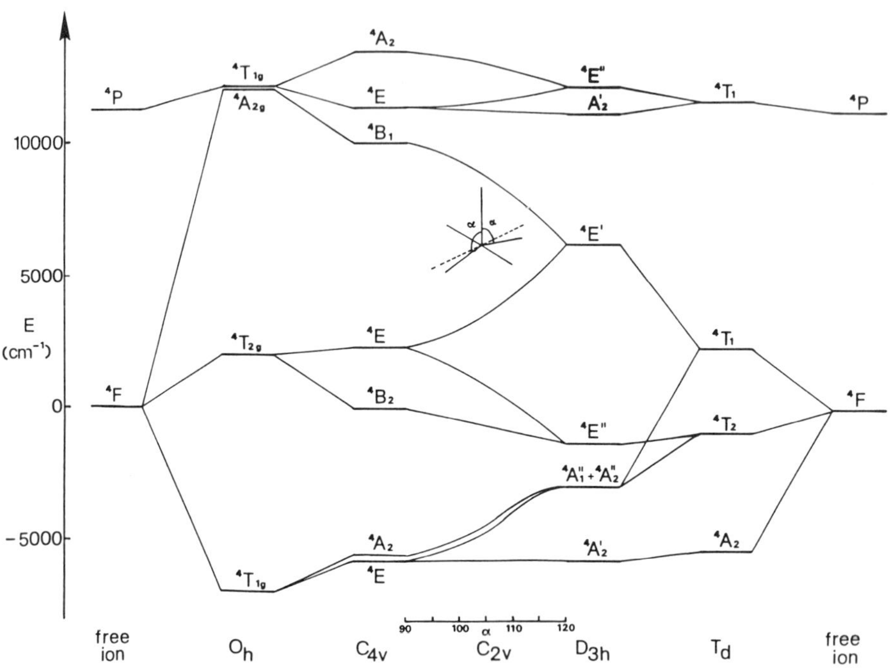

Figure 9.1 Splitting of the ^4F and ^4P free ion terms of high spin cobalt(II) in octahedral (O_h), square pyramidal (C_{4v}), trigonal bipyramidal (D_{3h}), and tetrahedral (T_d) geometries. The energies are calculated assuming equivalent ligands, $e_\sigma = 3{,}333\ cm^{-1}$, $e_\pi = 0$, and $10\ D_q = 10{,}000\ cm^{-1}$. For geometries intermediate between C_{4v} and D_{3h}, the energies are calculated, assuming C_{2v} symmetry, as a function of the angle α [1].

The similar coordination chemistry of cobalt(II) and zinc(II) makes cobalt(II) an ideal substituent for zinc(II) in many zinc proteins [1, 2] because the proteins maintain most of their native properties. Cobalt(II) has also been substituted for native iron(III) [3] in the metal site of transferrins, and has been found to specifically interact with lysozyme [4].

9.2.2 Water ^{1}H Relaxation Properties of High Spin Cobalt(II)

The hexaaqua cobalt(II) ion provides almost field-independent water proton T_1^{-1} values in the 0.01 to 60 MHz region [5] (Figure 9.2). By assuming the validity of the Solomon Equation (3.3), both the $\omega_s\tau_c = 1$ and $\omega_I\tau_c = 1$ dispersions can be placed at fields higher than 60 MHz, and therefore an upper limit for τ_c equal to 10^{-12} s can be set. Since the rotational correlation time, τ_r, is likely to be very similar to that of $Cu(OH_2)_6^{2+}$ (Section 6.4) and $Mn(OH_2)_6^{2+}$ (Section 8.4)—about 3×10^{-11} s—a shorter correlation time must dominate the interaction; thus the latter has to be ascribed to the electronic relaxation time, τ_s. Such a low τ_s value is consistent with the low water proton T_1^{-1} values. When measurements are performed in high viscous ethyleneglycol the data are nearly the same. This indicates that τ_s is rotation-independent [5].

Water ^{1}H T_1^{-1} measurements of the high pH species of cobalt(II)-substituted carbonic anhydrase, which is tetrahedral with three coordinated histidines,

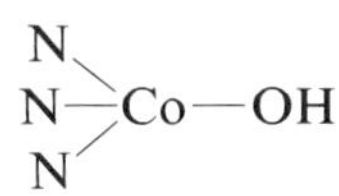

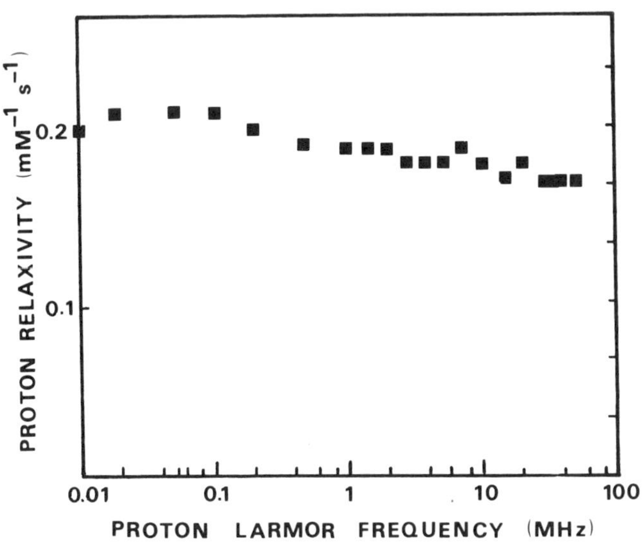

Figure 9.2 Water proton NMRD data for hexaaqua cobalt(II) solutions at 25 °C [5].

reveal the coordinated hydroxide ion [6]. Furthermore, the NMRD profile (Figure 9.3) shows a $\omega_s \tau_c = 1$ dispersion centered around 10 MHz, which qualitatively sets the τ_c value around 10^{-11} s. As the rotational correlation time of the molecule is much longer, this value is again therefore a measure of the electronic relaxation time. Compared to the hexaaqua complex, the nuclear-relaxing capability of this pseudotetrahedral cobalt(II) protein complex is thus relatively large.

A more quantitative analysis of the data would require consideration of the possible effects of zero field splitting (see Section 3.5), which is known to be sizeable in cobalt(II) complexes. Such effects have recently been taken into account to explain the smoother shape of the dispersion curve with respect to what is predicted by the Solomon equation, and reasonable values of the geometric factor G and τ_s have been obtained [7].

When the species interacting with the paramagnetic center is a hydroxide ion, water proton relaxation enhancement may only be detected through water exchange [8]. In fact, around neutral pH, the exchange rate of the hydroxide species cannot exceed the limit imposed by diffusion [6, 8–12]; the latter can be estimated by multiplying the diffusion coefficient of the hydroxide ion ($\sim 10^{10}$ M^{-1} s^{-1}) by its concentration. Thus, at pH 8 $\tau_M^{-1} < 10^{10} \times 10^{-6} = 10^4$ s^{-1}; but the observed f_M^{-1} T_{1p}^{-1} (Section 4.2.2) is about 10^5 s^{-1}, and the actual value of τ_M is much shorter than this.

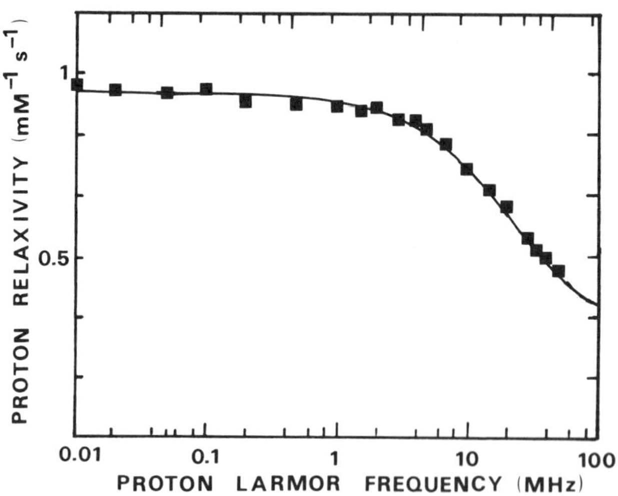

Figure 9.3 Water proton NMRD data for cobalt(II) human carbonic anhydrase I at pH 9.9 and 25 °C. Reproduced from [6], with permission; copyright © 1983 by Biophysical Society. The curve represents the best fitting of the data to a modified Solomon equation, including the effect of zero field splitting [7]. The best fit parameters are: $G = 5.0 \times 10^{-15}$ pm^{-6}, $\tau_s = 3.3 \times 10^{-11}$ s, and $\alpha = 72°$, where α is the average angle between the metal-proton vector and the z axis of the axial zero field splitting tensor.

Hydrogen exchange may occur through a hydrogen bond array of the type

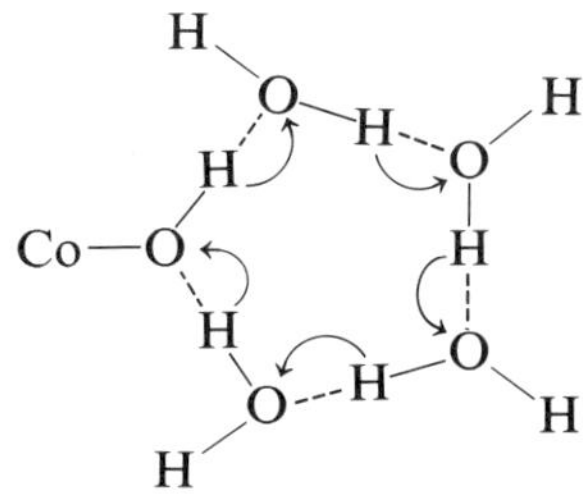

Here, covalent and hydrogen bonds can interchange, and a water molecule may diffuse away removing the hydrogen ion formerly bound to the hydroxide oxygen. In the case of cobalt(II) carbonic anhydrase, an exchange mechanism has been proposed to overcome the problem of diffusion based on the fact that the metalloenzyme can switch between four and five coordination [6]:

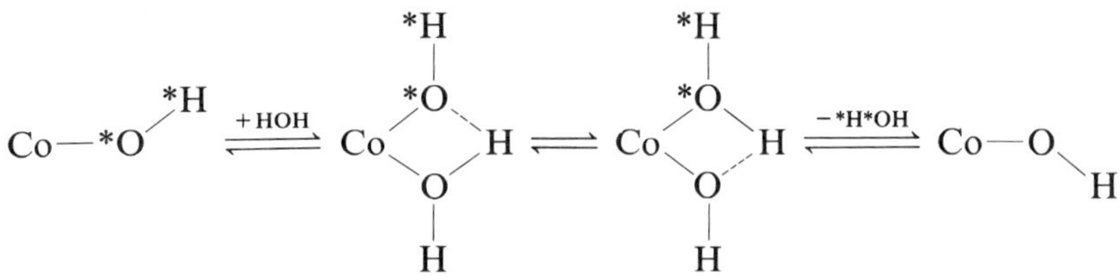

This mechanism, which can be quite general, could also account for the fast exchange of the solvent ^{17}O observed in the copper-substituted enzyme at high pH [13] (Section 6.5.2), when the latter derivative also contains a coordinated hydroxide ion.

Five-coordinated inhibited derivatives of the type [14]

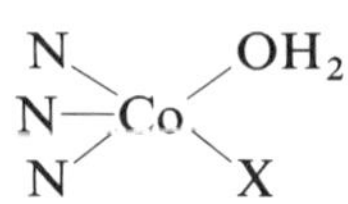

where X = I^-, NCS^-, or CH_3COO^-, display water ^1H T_1^{-1} values considerably lower than those of the tetrahedral species discussed [5, 15] (Figure 9.4); they are thought to have a τ_s of the order of 3 to 5×10^{-12} s. On the basis of water ^1H T_1^{-1} values, the low pH species of human carbonic anhydrase isoenzyme I has been assigned as largely five coordinate [6, 16]:

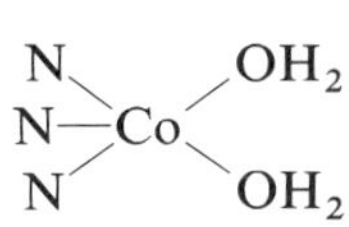

Therefore, present data indicate that the six-coordinated $Co(OH_2)_6^{2+}$ chromophore has a τ_s of the order of 10^{-12} s, the tetrahedral CoN_3O chromophore τ_s values of the order of 10^{-11} s, and the five-coordinated chromophores a τ_s of

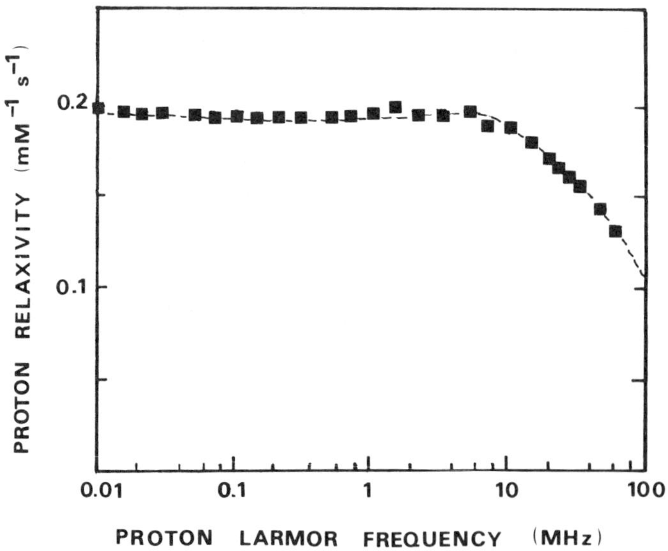

Figure 9.4 Water proton NMRD data for solutions of the nitrate adduct of cobalt(II) bovine carbonic anhydrase II at pH 6.0 and 25°C [1].

about 3 to 5 × 10^{-12} s. These values of τ_s follow the order of availability of low-lying excited states (Figure 9.1): as outlined in Section 3.7, the closer the excited states, the shorter τ_s.

9.2.3 Proton Isotropic Shifts of High Spin Cobalt(II)

From the foregoing discussion, six- and five-coordinated cobalt(II) chromophores are expected to display relatively narrow proton NMR signals. The spectrum of the five-coordinated iodide adduct of cobalt(II) bovine carbonic anhydrase II [17] is shown in Figure 9.5 together with that of the six-coordinated cobalt(II) conalbumin [3], the four-coordinated pseudotetrahedral high pH form of cobalt(II) bovine carbonic anhydrase II [17], and the cobalt(II)-substituted liver alcohol dehydrogenase [18]. Detection of the histidine N-H signals has been achieved through deuteration. The signal assignments are based both on estimates of the contact contributions from model complexes with imidazole and phenolate-like ligands, and on T_1^{-1} values, some of which are also shown in Figure 9.5. It is assumed that the largest contribution to T_1^{-1} is the point dipole relaxation mechanism. In the protein systems investigated the signals from the histidine N-H and the ring protons in *meta*-like position with respect to the coordinating nitrogen are always downfield, between 30 and 80 ppm. The isotropic shifts appear to be due to both contact and dipolar shifts, the latter being larger in five- and six-coordinated chromophores. In cobalt(II) conalbumin, the large upfield bias of the shifts arising from phenolate groups is attributed to large dipolar contributions,

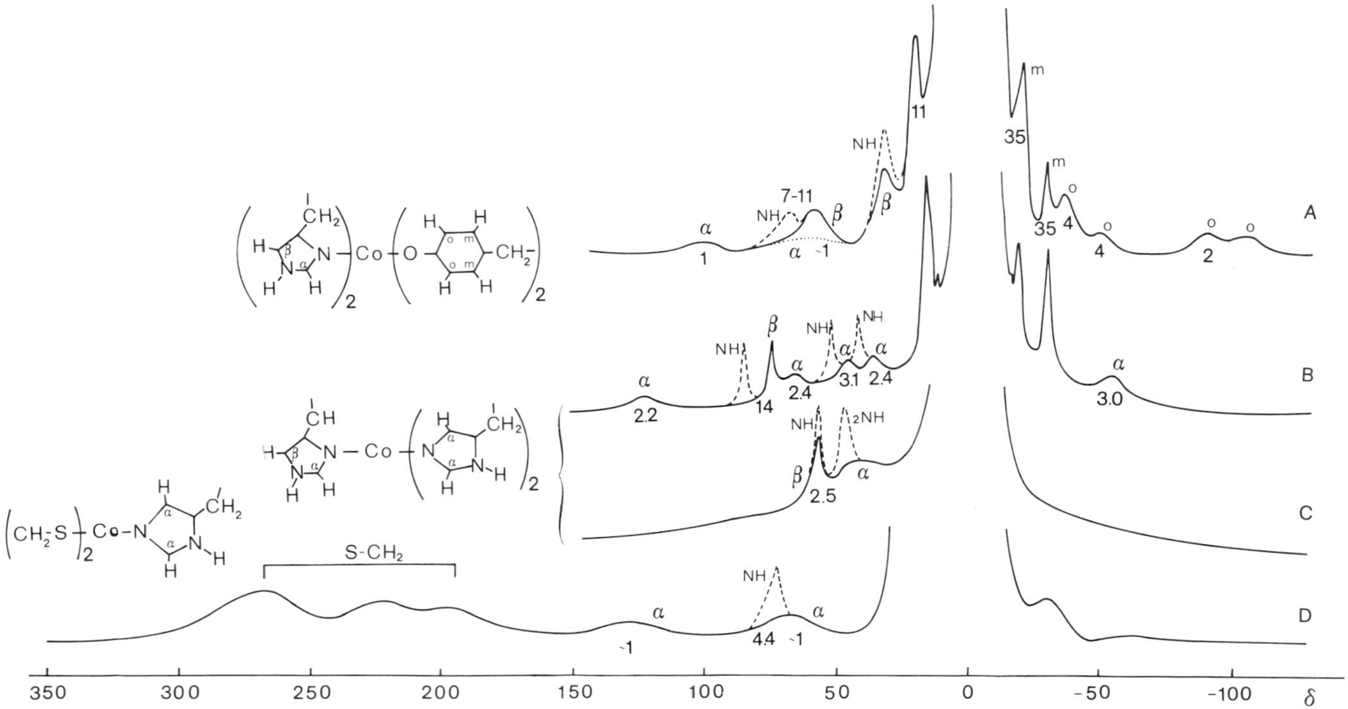

Figure 9.5 60 MHz ^{1}H NMR spectra of water solutions of: **A**: *bis*-cobalt(II) conalbumin [3]; **B**: iodide adduct of cobalt(II) bovine carbonic anhydrase II [17]; **C**: cobalt(II) bovine carbonic anhydrase II at pH 10.0 [17, 21]; **D**: *bis*-cobalt(II) liver alcohol dehydrogenase (catalytic sites) [18]. Chemical shifts are in ppm from TMS. The protein ligands are shown on the left-hand side. The signal assignment and the T_1^{-1} values (ms) are also reported. The dashed lines disappear when the spectra are recorded in D_2O.

which have been found to be operative in the axial position of elongated tetragonal model complexes [19]. The ^{1}H NMR spectra of cobalt(II) proteins thus provide direct information about the groups coordinated to the metal ion and enable subtle changes occurring upon inhibition, substrate binding, or pH changes to be detected at the active site [3, 17, 18, 20].

Ligand ^{1}H T_1^{-1} values are larger for four-coordinated molecules than for five- and six-coordinated molecules if the distance and overall shift are the same. Therefore, in the absence of water in the coordination sphere of the metal, the same conclusion drawn from NMRD measurements can also be reached by measuring the T_1^{-1} values of signals of metal ligands at a single frequency [21]. ^{13}C NMR studies on cobalt(II) human carbonic anhydrase I carboxymethylated at His 200 have revealed $^{13}COO^-$ isotropic shifts in five-coordinated complexes [22–24].

9.2.4 Low Spin Cobalt(II)

Low spin cobalt(II) chromophores occur under strong-ligand field conditions, and the ground level, which is characterized by $S = \frac{1}{2}$, is well separated from the first excited level. Although these compounds are most often investigated by EPR [25], cobalt(II) porphyrins have been extensively studied by NMR spectroscopy, despite their unfavorable electronic relaxation times (10^{-9} to 10^{-10} s). The ^{1}H NMR spectrum of cobalt(II) tetraphenylporphyrin, which is reported in Figure 9.6 [26], is consistent with a mainly dipolar nature of the isotropic shifts. This finding is expected in view of the slight tendency of the unpaired electron in the d_{z^2} orbital to delocalize into the ligand xy plane. ^{13}C NMR spectra [26], however, show that there is also some spin density on the carbon nuclei of the porphyrin ring. Nitrogen-containing ligands coordinate in the axial position and give rise to five-coordinated

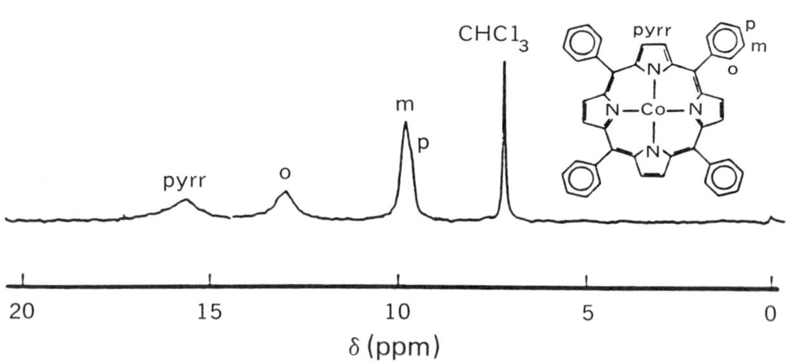

Figure 9.6 90 MHz ^{1}H NMR spectrum of tetraphenylporphynato cobalt(II) in $CDCl_3$ at 25 °C. Chemical shifts are in ppm from TMS. The signal assignment is also shown [26]. Reproduced from [26b], with permission; copyright © 1982 by American Chemical Society.

adducts [26]: the quality of the spectra for such compounds decreases because the electronic relaxation times become longer. Indeed, the first excited state contains two electrons in the d_{z^2} orbital, which is destabilized by the fifth ligand.

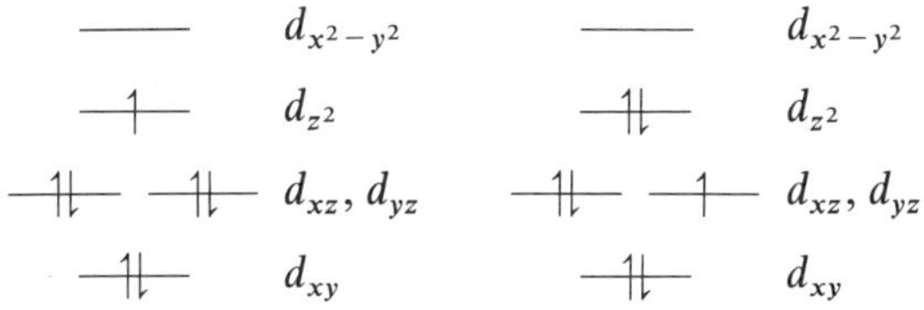

Ground state ${}^2A_1$ 1st excited state ${}^2E$

Therefore, the first excited state becomes higher in energy upon base binding to cobalt(II) porphyrins and the electronic relaxation times may become longer. Axial ligands capable of acting as π acceptors have been found to interact with the porphyrin plane and to experience sizeable isotropic shifts [26].

Studies have also been performed in which cobalt(II) has been substituted for the native metal in myoglobin (Mb) and hemoglobin (Hb). The signals at 62.2 ppm in Co(II)Mb and at 58.4 and 53.8 ppm in Co(II)(α)Co(II)(β)Hb (Figure 9.7) have been assigned to the N-H protons of the proximal histidine [27], and the shifts have been proposed to be largely contact in origin. The signal at 25 to 30 ppm has been assigned to the meso protons [27, 28]. The two N-H signals in Co(II)(α)Co(II)(β)Hb have also been specifically assigned to the α and β subunits as a result of investigations of the hybrid derivatives Co(II)(α)Fe(II)(β)Hb and Fe(II)(α)Co(II)(β)Hb (Figure 9.7).

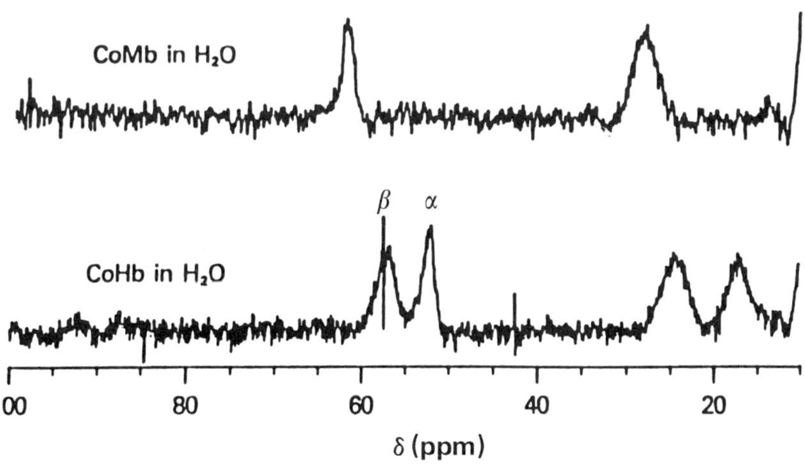

Figure 9.7 270 MHz ^1H NMR spectra of cobalt(II)-substituted myoglobin (CoMb) and hemoglobin (CoHb) in H_2O at pH 7. Chemical shifts are in ppm from DSS. Reproduced from [27], with permission; copyright © 1983 by American Chemical Society.

9.3 NICKEL

Nickel is a $3d^84s^2$ transition metal with oxidation numbers +1, +2, and +3, of which the most common is +2. This $3d^8$ ion has two unpaired electrons when it is six coordinated or four coordinated pseudotetrahedrally. When it is planar four coordinated it is always low spin diamagnetic.

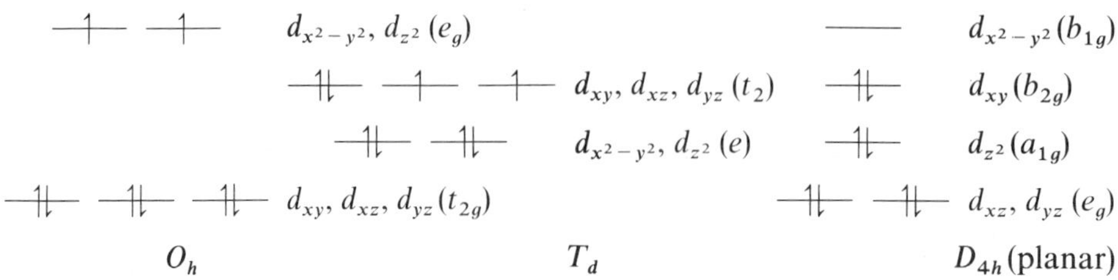

Five-coordinated nickel(II) complexes can either be high or low spin depending on the nature of the donor atoms.

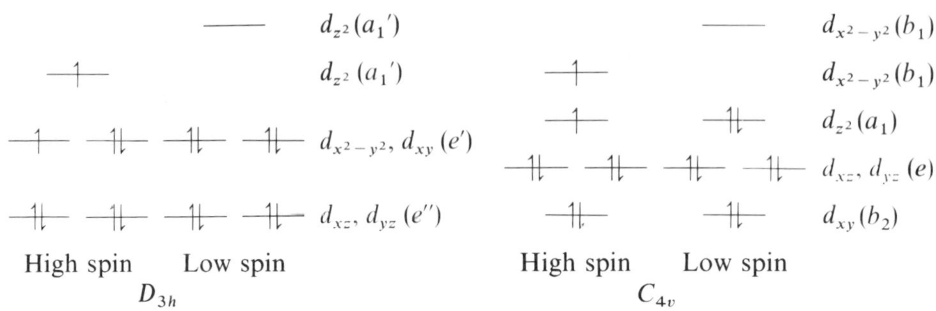

Nickel(I) and nickel(III) are rather rare; the former should have properties similar to copper(II) and the latter to low spin cobalt(II). Nickel(III) is possibly present in some dehydrogenases [29].

Nickel(II) has been introduced into the metal site of several metalloproteins and its NMR spectra recorded. When the site is pseudotetrahedral, sharp proton NMR signals are obtained.

9.3.1 Water Proton Relaxation Properties of High Spin Nickel(II)

The water proton T_1^{-1} values of the hexaaqua complex, which are shown in Figure 9.8 [30, 31], are relatively low and field independent up to 140 MHz. The increase noted at 270 MHz [31] is presumably due to a field dependence of τ_s of the type predicted by Bloembergen and Morgan—see Equation (3.27). Sophisti-

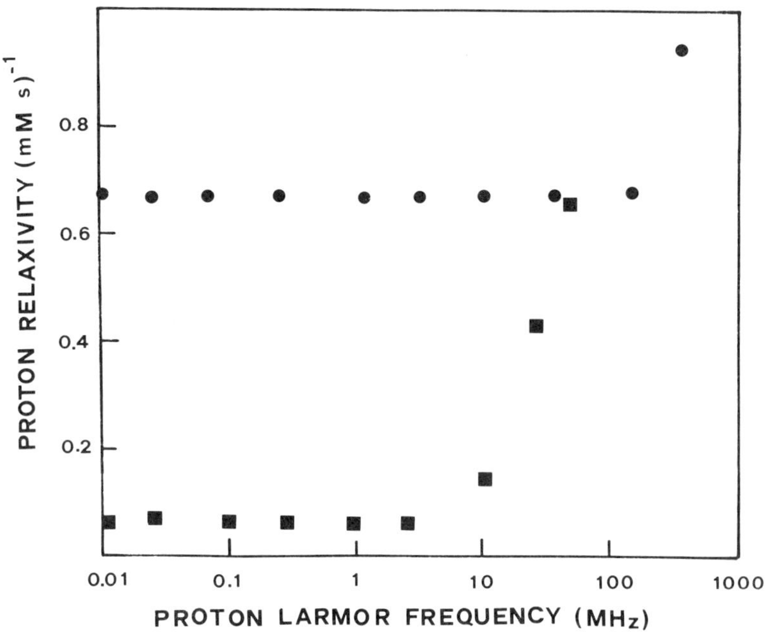

Figure 9.8 Water proton NMRD data for hexaaqua nickel(II) solutions at 25 °C (●) [30, 31] (see also note [64]) and nickel(II) bovine carbonic anhydrase II at pH 6.0 and 25 °C (■) [5].

cated analysis of the data seems to confirm that electron relaxation is due to fluctuations of the quadratic zero field splitting, and according to this analysis, the low field value of τ_s is estimated to be 3×10^{-12} s [31]. Furthermore, as predicted by theory (Section 3.5), the zero field splitting of the $S = 1$ manifold, depicted in Figure 3.12, provides large $\omega_s\tau_s$ values and therefore no $\omega_s\tau_s = 1$ dispersion is expected (Section 3.5) [32–35]. The $\omega_I\tau_s = 1$ dispersion presumably occurs at higher fields.

Water ^{1}H T_1^{-1} values have also been measured for nickel(II)-substituted bovine carbonic anhydrase II [33] (Figure 9.8), for which the following coordination polyhedron has been proposed [36]:

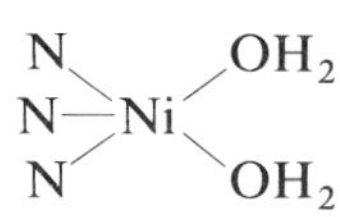

Although the low field profile is flat, a sharp rise occurs at high field. The absence of the $\omega_s\tau_s = 1$ dispersion is accounted for by the presence of large static zero field splitting [32, 33]. As before, the increase in T_{1p}^{-1} is presumably due to the field dependence of τ_s. The correlation time, τ_v, for the modulation of the zero field splitting is apparently longer than that in the aquaion.

9.3.2 Proton NMR Spectra of High Spin Nickel(II)

A change in coordination number from six to four perturbs the electronic relaxation times of nickel(II) complexes, in a sense opposite to that observed in cobalt(II) complexes. Nickel(II) complexes, when six coordinated, have an orbitally nondegenerate ground state labeled $^3A_2$ in O_h symmetry (Figure 1.1), with the first excited state at about a few thousand wave numbers; in pseudotetrahedral symmetry, on the contrary, there are always closely spaced levels originating from the threefold orbitally degenerate $^3T_1$ ground state of pure T_d symmetry—such levels cause a shortening of the electronic relaxation times.

An excellent example of the sharp NMR signals observed for pseudotetrahedral nickel(II) complexes is the ^{1}H NMR spectrum of nickel(II)-substituted azurin [37]. Azurin is a small protein (MW 14,000) containing a single type I copper (see Section 6.3) forced into a pseudotetrahedral environment, as shown in Figure 9.9. A similar environment is apparently maintained when nickel(II) is substituted for the native copper(II) ion. The spectrum in D_2O shows nine well-resolved signals (Figure 9.9) integrating to 12 to 13 protons. These signals arise from the methyl group of Met 121 (signal 4), which is the only assigned peak in the spectrum, the two pairs of protons in the α and β positions with respect to the coordinating nitrogens of His 117 and 46 (possibly signals 3 and 5), and probably the β or γ methylene groups of some of the four coordinating residues. The pH dependence

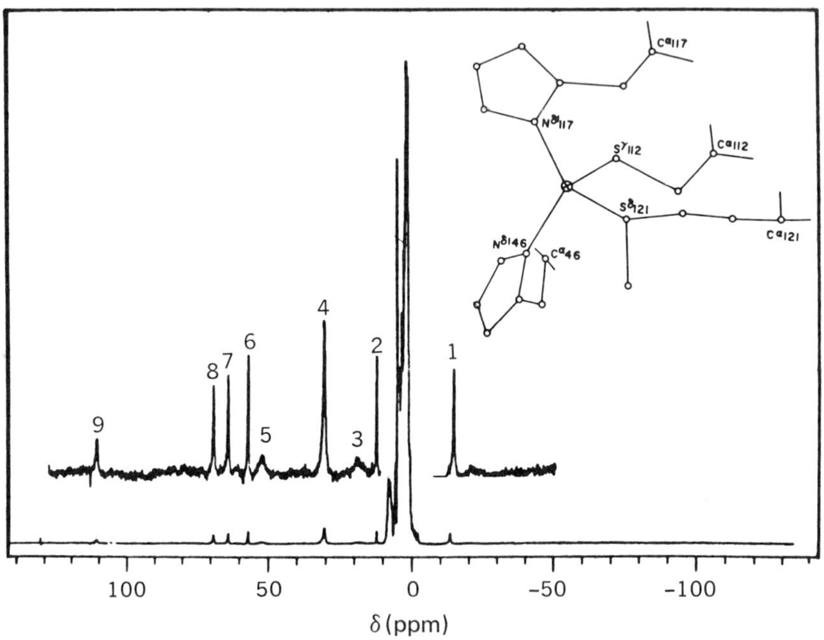

Figure 9.9 360 MHz ^{1}H NMR spectrum of nickel(II)-substituted azurin in D_2O at pH* 9.0 and 25 °C, and a schematic drawing of the metal coordination polyhedron. Chemical shifts are in ppm from DSS. Reproduced from [37], with permission; copyright © 1982 by American Chemical Society.

of the spectrum permits the detection of an acid-base equilibrium perturbing the active site and a slow conformational change [37]. Furthermore, due to the low molecular weight of the protein, the spectrum can be recorded at very high magnetic field (360 MHz) without excessive line broadening due to Curie spin relaxation (Section 3.4).

9.4 VANADIUM

Although vanadium has long been known to be present in sizeable quantities in vanadocytes—cells showing oxygen transport activity—of certain molluscs, studies have failed to reveal the presence of a specific vanadium-binding protein, mainly because purification procedures always result in the recovery of free oxovanadium(IV) ions. However, EPR investigations on intact cells have shown that only a minor fraction of the total vanadium content is in the form of the EPR-detectable oxovanadium(IV) [38, 39], suggesting the presence of vanadium(III). As vanadium(III) has a d^2 electron configuration with a $^3T_1$ ground state in O_h symmetry (Figure 1.1), pseudooctahedral vanadium(III), in analogy to pseudotetrahedral nickel(II), is expected to display very short electronic relaxation times. Thus, NMR experiments have been performed [39, 40] on intact vanadocytes with the aim of revealing isotropically shifted signals from possible organic donor groups. However, only a single signal around 21 ppm downfield could be detected. Furthermore, the shift value was somewhat variable and found to depend on the metal/water ratio in the cell preparation, which is consistent with the idea that water protons in fast exchange with the simple hexaaqua vanadium(III) species are observed [39, 40]. As the latter species is very sensitive to oxidation, especially at neutral or alkaline pH, it has been proposed that vanadocytes contain low pH compartments inaccessible to molecular oxygen. EXAFS [41] studies have subsequently indicated the ion to be in the +3 oxidation state and regularly surrounded by six second period atoms, most probably oxygens.

Oxovanadium(IV) has not been found to have any biological role. However, it has been introduced into the active site of many metalloproteins, including several zinc(II) proteins and transferrins. Being a d^1 ion, its electronic relaxation times are long, of the order of 10^{-8} to 10^{-9} s. Furthermore, its EPR spectra are very sharp and are characterized by a well-resolved hyperfine structure. The predominant isotope is ^{51}V, which has a natural abundance of 99.8% and $I = \frac{7}{2}$ (Table 1.1). A complete NMRD profile for *bis*-oxovanadium(IV) transferrin is shown in Figure 9.10 [42]. Despite the presence of the oxygen atom, which is likely to occupy the only solvent-accessible coordination position, the water proton T_1^{-1} values are still sensitive to the paramagnetic metal, probably through second sphere water. The overall NMRD profile is quite similar to that of *bis*-copper(II) transferrin and other copper(II) proteins (Figure 6.4). The first dispersion, qualitatively ascribed to the $\omega_S\tau_c = 1$ dispersion, is very small, even smaller than for copper(II); the second is easily attributed to the $\omega_I\tau_c = 1$ dispersion. From the latter, a τ_c of about 2×10^{-8} s is estimated; since the rotational correlation time of the protein is of the order of 2 to 3×10^{-7} s [43], τ_c is mainly determined by τ_s, with possible

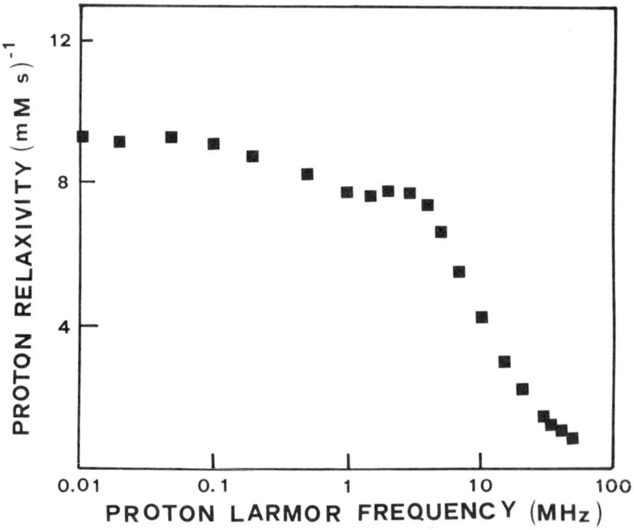

Figure 9.10 Water proton NMRD data for 1.6×10^{-3} M *bis*-oxovanadium(IV) transferrin at pH 8 and 25 °C [42].

contributions by τ_M. Analogous to the case for copper(II), the $\omega_s\tau_c = 1$ dispersion may be substantially lower than that predicted by the Solomon equation because of the effect of hyperfine coupling with the metal nucleus [42] (see Section 6.5.1).

9.5 CHROMIUM

Although chromium is an essential trace element, no naturally occurring chromium(III) compounds have been definitely identified. Nevertheless, the chemical properties of chromium(III) have stimulated its use as a substitution probe of other biochemically relevant metal ions.

Chromium(III) is generally six coordinated with a $^4A_{2g}$ (O_h symmetry) ground state (Figure 1.1); its electronic relaxation times are of the order of 10^{-9} to 10^{-10} s (Table 3.1). It is kinetically inert as far as ligand exchange is concerned, and the water exchange in the hexaaqua complex or in any water-containing complex is very slow ($\tau_M = 10^5$ to 10^6 s), whereas proton exchange is relatively fast [$\tau_M = 2$ to $5 \cdot 10^{-6}$ s in $Cr(OH_2)_6^{3+}$]. The NMRD profiles for hexaaqua chromium(III) are shown in Figure 9.11 [44]. Above 25 °C the proton exchange rate is substantially fast, so that the measured T_1^{-1} is not affected by τ_M. As the low field $\omega_s\tau_c = 1$ dispersion has a τ_c of $\sim 5 \times 10^{-10}$ s, it is too long to be the rotational time of the complex and has to be ascribed to the electronic relaxation time, τ_s. Furthermore, as discussed for manganese(II) (Section 8.4), such a dispersion can only originate from a contact contribution. The high field dispersion corresponds to another $\omega_s\tau_c = 1$ dispersion and yields a τ_c of $\sim 3 \times 10^{-11}$ s, which is the expected τ_r value. Comparison of the data at 25 °C and 65 °C [44], both already in the fast-exchange

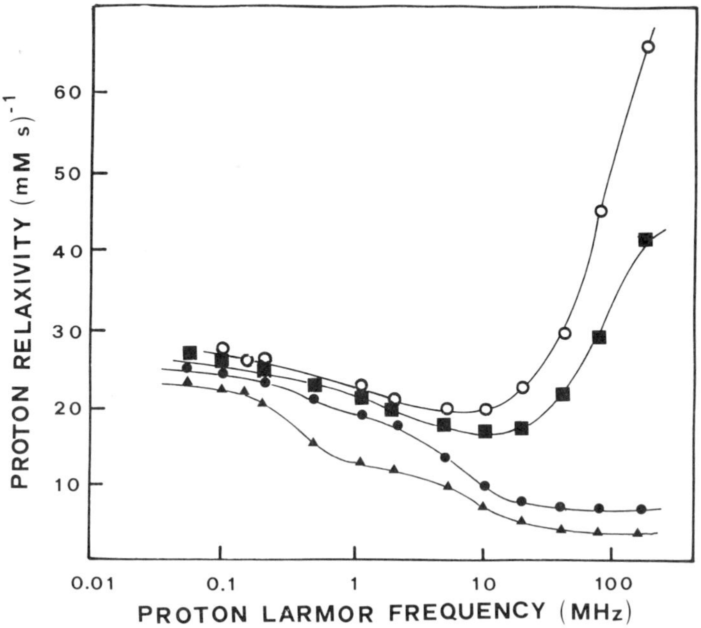

Figure 9.11 Water proton T_1^{-1} and T_2^{-1} data of hexaaqua chromium(III) solutions at 25 °C (● = T_1^{-1}, ○ = T_2^{-1}) and 65 °C (▲ = T_1^{-1}, ■ = T_2^{-1}) [44].

limit, shows that the two inflections move in opposite directions with increasing temperature. The shift of the high field dispersion is consistent with the expected temperature dependence of τ_r; the shift of the low field dispersion indicates that τ_s increases with increasing temperature.

Water proton relaxation measurements have been performed on a variety of mixed complexes [45, 46], where both τ_r and τ_M of the water protons are influenced by the nature of the other ligands. Qualitatively, the water proton relaxation rate has been found to be related to the acidity of the water protons, the net charge of the complex, and, of course, the number of coordinated water molecules.

When chromium(III) binds to a macromolecule, τ_s becomes the only correlation time and proton relaxation is further enhanced due to an increase of the dipolar term. The resulting T_{1M}^{-1} is therefore mainly dipolar in origin.

As chromium(III) is substitutionally inert, chromium(III) ATP and ADP complexes have been extensively studied [47, 48], in addition to ternary (enzyme) complexes. The proposed structures for the Cr(III)-ATP complex [48] (Figure 9.12) suggest that ATP, even when complexed to chromium(III), is still able to interact with its binding site in many ATP-dependent enzymes. For example, both ATP and Cr(III)-ATP are capable of interacting with the enzyme RNA polymerase in the presence of a sample nucleotide substrate cytidyl triphosphate (CTP), but the polymerization is blocked at the first step. By analyzing the CTP proton relaxation enhancements in the presence of Cr(III)-ATP, and by using a τ_c obtained through

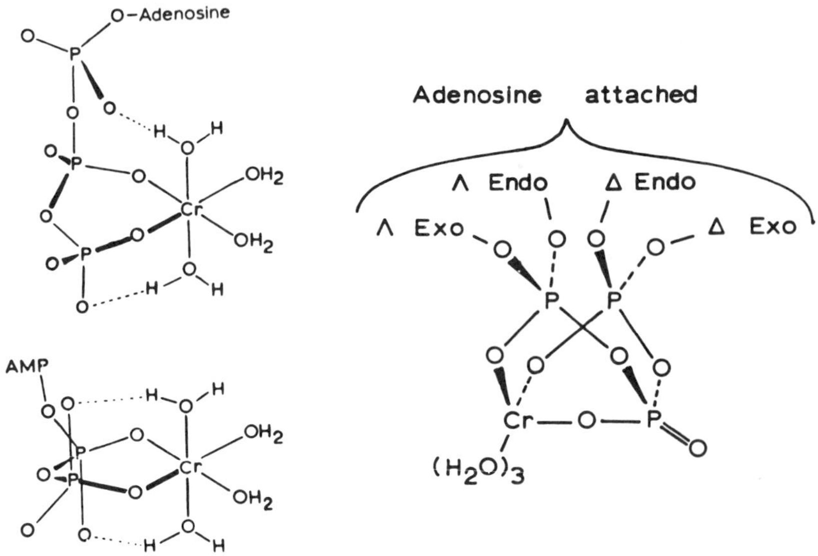

Figure 9.12 Possible bidentate (left) and tridentate (right) binding modes of ATP to chromium(III) ions. Reproduced from [48], with permission, copyright © 1979 by Elsevier Science Publishing Co., Inc.

water proton NMRD data, initiator-to-substrate distances have been proposed [49, 50].

Using a similar procedure, pyruvate has been mapped with respect to Cr(III)-ATP in the pyruvate kinase-Cr(III)-ATP-pyruvate ternary system [51, 52]; Cr(III)-ATP has been shown to bind 1.1 nm from the magnesium(II) (or manganese(II)) binding site of pyruvate kinase. Therefore, the divalent cation binds at a site near to, but different from, the nucleotide site.

Hexacyanochromate(III) has been used as a relaxation reagent to probe the specific binding site of hexacyanoferrate(III) in azurin. From line-broadening effects on assigned signals in the reduced protein upon addition of hexacyanochromate(III) the binding site has been proposed to be midway between Lys 85 and Lis 92 [53].

^2H NMR spectroscopy has been recently used in the study of small chromium(III) complexes through isotropic shifts. This is a promising technique also for the investigation of suitable problems in biochemical systems [54, 55].

9.6 RUTHENIUM

Pentaamineruthenium(II) complexes generated *in situ* have been recently used as probes for electron transfer processes from the surface of metalloproteins to the redox metal in the active center. The electron is transferred to the redox center, leaving pentaamineruthenium(III) coordinated to a surface histidine. Sperm-whale myoglobin can be labeled with ruthenium at His 12, 81, and 113. Isotropically shifted NMR signals arising from the ring protons have been observed at −36 ppm and assigned to the 2-H protons of the above histidines. The isotropic shifts of heme proton signals are unaffected by the presence of ruthenium [56].

9.7 SPIN LABELS

Spin labels are stable free radicals capable of reacting with a macromolecule. The most common spin labels are nitroxide-containing compounds, which are based on molecular skeletons of the type

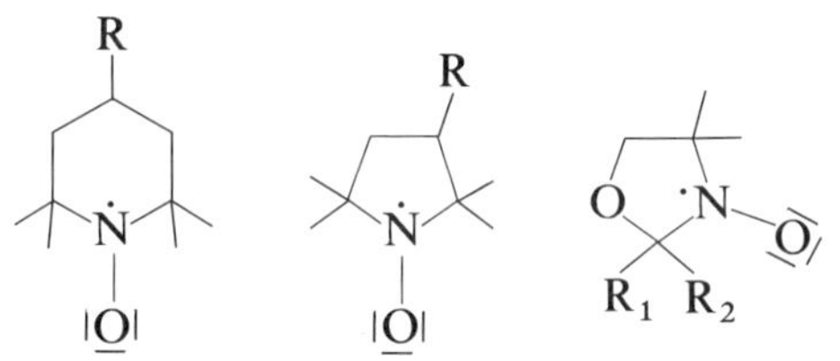

where the unpaired electron is localized on the N-O moiety. Free radicals of this type have electronic relaxation times of about 10^{-7} s, which are the largest among paramagnetic compounds. The correlation time for the NMR experiment when the radical is free in solution is the rotational correlation time; when the radical interacts rigidly with a macromolecule its rotational correlation time increases until, only for very large macromolecules, the correlation time for the NMR experiment eventually becomes the electronic relaxation time. If the bound radical still has motional freedom, this motion may be faster than the rotational time of the whole molecule and thus determines the correlation time. Of course, if the system is in chemical exchange, the exchange time may also be shorter than the electronic relaxation time and so represents the limiting NMR correlation time. An immobilized radical produces a very large broadening effect on the NMR line of nearby nuclei, which is probably the largest effect that a paramagnetic species can cause.

As the nuclear relaxation mechanism is dipolar in origin, nuclear T_1^{-1} or T_2^{-1} enhancements experienced by nuclei of a macromolecule interacting with a radical contain structural information. Owing to the long correlation times, the effects on transverse relaxation are detectable on nuclei that are as far as 1.5 to 2 nm away from the paramagnetic probe. On the other hand, nuclei that are much closer may not be detectable because of the excessive line broadening. Therefore, radicals are best used as long-range relaxing probes. By attaching a radical to a specific active site, water ^{1}H T_1^{-1} enhancements have also been detected in diamagnetic proteins [57], and substrates have been mapped [58].

Spin labels in solution also give rise to measurable isotropic shifts on solvent or interacting molecules [59]. Although the nature of the shift is mainly contact in origin [60], dipolar contributions have also been suggested to be operative [61].

^1H and ^2H NMR studies are available for ruthenium(III) and cobalt(III) porphyrin π-cations. Such systems show relatively sharp NMR signals [62]. Small molecules are at the limit of detectability as far as proton NMR is concerned—since the correlation time is determined by the rotational correlation time—and are easily detected through ^2H NMR. If other factors concur favorably (interactions between radicals at high concentration [63], tendency to aggregate, availability of low-lying excited states), the proton NMR signals can also be detected for radicals.

REFERENCES

1. Bertini, I., and Luchinat, C. *Adv. Inorg. Biochem.* **6** (1984) 71.
2. Bertini, I., and Luchinat, C. In *Metal Ions in Biological Systems*, H. Sigel (ed.), Marcel Dekker, New York, 1983, Volume 15, p. 101.
3. Bertini, I., Luchinat, C., Messori, L., and Scozzafava, A. *Eur. J. Biochem.* **141** (1984) 375.
4. McDonald, C. C., and Phillips, W. D. *Biochem. Biophys. Res. Commun.* **35** (1969) 43.
5. Banci, L., Bertini, I., and Luchinat, C. *Inorg. Chim. Acta* **100** (1985) 173.
6. Bertini, I., Brown, R. D. III., Koenig, S. H., and Luchinat, C. *Biophys. J.* **41** (1983) 179.
7. Bertini, I., Luchinat, C., Mancini, M., and Spina, G. *J. Magn. Reson.* **59** (1984) 213.
8. Bertini, I., Canti, G., Luchinat, C., and Messori, L. *Inorg. Chem.* **21** (1982) 3426.
9. Eigen, M., and Hammes, G. G. *Adv. Enzymol.* **25** (1963) 1.
10. Fabry, M. E., Koenig, S. H., and Schillinger, W. E. *J. Biol. Chem.* **245** (1970) 4256.
11. Melton, B. F., and Pollak, V. L. *J. Phys. Chem.* **73** (1969) 3669.
12. Grunwald, E., and Fong, D. W. *J. Am. Chem. Soc.* **94** (1972) 7371.
13. Bertini, I., Canti, G., and Luchinat, C. *Inorg. Chim. Acta* **56** (1981) 1.
14. Bertini, I., Canti, G., Luchinat, C., and Scozzafava, A. *J. Am. Chem. Soc.* **100** (1978) 4873.
15. Bertini, I., Canti, G., and Luchinat, C. *Inorg. Chim. Acta* **56** (1981) 99.
16. Bertini, I., and Luchinat, C. *Acc. Chem. Res.* **16** (1983) 272.
17. Bertini, I., Canti, G., Luchinat, C., and Mani, F. *J. Am. Chem. Soc.* **103** (1981) 7784.
18. Bertini, I., Gerber, M., Lanini, G., Luchinat, C., Maret, W., Rawer, S., and Zeppezauer, M. *J. Am. Chem. Soc.* **106** (1984) 1826.
19. Horrocks, W. DeW. Jr., Taylor, R. C., and La Mar, G. N. *J. Am. Chem. Soc.* **86** (1964) 3031. Kluiber, R. W., and Horrocks, W. DeW. Jr. *J. Am. Chem. Soc.* **87** (1965) 5350; *J. Am. Chem. Soc.* **88** (1966) 1399.
20. Hill, H. A. O., Smith, B. E., Storm, C. B., and Ambler, R. P. *Biochem. Biophys. Res. Commun.* **70** (1976) 783.
21. Bertini, I., Lanini, G., and Luchinat, C. *J. Am. Chem. Soc.* **105** (1983) 5116.
22. Khalifah, R. G., and Morley, P. J. *Arch. Biochem. Biophys.* **232** (1984) 632.
23. Khalifah, R. G., Rogers, J. I., Harmon, P., Morley, P. J., and Carroll, S. B. *Biochemistry* **23** (1984) 3129.
24. Morley, P. J., Carroll, S. B., Harmon. P., and Khalifah, R. G. *J. Am. Chem. Soc.* **106** (1984) 1830.
25. Haffner, P. H., and Coleman, J. E. *J. Biol. Chem.* **248** (1973) 6630.

26. Fulton, G. P., and La Mar, G. N. *J. Am. Chem. Soc.* **98** (1976) 2119. Shirazi, A., and Goff, H. M. *Inorg. Chem.* **21** (1982) 3420.
27. Inubushi, T., Ikeda-Saito, M., and Yonetani, T. *Biochemistry* **22** (1983) 2904.
28. Hill, H. A. O., Sadler, J. P., and Williams, R. J. P. *J. Chem. Soc.* **A** (1973) 1663.
29. LeGall, J., Ljungdahl, P. O., Moura, I., Peck, H. D., Xavier, A. V., Moura, J. J. G., Teixera, M., Huynh, B. H., and DerVartanian, D. V. *Biochem. Biophys. Res. Commun.* **106** (1982) 610.
30. Morgan, L. O., and Nolle, A. W. *J. Chem. Phys.* **31** (1959) 365.
31. Friedman, H. L., Holz, M., and Hertz, H. G. *J. Chem. Phys.* **70** (1979) 3369.
32. Lindner, U. *Ann. Phys.* **16** (1965) 319.
33. Bertini, I., Luchinat, C., Mancini, M., and Spina, G. In *Magneto-Structural Correlations in Exchange Coupled Systems*, R. D. Willett, D. Gatteschi, and O. Kahn (eds.), Reidel, Dordrecht, 1985, p. 421.
34. Kowalewski, J., Nordenskiöld, L., Benetis, N., and Westlund, P.-O. *Progr. NMR Spectr.* **17** (1985) 141.
35. Benetis, N., Kowalewski, J., Nordenskiöld, L., Wennerström, H., and Westlund, P.-O. *Mol. Phys.* **48** (1983) 329.
36. Bertini, I., Borghi, E., Luchinat, C., and Monnanni, R. *Inorg, Chim. Acta* **67** (1982) 99.
37. Blaszak, J. A., Ulrich, E. L., Markley, J. L., and McMillin, D. R. *Biochemistry* **21** (1982) 6253.
38. Swinehart, J. H., Biggs, W. R., Halko, D. J., and Schroeder, N. C. *Biol. Bull.* **146** (1974) 302.
39. Kustin, K., Levine, D. S., McLeod, G. C., and Curby, W. A. *Biol. Bull.* **150** (1976) 426.
40. Carlson, R. M. K. *Proc. Natl. Acad. Sci. U.S.A.* **72** (1975) 2217.
41. Tullius, T. D., Gillum, W. O., Carlson, R. M. K., and Hodgson, K. O. *J. Am. Chem. Soc.* **102** (1980) 5670.
42. Bertini, I., Briganti, F., Koenig, S. H., and Luchinat, C., *Biochemistry*, in press.
43. Koenig, S. H., and Schillinger, W. E. *J. Biol. Chem.* **244** (1969) 3283.
44. Hausser, R., and Noack, F. *Z. Phys.* **153** (1964) 394.
45. De Pamphilis, M. L., and Cleland, W. W. *Biochemistry* **12** (1973) 3714.
46. Foster, D. M., and Mildvan, A. S. *Bioinorg. Chem.* **1** (1972) 133.
47. Villafranca, J. J. *Methods Enzymol.* **87** (1982) 180.
48. Cleland, W. W., and Mildvan, A. S. *Adv. Inorg. Biochem.* **1** (1979) 163.
49. Bean, B. L., Koren, R., and Mildvan, A. S. *Biochemistry* **16** (1977) 3322.
50. Stein, P. J., and Mildvan, A. S. *Biochemistry* **17** (1978) 2675.
51. Gupta, R. K., Fung, C. H., and Mildvan, A. S. *J. Biol. Chem.* **251** (1976) 2421.
52. Mildvan A. S., Sloan, D. L., Fung, C. H., Gupta, R. K., and Melamud, E. *J. Biol. Chem.* **251** (1976) 2431.
53. Cho, K. C., Blair, D. F., Banerjee, U., Hopfield, J. J., Gray, H. B., Pecht, I., and Chan, S. I. *Biochemistry* **23** (1984) 1858.
54. Wheeler, W. D., Kaizaki, S., and Legg, J. I. *Inorg. Chem.* **21** (1982) 3248.
55. Wheeler, W. D., and Legg, J. I. *Inorg. Chem.* **23** (1984) 3798.
56. Toi, H., La Mar, G. N., Margalit, R., Che, C.-M., and Gray, H. B. *J. Am. Chem. Soc.* **106** (1984) 6213.
57. Mildvan, A. S., and Weiner, H. *Biochemistry* **8** (1969) 552.
58. Mildvan, A. S., and Weiner, H. *J. Biol. Chem.* **244** (1969) 2465.
59. Screttas, C. G., and Micha-Screttas, M. *J. Chem. Soc., Chem. Commun.* (1982) 1168.

60. Kreilick, R. W. *Adv. Magn. Reson.* **6** (1973) 141.
61. Waterton, J. C., and Sanders, J. K. M. *J. Am. Chem. Soc.* **100** (1978) 1295.
62. Morishima, I., Takamuri, Y., Shiro, Y. *J. Am. Chem. Soc.* **106** (1984) 7666.
63. Kreilick, R. W., *J. Am. Chem. Soc.* **90** (1968) 2711.
64. Note added in proof. Recent NMRD data in the 2-88 MHz range on water solutions of hexaaqua nickel(II) have shown some field dependence of relaxivity that requires further theoretical analysis (Hertz, H. G., and Holz, M. *J. Magn. Reson.* **63** (1985) 64).

CHAPTER 10

Lanthanides

10.1 LANTHANIDES IN BIOLOGICAL SYSTEMS

Owing to the special NMR properties of lanthanide-containing compounds, lanthanide ions have been used to study biological systems through NMR. Hydrated lanthanides have a great affinity for oxygen and nitrogen donors of protein molecules, and, even when complexed by chelating ligands, they may interact with further donors, possibly of a protein molecule. Furthermore, lanthanides can replace calcium in calcium-containing proteins, since the ionic radii of calcium(II) and lanthanides(III) are similar. This property has been used, for example, to investigate the metal binding site of lysozyme, the calcium-binding protein parvalbumin, and the calcium-activated protein trypsin. Lanthanides have also been introduced into the metal site of transferrins and the calcium sites of thermolysin. Finally, the NMR properties of lanthanides have been exploited to investigate the transport of ions and molecules through membranes.

Recently, several excellent reviews on the use of lanthanides as shift reagents and relaxation reagents have been published [1–8], some of which also discuss biochemical applications.

10.2 ELECTRONIC PROPERTIES OF LANTHANIDES

The electronic properties of lanthanides are peculiar in that spin-orbit interactions are very large, larger in fact than ligand field effects. Consequently, the description given in Section 1.2 is no longer adequate.

Spin-orbit interactions couple the orbital, L, and spin, S, angular momenta. The resulting free ion terms are characterized by J values which are vector combinations of L and S. Again, the level of interest is the ground level, which is that with largest S, largest L with equally large S, smallest J for f^1 to f^6 lanthanides, and largest J for f^8 to f^{13} lanthanides. For example, Ce(III) has a ground state with

Table 10.1 Some Electronic Properties of Lanthanide Ions

Ion	Configuration	$^{2S+1}L_J$ of Ground State/(Multiplicity)	g_J^a	$\langle S_z \rangle_J^b$	$\langle S_z \rangle_J^c$	$\left(\frac{\Delta\nu}{\nu_0}\right)^{dip}$ (ppm)[d]
Ce^{3+}	$4f^1$	$^2F_{5/2}$ (6)	$\frac{6}{7}$	1.07	0.98	+1.6
Pr^{3+}	$4f^2$	$^3H_4$ (9)	$\frac{4}{5}$	3.20	2.97	+2.7
Nd^{3+}	$4f^3$	$^4I_{9/2}$ (10)	$\frac{8}{11}$	4.91	4.49	+1.05
Pm^{3+}	$4f^4$	$^5I_4$ (9)	$\frac{3}{5}$	4.80	4.01	−0.6
Sm^{3+}	$4f^5$	$^6H_{5/2}$ (6)	$\frac{2}{7}$	1.79	−0.06	+0.17[e]
Eu^{3+} (Sm^{2+})	$4f^6$	$^7F_0$ (1)	—	—	−10.68	~−1.0[e]
Gd^{3+} (Eu^{2+})	$4f^7$	$^8S_{7/2}$ (8)	2	−31.5	−31.50	~0
Tb^{3+}	$4f^8$	$^7F_6$ (13)	$\frac{3}{2}$	−31.5	−31.82	+20.7
Dy^{3+}	$4f^9$	$^6H_{15/2}$ (16)	$\frac{4}{3}$	−28.30	−28.54	+23.8
Ho^{3+}	$4f^{10}$	$^5I_8$ (17)	$\frac{5}{4}$	−22.50	−22.63	+9.4
Er^{3+}	$4f^{11}$	$^4I_{15/2}$ (16)	$\frac{6}{5}$	−15.30	−15.37	−7.7
Tm^{3+}	$4f^{12}$	$^3H_6$ (13)	$\frac{7}{6}$	−8.17	−8.21	−12.7
Yb^{3+}	$4f^{13}$	$^2F_{7/2}$ (8)	$\frac{8}{7}$	−2.57	−2.59	−5.2

[a] Calculated from $g_J = 1 + \frac{J(J+1) - L(L+1) + S(S+1)}{2J(J+1)}$; the formula does not hold for f^6 ions.

[b] Calculated according to Equation (10.4).

[c] Calculated by inclusion of the excited states [12].

[d] Dipolar shift predicted from Equation (10.3) for $r = 300$ pm, $3\cos^2\theta - 1 = 1$, $T = 300$ K, and D_z values for each lanthanide estimated from reference [9].

[e] Including contributions from excited J manifolds [9]. For f^6 ions, the contribution of the ground state manifold is zero.

$J = \frac{5}{2}$, since one electron in an f orbital gives rise to a free ion term ^2F with $S = \frac{1}{2}$ and $L = 3$. Two terms arise through spin-orbit coupling: one with $J = 3 - \frac{1}{2} = \frac{5}{2}$ and the other with $J = 3 + \frac{1}{2} = \frac{7}{2}$. The former term is the ground state as shown in Table 10.1. The J values of the ground state vary from $\frac{5}{2}$ to 8, including $J = 0$ for Eu(III) and Sm(II). In general, it can be assumed that the ground term is the only one of interest; however, this is not true for Sm(III), which has an excited J manifold at about 1,000 cm^{-1}, and Eu(III), whose ground term has $J = 0$ and would be diamagnetic if excited terms were not taken into consideration.

In actinides, the J terms are better described as linear combinations of ground and excited terms, owing to even larger spin-orbit coupling effects and the vicinity of free ion terms.

Ligand field effects split the J manifold in a way that is not easily predicted without specific calculations. However, the overall splitting is such that many of the levels are appreciably populated at room temperature. An elegant procedure that takes such effects into account in a general way with respect to dipolar shifts has been provided by Bleaney [9].

10.3 THE DIPOLAR CONTRIBUTION TO THE ISOTROPIC SHIFTS

Ligand field effects remove the spherical symmetry around the metal ion and cause magnetic susceptibility anisotropy. The dipolar shifts are then given by Equation (2.18). As already mentioned in Section 2.4 for high spin iron(III), the susceptibility can be factorized into two contributions: one is given by a Zeeman term of the type $g\mu_B B_0 S_z$, and the other is given by the splitting of the magnetic manifold, which provides anisotropy. Because the orbital and spin angular momenta are strongly coupled, the g factor in lanthanides is a function of both L and S and of their vector combination J; it is given by

$$g_J = 1 + \frac{J(J+1) - L(L+1) + S(S+1)}{2J(J+1)}$$

and the Zeeman term then becomes $g_J \mu_B B_0 J_z$. The Hamiltonian for the two contributions is [9]

$$\hat{\mathscr{H}} = g_J \mu_B (B_x \hat{J}_x + B_y \hat{J}_y + B_z \hat{J}_z) + D_x \hat{J}_x^2 + D_y \hat{J}_y^2 + D_z \hat{J}_z^2 \tag{10.1}$$

where $D_{x,y,z}$ are parameters that take into account the ligand field splitting. In analogy with Equation (2.22), the former contribution provides a term in T^{-1} and the latter a term in T^{-2} if D is larger than the Zeeman energy and smaller than kT. The general expression for isotropic g_J is [9]

$$\left(\frac{\Delta\nu}{\nu_0}\right)^{\text{dip}} = -\frac{\mu_0}{4\pi} \frac{g_J^2 \mu_B^2 J(J+1)(2J-1)(2J+3)}{60(kT)^2} \times \frac{D_z(3\cos^2\theta - 1) + (D_x - D_y)\sin^2\theta\cos 2\Omega}{r^3} \tag{10.2}$$

where r, θ, and Ω are polar coordinates with respect to the principal directions of the **D** tensor, and g_J is given in Table 10.1. This equation differs from Equation (2.22) in that only the term in T^{-2} has remained.

For axial symmetry, Equation (10.2) becomes

$$\left(\frac{\Delta\nu}{\nu_0}\right)^{\text{dip}} = -\frac{\mu_0}{4\pi}\frac{g_J^2\mu_B^2 J(J+1)(2J-1)(2J+3)}{60(kT)^2}\frac{D_z(3\cos^2\theta - 1)}{r^3} \tag{10.3}$$

The parameter D_z accounts for crystal field effects, assuming that the splitting of the $2J+1$ levels does not exceed kT in energy. Although Horrocks [10] has pointed out that such an assumption is not correct, a more thorough treatment by McGarvey [11] has shown that Bleaney's treatment is capable of accounting for dipolar shifts at or near room temperature within an accuracy of 10 to 20%. The shifts at 300 K for a nucleus at 300 pm and $3\cos^2\theta - 1 = 1$, calculated following Bleaney's treatment and using a plausible value for the ligand field parameter to calculate D_z for each lanthanide [9], are reported in the last column of Table 10.1 and in Figure 10.1. For Sm(III), the shift value has been obtained by also considering the first excited $J = \frac{7}{2}$ manifold. For Eu(III), the ground state $J = 0$ does not contribute to the shift; instead, the excited levels with $J = 1$ and $J = 2$, which lie at about 400 and 1200 cm^{-1} above it, provide the dipolar shift, which has a rather complex temperature dependence.

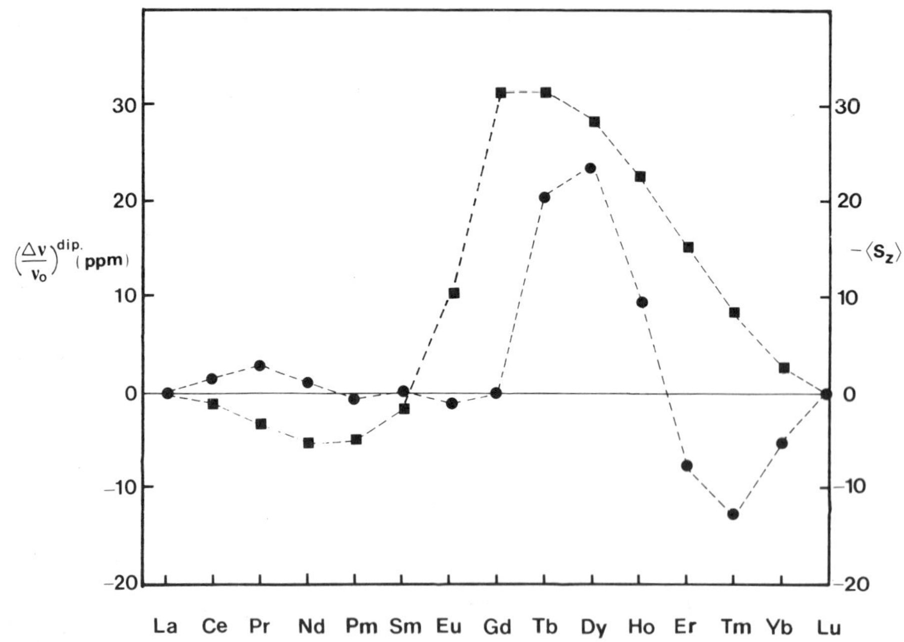

Figure 10.1 Calculated patterns of: (●) dipolar shifts [9] and (■) $-\langle S_z\rangle$ values [12] (proportional to the contact shifts) induced by lanthanide ions.

For f^7 ions, the ligand field splitting is zero under first-order conditions, whereas higher order effects account for the splitting of the $J = \frac{7}{2}$ level and a small dipolar shift. However, f^7 ions, such as Gd(III) and Eu(II), are of no use as shift reagents, rather are used as nuclear relaxing probes.

The accuracy of the present point dipole approximation with respect to a model with delocalized unpaired electrons (Section 2.2) has also been recently evaluated for f^n systems [13]. The deviations have been found to be substantially smaller than those estimated for 3d metal ions.

10.4 THE CONTACT CONTRIBUTION TO THE ISOTROPIC SHIFTS

Although the contact contribution is often assumed to be negligible for nuclei a few bonds apart from the metal ion, unless efficient spin delocalization mechanisms are operative [14], it can be sizeable and sometimes dominant for ^{17}O and ^{14}N nuclei directly coordinated to the metal [12]. The contact shift can be evaluated—see Equation (2.2)—through the average value, $\langle S_z \rangle$, on the ground J manifold, the Zeeman splitting of which is less than kT [15]:

$$\langle S_z \rangle_J = -g_J(g_J - 1)J(J+1)\frac{\mu_B B_0}{3kT} \tag{10.4}$$

where g_J is given in Table 10.1. The contact contribution is therefore [15]

$$\left(\frac{\Delta\nu}{\nu_0}\right)^{\text{con}} = \left(\frac{A_c}{\hbar}\right)\frac{g_J(g_J - 1)\mu_B J(J+1)}{3\gamma_N kT} \tag{10.5}$$

or

$$\left(\frac{\Delta\nu}{\nu_0}\right)^{\text{con}} = \frac{1}{\mu_0}\frac{A_c}{\hbar\gamma_N\mu_B}\frac{g_J - 1}{g_J}\chi \tag{10.6}$$

Although this simple treatment does not take into account the ligand field splitting of the J manifold, it provides a useful criterion for predicting the general trend of contact shifts along the series. Note that this method does not apply to Sm(III) and Eu(III) salts for the reason already outlined [12, 15]. The values of $\langle S_z \rangle$, as calculated by Golding and Halton [12] with or without inclusion of the excited states, are reported in Table 10.1 and Figure 10.1. It is noteworthy that $\langle S_z \rangle_J$ in lanthanides can be either negative or positive owing to the large orbital contribution to the determination of the energy levels. Since orbital contributions are larger than spin contributions, $\langle S_z \rangle_J$ can also differ in sign with respect to the spin-only value.

It is also often assumed that A_c is constant along a series of lanthanide complexes with the same ligand. Values of $A_c/\hbar$ for ^{17}O in a series of hydrated lanthanides [16] are reported in Table 10.2. The mechanism providing spin density delocalization from the metal to the ligands is due to a weak covalent bonding involving the 6s metal orbital, which, in turn, can transfer unpaired spin density on nearby nuclei through spin polarization from 4f orbitals [15].

Table 10.2 ^{17}O Hyperfine Coupling Constants for Some Lanthanide(III) Aquaions. [16]

Lanthanide Ion	$\frac{A_c}{h}$ (MHz)
Pr^{3+}	0.84
Nd^{3+}	0.69
Gd^{3+}	0.71
Tb^{3+}	0.75
Dy^{3+}	0.65
Ho^{3+}	0.69
Er^{3+}	0.67
Tm^{3+}	0.66
Yb^{3+}	0.57

10.5 SEPARATION OF DIPOLAR AND CONTACT CONTRIBUTIONS

The following method for separation of dipolar and contact shifts, proposed by Dobson, Williams, and Xavier [17], is valid for axial systems assuming that: (1) there is a nucleus in such a position in the molecule that the contact contribution is zero because it is far from the paramagnetic center; and (2) the contact hyperfine coupling is constant ($\pm 10\%$) for Nd-Tm.

For nuclei i and j

$$\frac{(\Delta\nu_i)^{\text{dip}}}{(\Delta\nu_j)^{\text{dip}}} = R_{ij}$$

If $(\Delta\nu_j)^{\text{con}} = 0$, then

$$\frac{(\Delta\nu_i)^{\text{iso}}}{(\Delta\nu_j)^{\text{iso}}} = R_{ij} + \frac{(\Delta\nu_i)^{\text{con}}}{(\Delta\nu_j)^{\text{iso}}} = R_{ij} + \frac{A_{c(i)}}{\hbar}\,\frac{g_J(g_J - 1)\mu_B J(J+1)}{3\gamma_N kT(\Delta\nu_j)^{\text{iso}}}\,\nu_0$$

By plotting $(\Delta\nu_i)^{\text{iso}}/(\Delta\nu_j)^{\text{iso}}$ versus $g_J(g_J - 1)J(J+1)/(\Delta\nu_j)^{\text{iso}}$ for all the lanthanides, the intercept gives R_{ij} and hence the ratio between pairs of dipolar shifts (Figure 10.2). The requirement for axial symmetry apparently holds for many systems, even with low symmetry, possibly as a result of ligand rearrangement which might be fast on the NMR time scale [18–20].

Separation of dipolar and contact shifts in axial systems can also be achieved, again by using a variety of lanthanides, without assuming that for a given signal the contact shift is zero [21], as long as the patterns of both dipolar and contact shifts (Figure 10.1) hold. Here, the isotropic shift can be expressed as the sum of Equations (10.3) and (10.5):

$$\left(\frac{\Delta\nu}{\nu_0}\right)^{\text{iso}}_{ij} = \left(\frac{\Delta\nu}{\nu_0}\right)^{\text{con}}_{ij} + \left(\frac{\Delta\nu}{\nu_0}\right)^{\text{dip}}_{ij} = KA_{c(i)}\langle S_z\rangle_j + G_i C_j$$

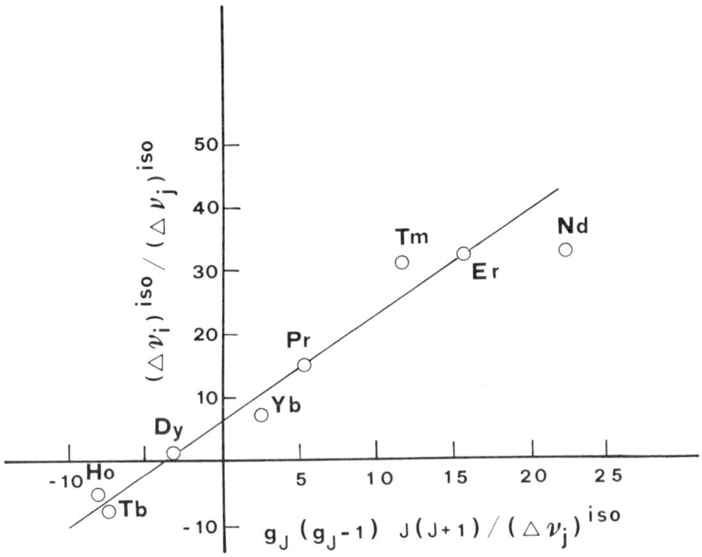

Figure 10.2 Observed ratios between the isotropic shifts of the ^{31}P [$(\Delta\nu_i)^{iso}$] and ^{1}H [$(\Delta\nu_j)^{iso}$] resonances of cytidine monophosphate interacting with lanthanide ions, as a function of $g_J(g_J - 1)J(J + 1)/(\Delta\nu_j)^{iso}$ for each lanthanide. Reproduced from [17], with permission; copyright © 1973 by The Royal Society of Chemistry.

where the subscript i refers to the observed nucleus and j to the lanthanide employed. $\langle S_z \rangle_j$ and C_j depend only on the lanthanide, whereas $A_{c(i)}$ and G_i are assumed to depend only on the nucleus and not on the lanthanide. G_i is given by $(3\cos^2\theta - 1)/r_i^3$ of equation (10.3). The unknowns for a single shift are $A_{c(i)}$ and G_i. If n signals are observed, the unknowns are $2n$; that is, $A_{c(i)}$ and G_i for the n nuclei. For a second lanthanide derivative, $A_{c(i)}$ and G_i are the same, and the experimental data are now $2n$—that is, as many as the unknowns. Using a variety of lanthanides, it is possible, in general, to set up and solve the appropriate number of simultaneous equations and thus evaluate the terms $A_{c(i)}$ and G_i for each nucleus [21]. This procedure is again valid as long as $A_{c(i)}$ and G_i depend only on the nucleus, and $\langle S_z \rangle_j$ and C_j depend only on the lanthanide in a predictable way and not on the nucleus under observation. The accuracy of the results depends on how precisely the patterns of Figure 10.1 are capable of reproducing the contact and dipolar shifts within the lanthanide series.

A method has recently been developed for factoring out dipolar and contact shifts in nonaxial cases [22]. It is, however, assumed that the in-plane anisotropy plays a minor role.

When contact contributions can be neglected, the dipolar shifts for ligand protons in series of lanthanide complexes can be reproduced by using the complete equation for nonaxial cases—Equation 10.2—and the geometric factors calculated from the X-ray structure [23–25]. The values for the D_z and $D_x - D_y$ parameters obtained through best fitting procedures are in fair to good agreement with the single-crystal anisotropy data [25].

10.6 LANTHANIDES AS DIPOLAR SHIFT REAGENTS

The f^n configurations give rise to several free ion terms that upon spin-orbit coupling give rise, except for f^7 configurations, to several closely spaced energy levels. The multiplicity of the ground levels varies from 6 to 17 (Table 10.1), and is further split by crystal field effects. According to the electronic relaxation mechanisms discussed in Section 3.7, the availability of low-lying excited levels accounts for short electronic relaxation times, again with the exception of metal ions with an f^7 configuration. For these fast-relaxing systems, the correlation times are determined by the electronic relaxation times. Values for the latter have been estimated from nuclear relaxation measurements assuming that the Solomon Equation (3.8), written in the following form, holds:

$$T_{1M}^{-1} = \frac{2}{15}\left(\frac{\mu_0}{4\pi}\right)^2 \frac{\gamma_N^2 g_J^2 \mu_B^2 J(J+1)}{r^6}\left(\frac{7\tau_s}{1+\omega_s^2\tau_s^2} + \frac{3\tau_s}{1+\omega_I^2\tau_s^2}\right) \tag{10.7}$$

where $g_J^2\mu_B^2 J(J+1)$ is more properly replaced by the experimental μ_{eff}^2 value. The τ_s values obtained for the aquaions are reported in Figure 10.3 [26], from which

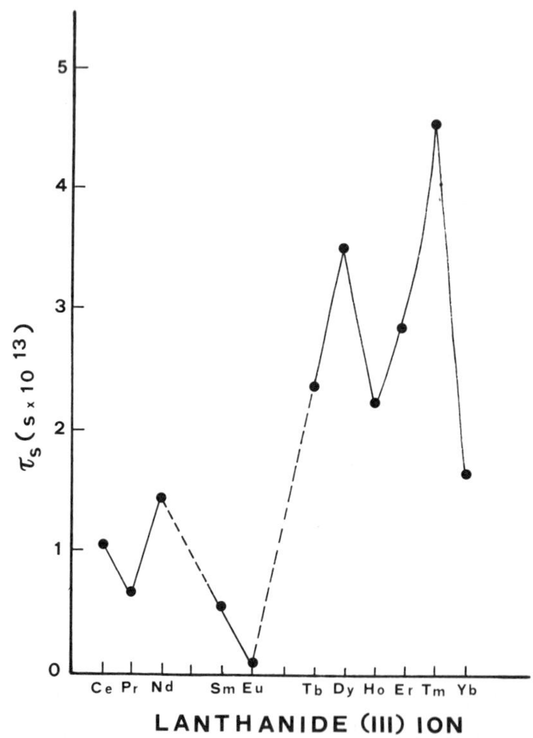

Figure 10.3 τ_s values for lanthanide(III) aquaions estimated from Equation (10.7). Reproduced from [26], with permission; copyright © 1980 by The Royal Society of Chemistry.

it can be seen that Pr(III), Sm(III), and Eu(III) display the shortest relaxation times. Information on trends in τ_s for a range of lanthanides have been obtained from T_1 values of protons of lanthanide complexes as a function of the effective magnetic moment of the complex [27]. The plot of T_{1M}^{-1} versus μ_{eff}^2 (Figure 10.4) is a straight line, indicating that τ_s is essentially independent of the lanthanide ion [apart from Gd(III)]. The τ_s values are in the range 7 to 10×10^{-13} s [27].

Such short electronic relaxation times result in very little line broadening, besides Curie relaxation [27], in NMR signals. The isotropic shifts are mainly dipolar in origin and the contact contribution, which arises from covalent bonds, is small owing to the strong radial contraction of the *f* orbitals. The hyperfine coupling constant presumably has a lower value than that of the 3*d* transition metal ions, for the same type of nuclei. Although some ions have been found to give rise to larger contact contributions than other ions of the series, such contributions are generally neglected for nuclei not directly coordinated to the metal. Factorization of the isotropic shifts can in principle be successfully achieved by utilizing different lanthanides and assuming that the dipolar shifts have axial symmetry [17, 21]. The combined effects of negligible line broadening and dominant dipolar coupling, which depends on the geometrical coordinates described in Equations (10.2) and (10.3), justify the use of these ions and their complexes as shift

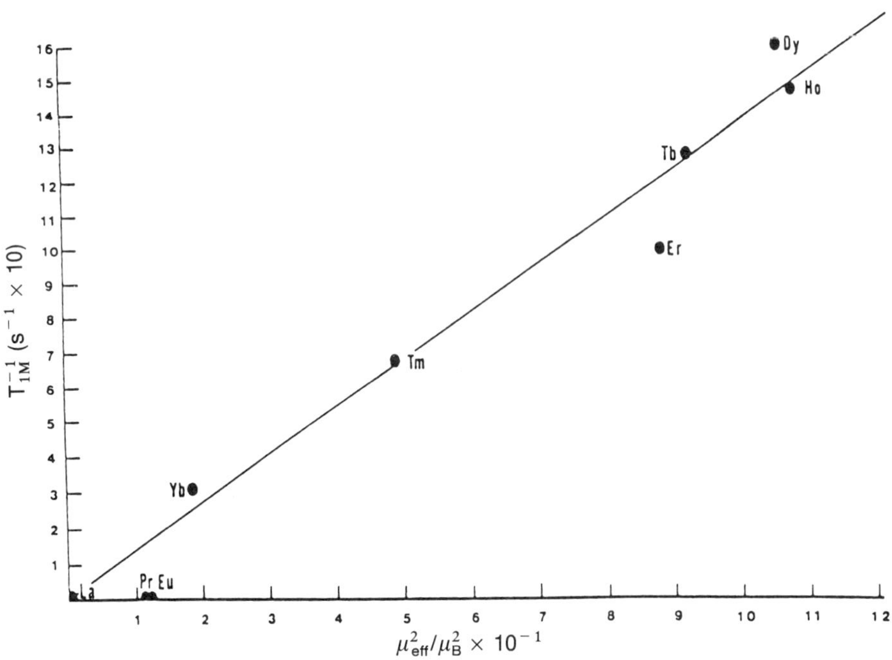

Figure 10.4 Plot of the paramagnetic contribution to the longitudinal relaxation rate at 100 MHz of the *t*-butyl protons in Ln(dipivaloylmethanate)$_3$(3,5-lutidine)$_2$ versus the square of the effective magnetic moment, μ_{eff}^2, of the various lanthanides. Reproduced from [27], with permission; copyright © 1982 by Academic Press, Inc.

reagents, and they are frequently used in organic chemistry. Since these systems have fast electronic relaxation times, the availability of high magnetic fields enables exploitation of Curie relaxation (Section 3.4) for an independent estimate of the distance of the resonating nucleus from line broadening measurements.

The successful use of lanthanides as dipolar shift reagents is also due to their tendency to interact with molecules containing oxygen and nitrogen as donor atoms. For example, lanthanide complexes with ligands of the type shown in scheme I interact with organic molecules, possibly of biological interest. Dipolar shifts and relaxation measurements may then allow mapping of the nuclei in the molecule of interest according to the guidelines given in Section 4.6.

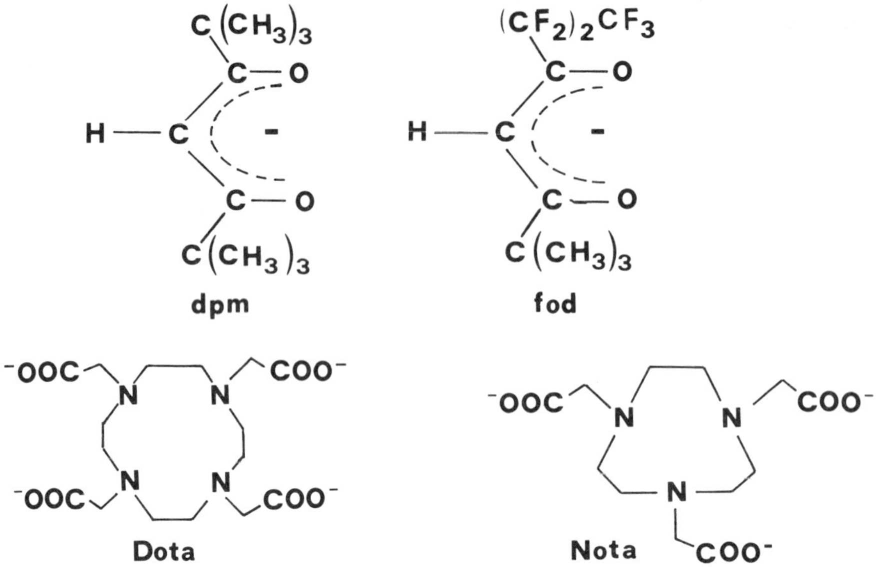

Scheme I

The use of lanthanides in biochemistry is relatively limited because the shifts are small as compared, for example, to those in 3*d* metal ion complexes for which the contact contribution is also large. In the authors' opinion, the lack of sizeable contact contributions removes a possible criterion for the assignment of the observed signals.

In particular, two systems—parvalbumin and lysozyme—have been intensively investigated, and in both cases agreement with expectations is not completely satisfactory. For both these systems, the X-ray structure of the protein was known, and the expected shift ratios were determined through experimental distances. The moderate agreement found for lysozyme [28, 29] might possibly be improved if dipolar contributions due to rhombic components in the magnetic anisotropy are also included [30]. Furthermore, the possibility that there is more than one metal binding site has also been considered [31, 32].

10.7 PARVALBUMIN

X-ray studies of parvalbumin, a calcium-binding protein isolated from carp, have revealed two calcium binding sites [33]. Furthermore, it appears that lanthanides can substitute for the calcium ion, presumably without major structural changes [34, 35]. Ytterbium(III) has been used preferentially as a lanthanide shift reagent to gain structural information in solution [36–41] because it has (1) the largest ratio between pseudocontact and contact shifts; (2) the smallest absolute value of the contact shift; and (3) a relatively large pseudocontact shift.

When the lanthanide is added in increasing concentrations to the protein, a set of signals develops up to a 1:1 ratio; upon further addition, some of these signals decrease and new ones appear (Figure 10.5). This behavior is typical of slow exchange of metal ions between two sites. The first set of signals has been assigned to protons of the first binding site; the second set has also been assigned to protons of the first site, but these protons are also believed to be influenced by the metal in the second site; the third set of signals is influenced by the second metal or both. Metal titration of parvalbumin demonstrates the inequivalence of the two sites [38]. The signals of the first site have been analyzed assuming that this site corresponds to the so-called EF site [33], according to the protein's tertiary structure.

By studying the temperature dependence of the observed shift in such a way as to determine the diamagnetic shift, the isotropic shifts have been evaluated, and have been found to follow a $T^{-2.8}$ to $T^{-3.2}$ dependence [39]. The geometrical factors were then calculated from the X-ray data for the four nuclei whose signals could be unambiguously assigned: these included the two ring protons of a histidine residue, the methyl group of the acetylated N-terminal residue, and the $^{113}Cd^{2+}$ ion purposely introduced in the other metal site (Table 10.3A) [40]. By using at least five isotropic shift values and geometrical factors, Equation (10.2) could have been used to evaluate A_1 and A_2, which are defined as follows:

$$A_1 = -\frac{\mu_0}{4\pi} \frac{g_J^2 \mu_B^2 J(J+1)(2J-1)(2J+3)}{60(kT)^2} D_z$$

$$A_2 = -\frac{\mu_0}{4\pi} \frac{g_J^2 \mu_B^2 J(J+1)(2J-1)(2J+3)}{60(kT)^2} (D_x - D_y)$$

and the three geometrical parameters, r, θ, and Ω in Equation 10.2, defining the orientation of the susceptibility tensor of the metal with respect to the X-ray coordinate system. Since only four signals were confidently assigned, six additional resonances arising from methyl protons (Table 10.3B) were considered in a best-fitting procedure. This approach gave a unique set of five parameters in reasonable agreement with the observed shifts and hence allowed the assignment of the other signals.

In principle, this treatment may be used to refine the structure on the sensitive r^3 scale of the NMR shifts, to follow conformational changes of the metal

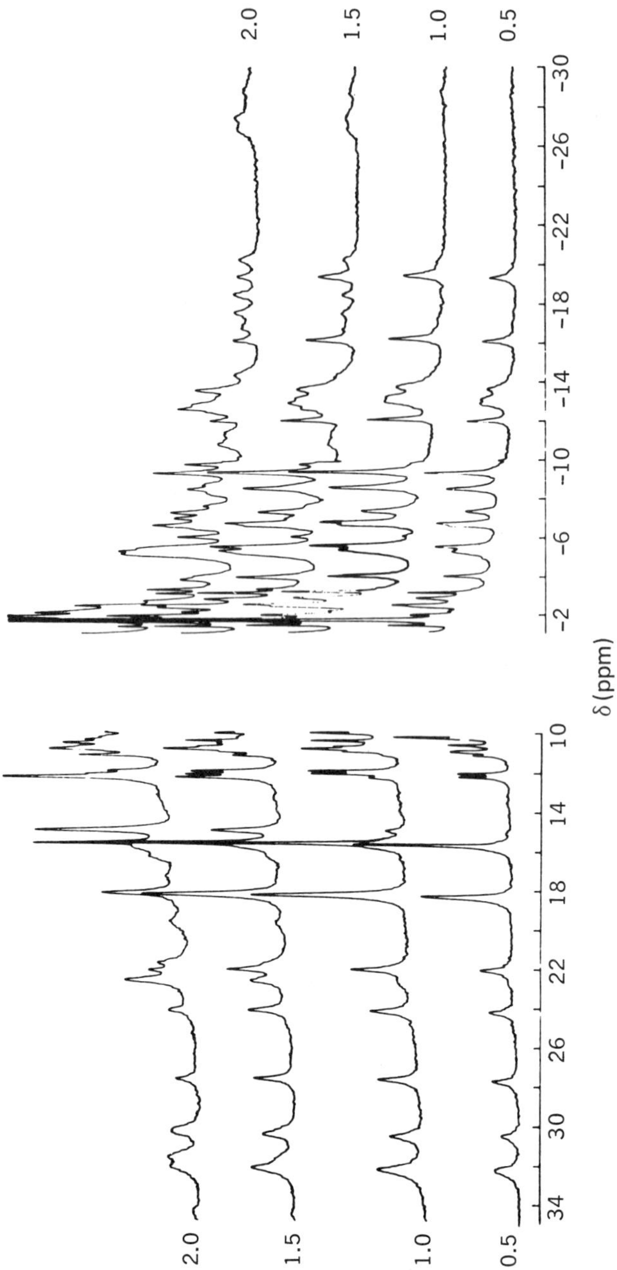

Figure 10.5 270 MHz ^{1}H NMR spectra of carp parvalbumin solutions upon addition of increasing amounts of ytterbium(III) ions (from 0.5 to 2.0 metal/protein ratios). Chemical shifts are in ppm from DSS. Reproduced from [41], with permission; copyright © 1980 by Elsevier Science Publishing Co., Inc.

Table 10.3 Observed and Calculated Isotropic Shifts for the Protons of Parvalbumin Substituted with Ytterbium(III) at the EF Site [40]

A. Assigned Resonances

Nucleus	$\left(\frac{\Delta\nu}{\nu_0}\right)^{iso}$ (ppm) Observed	Calculated	r (nm)[a]
^{1}H His 26 C2-H	0.485	0.475	1.36
^{1}H His 26 C4-H	0.343	0.332	1.51
^{1}H *N*-acetyl CH_3	0.033	0.082	2.00
^{113}Cd CD metal site	−0.270	−0.286	1.19
(^{13}C Arg 75 ζ-carbon	0.318	0.351	2.32)

B. Methyl Groups Within 1.5 nm of the EF Site

Observed $\left(\frac{\Delta\nu}{\nu_0}\right)^{iso}$ (ppm)	Observed r (nm)[a]	Calculated $\left(\frac{\Delta\nu}{\nu_0}\right)^{iso}$ (ppm)	Nucleus		r (nm)[b]
17.69	0.62	17.76	Leu 86	δ-1	0.614
14.17	0.79	10.17	Ile 97	δ-1	0.719
		6.13	Leu 86	δ-2	0.835
		5.82	Ile 97	γ-2	0.598
		4.39	Leu 63	δ-1	1.013
		3.50	Ile 58	γ-2	0.959
		2.87	Leu 63	δ-2	1.126
		1.87	Thr 82	γ-2	1.320
		1.30	Leu 105	δ-2	1.028
		0.94	Ile 50	γ-2	1.334
		0.60	Ile 58	δ-1	1.024
		0.58	Ile 50	δ-1	1.422
		0.28	Ala 88	β	1.173
		0.18	Leu 105	δ-1	1.118
		+0.17	Val 106	γ-2	1.345
		−0.50	Val 43	γ-1	1.453
−0.37		−0.65	Thr 103	γ-2	1.414
−1.44		−1.22	Ala 104	β	1.068
−1.62		−2.31	Val 99	γ-1	1.076
−1.90		−2.97	Val 99	γ-2	0.902

Table 10.3 *(continued)*

C. Most Downfield Shifted Signals

$\left(\frac{\Delta\nu}{\nu_0}\right)^{iso}$ (ppm)		Calculated	
Observed	Calculated	CH Nucleus	Relative Broadening[c]
	39.5	Ile 97 α	25.4
27.62	36.8	Asp 92 β-2	4.8
25.77	36.7	Ile 97 γ-1	3.6
22.26	28.3	Asp 92 β-1	3.8
22.09	25.3	Ile 97 γ-2	1.3
20.10	21.6	Asp 90 β-1	3.5
18.90	17.5	Lys 96 α	2.1

[a] r is the distance between the nucleus listed and the EF site.

[b] Calculated from the X-ray crystallographic coordinates.

[c] Predicted line broadening relative to the linewidth of the resonance with an observed chemical shift of 29.80 ppm (not shown) and $r = 0.585$ nm. The predicted line broadening is based on a r^{-6} dependence.

binding sites in solution, and to determine the structure of similar sites in other calcium-binding proteins.

Table 10.3 reports the observed and calculated signals [40]. Overall, the agreement is relatively good. Major discrepancies are only observed in the most shifted signals (Table 10.3C), which are expected to be more sensitive to inaccuracies of the X-ray coordinates. Furthermore, the most downfield signal predicted is not observed, possibly because it is broadened beyond detection.

The relaxation times, T_1 and T_2, which are determined by dipolar and Curie mechanisms (at least at 270 MHz), respectively, provide a further means of checking the assignment. For example, the assignment of two of the methyl signals has been checked through T_2 measurements (Table 10.3B): the resonance at 36.8 ppm, which has a T_2 of 2.3×10^{-3} s corresponding to a distance of 480 pm ($\tau_r = 1.2 \times 10^{-8}$ s), is assigned to the β-1 and β-2 protons of Asp 92 (Table 10.3C) [41].

10.8 GADOLINIUM(III) AS A RELAXATION PROBE

Gadolinium(III) is an f^7 ion with an $^8S_{7/2}$ ground term. The seven electrons occupy the seven f orbitals with one electron per orbital. In analogy to manganese(II), which has a ^6S ground term and displays the longest electronic relaxation time of the 3d metal ions, gadolinium(III) also has the longest relaxation times of all the lanthanides. Indeed, it is the only ion in the series that can be easily investigated through EPR. A lower limit to its electronic relaxation time can be set to be around 2×10^{-10} s from the room temperature X-band EPR linewidth of the hydrated ion [42], but this is probably longer in macromolecular [42, 43] complexes. Therefore, the ion can be used as a relaxing probe. Despite the similarity of its electronic relaxation time to that of copper(II), the high J value of gadolinium(III)—see Equation (10.5)—results in a nuclear relaxing capability closer to that of manganese(II). The formation of gadolinium(III) protein complexes allows the mapping of molecules interacting with the paramagnetic center in fast chemical exchange [44, 45] (see Section 4.6). A particularly useful application of this technique is the detection of water coordinated to gadolinium(III) in gadolinium(III) protein complexes through water ^{1}H NMR measurements, possibly at variable magnetic fields (see Section 10.8.2).

10.8.1 Water ^{1}H NMRD of Gadolinium(III) Solutions

The NMRD data for water solutions of gadolinium(III) are shown in Figure 10.6 [46]. The only dispersion observed, at around 5 MHz, is easily ascribed to the

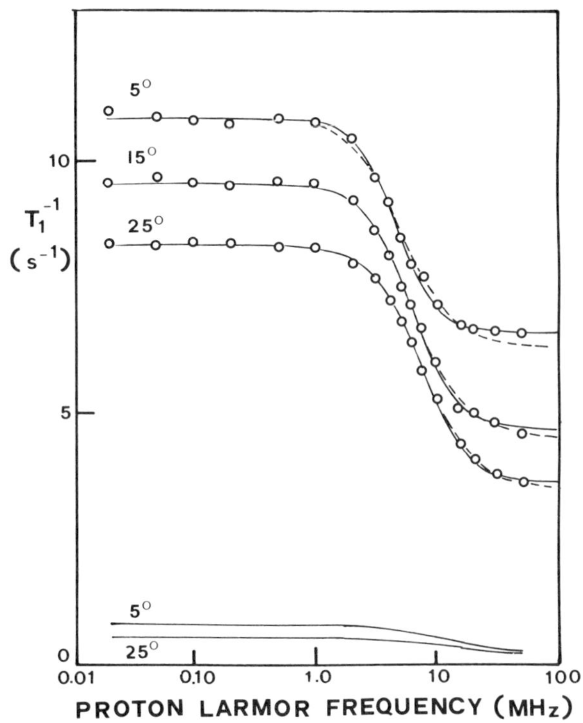

Figure 10.6 Water proton NMRD data for 3×10^{-4}M water solutions of gadolinium(III) aquaion. The solid lines represent fitting of the data assuming fast exchange and field-dependent τ_c; the dashed lines represent fitting of the data assuming field-independent τ_c and contribution of τ_M to T_1. The lower solid lines represent the estimated outer-sphere contribution [46].

"$\omega_s\tau_c$" = 1 dispersion, yielding an estimate for τ_c of about 5×10^{-11} s. Such a correlation time is largely determined by the rotational correlation time, τ_r. The observed drop in T_1^{-1} along this dispersion is definitely less than that expected from the Solomon equation. Therefore, assuming that the Solomon equation holds, either some contribution to T_{1M} from τ_M, or to τ_c from a field dependence of τ_s of the type shown in Equation (3.27), or both, have been invoked. The contribution from τ_M would be relatively more important at low field, where T_{1M} is shorter, than at high field, and would thus predominantly quench the low field plateau. On the other hand, if τ_s contributes to τ_c, its field dependence may account for a higher than expected high field plateau [46]. Independent data on the proton exchange time, τ_M, were needed to obtain a nonequivocal solution. ^{17}O and ^2H NMR data seem to indicate that τ_M is very short [47], thus suggesting a τ_s contribution to τ_c.

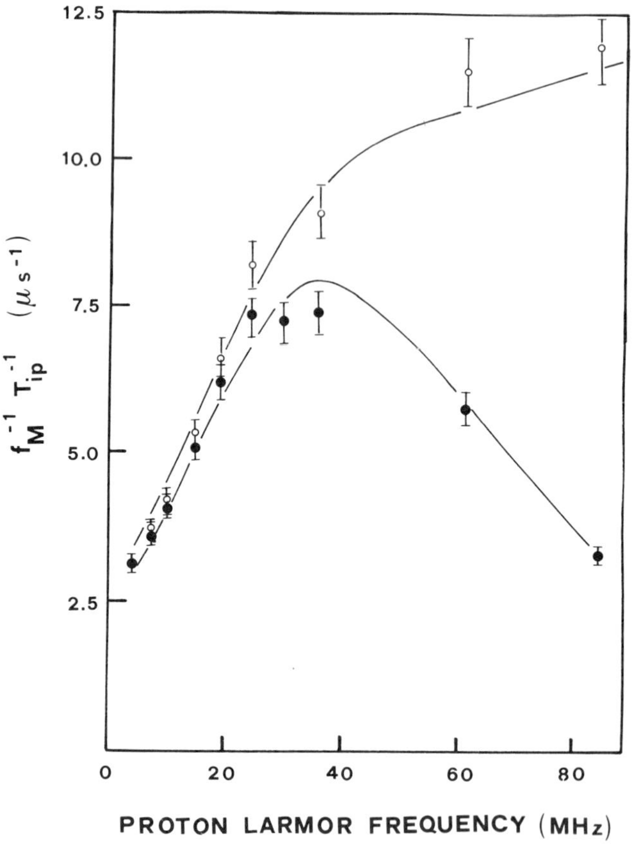

Figure 10.7 Normalized water proton relaxation rates in the gadolinium(III)-immunoglobulin G complex at 19 °C as a function of the proton Larmor frequency. (●) $f_M^{-1}T_{1p}^{-1}$; (○) $f_M^{-1}T_{2p}^{-1}$. Reproduced from [48], with permission; copyright © 1976 by Springer Verlag, N.Y.

10.8.2 Gadolinium(III)-Immunoglobulin G

The difficulties already underlined in interpreting water proton relaxation measurements in aqueous gadolinium(III) solutions are mirrored in a thorough analysis carried out by Dwek *et al.* [48, 49] on the gadolinium(III) complex of immunoglobulin G, the principal antibody in human serum. This protein has six gadolinium(III) binding sites, two of which have much higher binding constants than the other four [50]. Under the experimental conditions employed by Dwek *et al.*, about 90% saturation of the strong sites and negligible saturation of the others was obtained. T_1 and T_2 measurements were performed in the field range 4 to 84 MHz (Figure 10.7) and the temperature dependence of both relaxation rates was measured at two different frequencies (Figure 10.8). Some qualitative considerations can be readily made from this data: (1) the peak in T_1^{-1} around 30 to 40 MHz is indicative of a field dependence of τ_c, analogous to that observed in manganese (II) macromolecule complexes (see Section 8.5); (2) the higher values of T_2^{-1} compared to T_1^{-1} at high field rule out any important contribution of τ_M to T_1^{-1}; (3) the leveling off of T_2^{-1} at high field is, on the contrary, indicative of the slow-exchange limit being approached by the observed transverse relaxation rate at high field—this is confirmed by the temperature dependence of T_2^{-1}; and (4) the temperature dependence of T_1^{-1} which, taken alone, would be also indicative of slow exchange, can be accounted for by fast exchange and a large contribution of τ_r to τ_c at high field.

A computer analysis of the data and an accurate search for all the minima in the least-squares fit procedure, for the sharpness of the minima, and for the extent of covariance of the parameters, shows that most parameters are rather ill-defined. In particular, the best-fit value for the hydration number, 4.5, has an estimated lower limit of 1.8 and an upper limit of 8.0, showing how dangerous it can be to rely too much on a computer-fitting procedure without a deep understanding of all the possible sources of indetermination.

For the particular case of the gadolinium(III)-immunoglobulin G system, an alternative procedure has been developed [49] for obtaining chemically relevant information from relaxation data of water nuclei. This method will be described in detail since it is in principle very general, although it is probably only useful with metal ions such as gadolinium(III) and manganese(II) (cf. Section 8.5.5).

By combining Equations (3.3) and (4.4), in the limit $\omega_s\tau_c \gg 1$, the paramagnetic contribution to the nuclear relaxation time can be expressed as

$$T_{1p} = \frac{55}{[E]q}\left[\frac{K}{\gamma_N^2}\left(\frac{1 + \omega_I^2\tau_c^2}{\tau_c}\right) + \tau_M\right] \tag{10.8}$$

where the unknowns are q, τ_c, and τ_M. If T_1 is measured for protons and deuterons at the same magnetic field, Equation (10.8) can be used for both nuclei, provided the appropriate values for γ_N and ω_I are used. In this way, two equations are obtained with the same three unknowns, τ_c being identical at the same magnetic field.

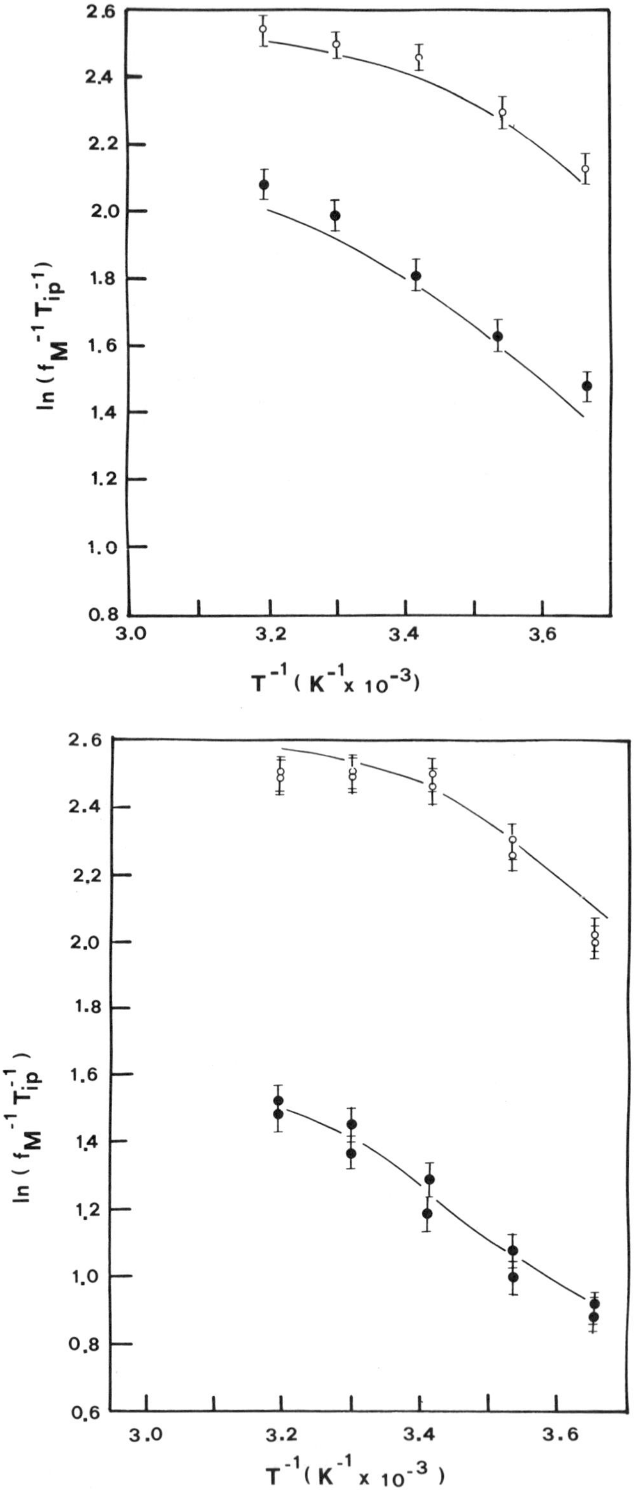

Figure 10.8 Temperature variation of the normalized water proton relaxation rates in the gadolinium(III)-immunoglobulin G complex at 61 MHz (**A**) and 84 MHz (**B**). (●) $f_M^{-1}T_{1p}^{-1}$; (○) $f_M^{-1}T_{2p}^{-1}$. Reproduced from [48], with permission; copyright © 1976 by Springer Verlag, N.Y.

When the measurements are subsequently repeated at a different magnetic field, two more equations and one further unknown—a new value for the correlation time, τ_c'—are obtained. The set of simultaneous equations can then be solved. Data at additional magnetic field values can also be added to improve the reliability of the results through a best-fit procedure. Application of this procedure to the gadolinium(III)-immunoglobulin G system has resulted in a much better definition of the q value of between 3.7 and 4.6; the possibility of the presence of four water molecules in the coordination sphere of the bound gadolinium is not surprising in the light of its tendency to exceed the coordination number six.

Although the foregoing method seems to overcome most of the problems connected with the use of nuclear relaxation rate measurements for obtaining chemical information, its applicability is limited by several factors, which have been discussed at length [49]. The first limitation arises from the intrinsically low relaxation enhancement experienced by deuteron nuclei, owing to their relatively small γ_N values (see Table 1.1 and Section 3.6), coupled with the high diamagnetic relaxation rates typical of quadrupolar nuclei. Therefore, a very efficient relaxing agent, such as gadolinium(III), is needed in a relatively high concentration, which is not always attainable with protein systems. Furthermore, the data have to be taken in a frequency range in which the $\omega_I\tau_c$ dispersion occurs, either for deuterons or, more likely, for protons. If the condition $\omega_I\tau_c \ll 1$ holds for both nuclei, then ω_I disappears from Equation (10.8) and large uncertainties can arise in the derived parameters.

10.9 THE INVESTIGATION OF LIPID BILAYERS

Lanthanides have been successfully used as shift reagents in the investigation of vesicles formed by lipid bilayers—these vesicles are obtained by sonicating a lipid dispersion. Lipid molecules have a hydrophilic head and a hydrophobic moiety composed of fatty acid chains. In water, hydrophobic interactions cause these lipid molecules to align in a head-to-tail fashion, forming a bilayer and eventually a vesicle (Figure 10.9). Phospholipids, such as phosphatidylcholine (see scheme II) have been widely used to produce vesicles of about 50 nm diameter. For vesicles of this size, the number of hydrophilic heads exposed to the outside is roughly twice that of the heads pointing inwards. Upon addition of lanthanides, interactions between the phosphate groups and the metal occur. When the lanthanide is paramagnetic, the signals of the nuclei influenced by the paramagnetic center are shifted; this shift increases with increasing lanthanide concentration, indicating that the system is in fast-exchange conditions [51, 52]. As the phosphate groups buried inside the vesicle are substantially unaffected by the outer paramagnetic ions, two signals are observed for a given nucleus with an intensity ratio of approximately 1:2 [53]. The ratios between the proton isotropic shifts are independent of the lanthanide used, since the shifts are dipolar in origin [54, 55]. This feature has been exploited to study the conformation of the choline head groups

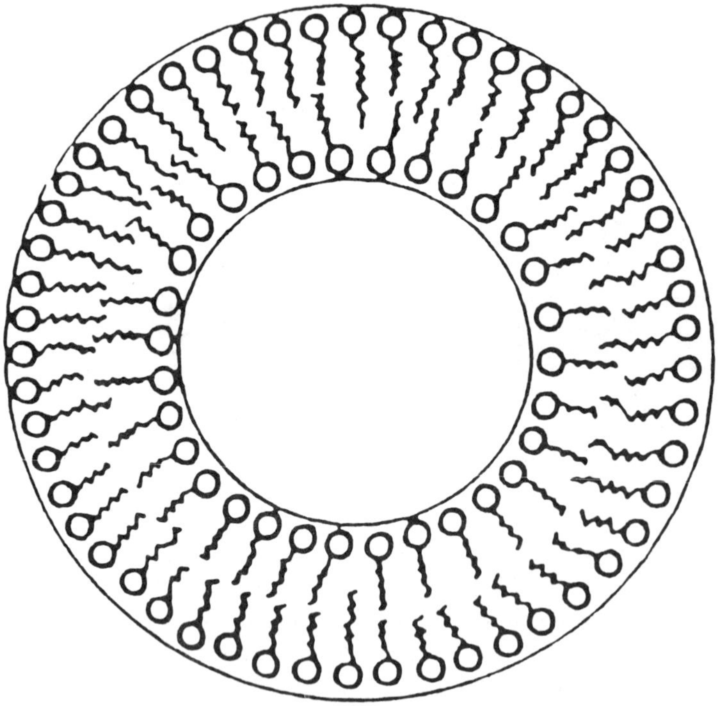

Figure 10.9 Schematic representation of a phosphatidylcholine vesicle. Open circles indicate the polar head groups.

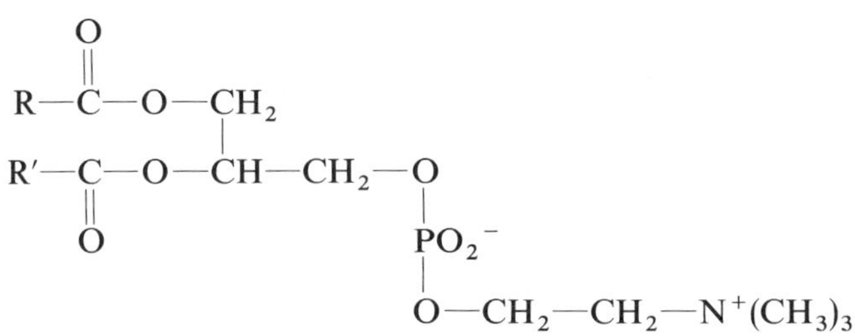

Scheme II

[55–57] (Figure 10.10). In addition, the use of gadolinium(III) as a nuclear relaxing probe [54, 55] provides further independent structural information. The isotropic shifts of the ^{31}P nuclei of the phosphate groups have large contact contributions [54], and the dipolar shift can be factorized out using the methods described in Section 10.5. The effects of lanthanide ions on the quadrupolar spectra of $I = 1$ nuclei— such as ^{2}H and ^{14}N—in lipid bilayers have also been recently analyzed in detail [59].

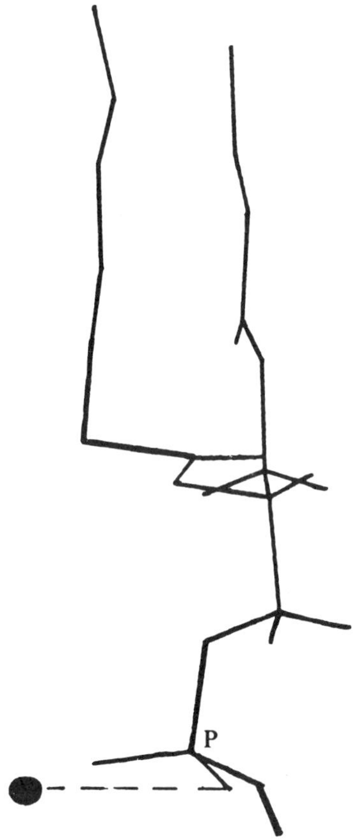

Figure 10.10 Conformation of the choline head group as determined through the use of lanthanide probes (●). Reproduced from [55], with permission; copyright © 1976 by Macmillan Journals Ltd.

In general, upon sonication of the vesicles, a single paramagnetically shifted signal is obtained, because the metal enters the vesicles [51]. The vesicles may also be made permeable to metallic cations by additions of a variety of molecules that partially substitute the starting lipid in the bilayer [58]; vesicles made from synthetic lipids with relatively short fatty chains are also permeable to metal cations [60]. The use of shift reagents thus provides a method to study the kinetics of metal transport across the membrane.

Anionic complexes of lanthanides have recently been found to induce sizeable isotropic shifts in the NMR signals of physiologically important group IA and IIA cations [61–64]. The use of such complexes seems very promising to resolve resonances from intra- and extracellular cations in intact cells and tissues [61, 65].

REFERENCES

1. Horrocks, W. DeW. Jr. *Adv. Inorg. Biochem.* **4** (1982) 201.
2. Ascenso, J. R., and Xavier, A. V. In *Systematic Properties of Lanthanides*, S. Sinha (ed.), Reidel, Dordrecht, 1983, p. 501.

3. Williams, R. J. P. *Struct. Bonding* **50** (1982) 79.
4. Swaninathan, V., and Sundaralingam, M. *CRC, Crit. Rev. Biochem.* **6** (1979) 245.
5. Reuben, J. *Prog. NMR Spectrosc.* **9** (1979) 1.
6. Inagaki, F., and Miyazawa, T. *Prog. NMR Spectrosc.* **47** (1981) 67.
7. Deslauriers, R., and Smith, I. C. P. In *Biological Magnetic Resonance*, L. J. Berliner and J. Reuben (eds.), Plenum Press, New York, 1980, Volume 2, p. 243.
8. Martin, R. B. In *Metal Ions in Biology*, T. G. Spiro (ed.), Volume 6, 1983, p. 235; Gribnau, M.C.M., Keijzers, C.P., and De Boer, E. *Magnetic Resonance Review* (1985); Banci, L., Bertini, I., and Luchinat, C. In *Rare Earths Spectroscopy*, B. Jeżowska-Trzebiatowska, J. Legendziewicz, and W. Stręk (eds.), World Scientific Publishing Co. Pte. Ltd., Singapore, 1985, p. 80.
9. Bleaney, B. *J. Magn. Reson.* **8** (1972) 91.
10. Horrocks, W. DeW. Jr., Sipe, J.P. III., and Sudwick, D. In *Nuclear Magnetic Resonance Shift Reagents*, R. E. Sievers (ed.), Academic Press, New York, 1973, p. 53; Horrocks, W. DeW. Jr. *J. Magn. Reson.* **26** (1977) 333.
11. McGarvey, B. R. *J. Magn. Reson.* **33** (1979) 445.
12. Golding, R. M., and Halton, M. P. *Aust. J. Chem.* **25** (1972) 2577.
13. Golding, R. M., Pascual, R. O., and McGarvey, B. R. *J. Magn. Reson.* **46** (1982) 30.
14. Wong, C.-P., and Horrocks, W. DeW. Jr. *J. Am. Chem. Soc.* **98** (1976) 7157.
15. Lewis, W. B., Jackson, J. A., Lemons, J. F., and Taube, H. *J. Chem. Phys.* **36** (1962) 694.
16. Reuben, J., and Fiat, D. *J. Chem. Phys.* **51** (1969) 4909.
17. Dobson, C. M., Williams, R. J. P., and Xavier, A. V. *J. Chem. Soc., Dalton Trans.* (1973) 2662.
18. Bleaney, B., Dobson, C. M., Levine, B. A., Martin, R. B., Williams, R. J. P., and Xavier, A. V. *J. Chem. Soc., Chem. Commun.* (1972) 791.
19. Horrocks, W. DeW. Jr. *J. Am. Chem. Soc.* **96** (1974) 3022.
20. Alsaadi, B. M., Rossotti, J. C., and Williams, R. J. P. *J. Chem. Soc., Dalton Trans.* (1980) 2151.
21. Reilley, C. N., Wood, B. W., and Desreux, J. F. *Anal. Chem.* **47** (1975) 2110.
22. Shelling, J. G., Bjornson, M. E., Hodges, R. S., Taneja, A. K., and Sykes, B. D. *J. Magn. Reson.* **57** (1984) 99.
23. Cramer, R. E., Dubois, R., and Furuike, C. K. *Inorg. Chem.* **14** (1975) 1005.
24. Cramer, R. E., Dubois, R., and Seff, K. *J. Am. Chem. Soc.* **96** (1974) 4125.
25. Cramer, R. E., and Maynard, R. B. *J. Magn. Reson.* **31** (1978) 295.
26. Alsaadi, B.M., Rossotti, F. J. C., and Williams, R. J. P. *J. Chem. Soc., Dalton Trans.* (1980) 2147.
27. Burns, P. D., and La Mar, G. N. *J. Magn. Reson.* **46** (1982) 61.
28. Campbell, I. D., Dobson, C. M., Williams, R. J. P., and Xavier, A. V. *Ann. N.Y. Acad. Sci.* **222** (1973) 163.
29. Campbell I. D., Dobson, C. M., and Williams, R. J. P. *Proc. Roy. Soc. London* **A345** (1975) 41.
30. Agresti, D. G., Lenkinski, R. E., and Glickson, J. D. *Biochem. Biophys. Res. Commun.* **76** (1977) 711.
31. Kurachi, K., Sicker, L. C., and Jensen, L. H. *J. Biol. Chem.* **250** (1975) 7663.
32. Perkins, S. J., Johnson, L. N., Machin, P. A., and Phillips, D. C. *Biochem. J.* **181** (1979) 21.
33. Kretsinger, R. H., and Nockolds, C. E. *J. Biol. Chem.* **248** (1973) 3313. Moews, P. C., and Kretsinger, R. H. *J. Mol. Biol.* **91** (1975) 201.

34. Sowadski, J., Cornick, G., and Kretsinger, R. H. *J. Mol. Biol.* **124** (1978) 123.
35. Horrocks, W. DeW. Jr., and Sudnick, D. R. *J. Am. Chem. Soc.* **101** (1979) 334.
36. Lee, L., and Sykes, B. D. *Biochemistry* **19** (1980) 3208.
37. Lee, L., and Sykes, B. D. *Biochemistry* **20** (1981) 1156.
38. Lee, L., and Sykes, B. D. *Biophys. J.* **32** (1980) 193.
39. Lee, L., and Sykes, B. D. *J. Magn. Reson.* **41** (1980) 512.
40. Lee, L., and Sykes, B. D. *Biochemistry* **22** (1983) 4366.
41. Lee, L., and Sykes, B. D. In *Advances in Inorganic Biochemistry*, D. W. Darnall and R. G. Wilkins (eds.), Elsevier North Holland, New York, 1980, Volume 2, p. 183.
42. Stephens, E. M., and Grisham, C. M. *Biochemistry* **18** (1979) 4876.
43. Epstein, M., Reuben, J., and Levitzki, A. *Biochemistry* **16** (1977) 2449.
44. Furie, B., Griffin, J. H., Feldmann, R. J., Sokoloski, E. A., and Schechter, A. N. *Proc. Natl. Acad. Sci. U.S.A.* **71** (1974) 2833.
45. Abbott, F., Gomez, J. E., Birnbaum, E. R., and Darnall, D. W. *Biochemistry* **14** (1975) 4935.
46. Koenig, S. H., and Epstein, M. *J. Chem. Phys.* **63** (1975) 2279.
47. Marianelli, R. Unpublished results. Reuben, J. *J. Chem. Phys.* **63** (1975) 5063.
48. Burton, D. R., Forsén, S., Karlström, G., Dwek, R. A., McLaughlin, A. C., and Wain-Hobson, S. *Eur. J. Biochem.* **71** (1976) 519.
49. Burton, D. R., Dwek, R. A., Forsén, S., and Karlström, G. *Biochemistry* **16** (1977) 250.
50. Dower, S. K., Dwek, R. A., McLaughlin, A. C., Mole, L. E., Press, E. M., and Sunderland, C. A. *Biochem. J.* **149** (1975) 73.
51. Bystrov, V. F., Dubrovina, N. I., Barsukov, L. I., and Bergelson, L. D. *Chem. Phys. Lipids* **6** (1971) 343.
52. Kostelnik, R. J., and Castellano, S. M. *J. Magn. Reson.* **7** (1972) 219.
53. Bystrov, V. F., Shapiro, Yu. E., Victorov, A. V., Barsukov, L. I., and Bergelson, L. D. *FEBS Lett.* **25** (1972) 337.
54. Hauser, H., Phillips, M. C., Levine, B. A., and Williams, R. J. P. *Eur. J. Biochem.* **58** (1975) 133.
55. Hauser, H., Phillips, M. C., Levine, B. A., and Williams, R. J. P. *Nature (London)* **261** (1976) 390.
56. Barry, C. D., Glasel, J. A., Williams, R. J. P., and Xavier, A. V. *J. Mol. Biol.* **84** (1974) 471.
57. Barry, C. D., Martin, D. R., Williams, R. J. P., and Xavier, A. V. *J. Mol. Biol.* **84** (1974) 491.
58. Barsukov, L. I., Parfen'eva, A. M., Viktorov, A. V., Shapiro, Yu. E., Bystrov, V. F., and Bergelson, L. D. *Biophys Engl. Transl.* **19** (1974) 463.
59. Siminovitch, D. J., Rance, M., and Jeffrey, K. R. *J. Magn. Reson.* **58** (1984) 62.
60. Hauser, H., and Barratt, M. D. *Biochem. Biophys. Res. Commun.* **53** (1973) 399.
61. Gupta, R. K., and Gupta, P. *J. Magn. Reson.* **47** (1982) 344.
62. Pike, M. M., and Springer, C. S., Jr. *J. Magn. Reson.* **46** (1982) 348.
63. Pike, M. M., Yarmush, D. M., Balschi, J. A., Lenkinski, R. E., and Springer, C. S., Jr. *Inorg. Chem.* **22** (1983) 2388.
64. Chu, S. C., Pike, M. M., Fossel, E. T., Smith, T. W., Balschi, J. A., and Springer, C. S., Jr. *J. Magn. Reson.* **56** (1984) 33.
65. Shinar, H., and Navon, G. *Biophys. Chem.* **20** (1984) 275.

Appendix I Glossary of Abbreviations

Abbreviation	Meaning
ADC	Analog-to-digital converter
ADP	Adenosine diphosphate
AMP	Adenosine monophosphate
AO	Atomic orbital
AP	Alkaline phosphatase
ATP	Adenosine triphosphate
Ac-py	Acetylpyridine
BCA II	Bovine carbonic anhydrase, isoenzyme II
CCP	Cytochrome *c* peroxidase
CD	Circular dichroism
CGS-emu	Unit system based on three basic units (electromagnetic units version)
CP	Cross-polarization
CTP	Cytidine triphosphate
CW	Continuous wave
Cyt	Cytochrome
DEFT	Driven equilibrium Fourier transform
DMSO	Dimethylsulphoxide
DNA	Deoxyribonucleic acid
DOTA	1,4,7,10 tetraazacyclododecane-N,N′,N″,N‴-tetraacetate
DPM	Dipyvaloylmethanate
DSS	2,2-Dimethyl-2-silapentane-5-sulphonate
Dacoda	1,5-Diazacyclooctane-N,N′-diacetate
EDTA	Ethylenediamine N,N,N′,N′-tetraacetate
ENDOR	Electron-nuclear double resonance
EP	Etioporphyrin
EPR	Electron paramagnetic resonance
ESR	Electron spin resonance
EXAFS	Extended X-ray absorption fine structure
FID	Free induction decay
FOD	1,1,1,2,2,3,3-heptafluoro-7,7-dimethyl-4,6-octanedione
FOZ	First-order Zeeman
FT	Fourier transform
GOase	Galactose oxidase
HCA I	Human carbonic anhydrase, isoenzyme I
HRP	Horseradish peroxidase
Hb	Hemoglobin
HiPIP	High-potential iron-sulfur protein
Im	Imidazole

(*continued*)

Appendix I (*continued*)

Abbreviation	Meaning
INDO	Intermediate neglect of differential overlap
IR	Infrared
LCAO	Linear combination of atomic orbitals
Lu	Lutidine
MAS	Magic angle spinning
MCD	Magnetic circular dichroism
MINDO	Modified intermediate neglect of differential overlap
MO	Molecular orbital
MPDME	Mesoporphyrin IX dimethylester
MW	Molecular weight
Mb	Myoglobin
Me	Methyl
NAD^+	Nicotinamide adenine dinucleotide (oxidized form)
NADH	Nicotinamide adenine dinucleotide (reduced form)
$NADP^+$	Nicotinamide adenine dinucleotide phosphate (oxidized form)
NADPH	Nicotinamide adenine dinucleotide phosphate (reduced form)
NMRD	Nuclear magnetic relaxation dispersion
NOE	Nuclear Overhauser effect
NOTA	1,4,7 triazacyclononane-N,N′,N″-triacetate
NTA	Nitrilotriacetate
OEP	Octaethylporphyrin
PTS	*p*-toluenesulphonamide
PhDTA	*o*-phenylenediamine-N,N,N′,N′-tetraacetate
P_i	Inorganic phosphate
Py	Pyridine
Pyrr	Pyrrole
RF	Radio-frequency
RNA	Ribonucleic acid
SALEN	1,2-ethylenebis (salicylideneiminate)
SI	International system of units
SOD	Superoxide dismutase
SOZ	Second-order Zeeman
T-n-PrP	Tetra-n-propyl porphyrin
TMC	1,4,8,11-tetramethyl-1,4,8,11-tetraaza cyclotetradecane
TMS	Tetramethylsilane
T-p-CH_3PP	Tetra-p-tolylporphyrin
TPP	Tetraphenylporphyrin

Appendix I *(continued)*

Abbreviation	Meaning	
TRN	Transferrin	
TSP	3-trimethylsilyl-[2,2,3,3,-$^2H_4$] propionate	
2D	Two-dimensional	
WEFT	Water-eliminated Fourier transform	
ZFS	Zero field splitting	
ppm	parts per million	
	Amino acids	R—
Ala	Alanine	CH_3—
Arg	Arginine	H_2N—C(NH)—NH—CH_2—CH_2—CH_2—
Asn	Asparagine	H_2N—CO—CH_2—
Asp	Aspartic acid	HOOC—CH_2—
Cys	Cysteine	HS—CH_2—
Gln	Glutamine	H_2N—CO—CH_2—CH_2—
Glu	Glutamic acid	HOOC—CH_2—CH_2—
Gly	Glycine	H—
His	Histidine	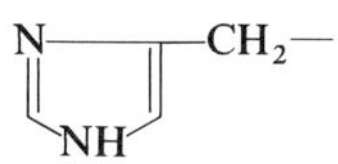
Ile	Isoleucine	CH_3—CH_2—$CH(CH_3)$—
Leu	Leucine	$(CH_3)_2CH$—CH_2—
Lys	Lysine	H_2N—CH_2—CH_2—CH_2—
Met	Methionine	CH_3—S—CH_2—CH_2—
Phe	Phenylalanine	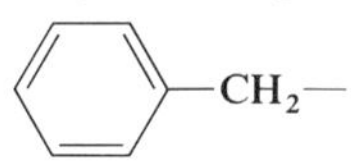
Pro	Proline	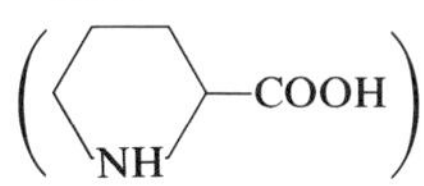
Ser	Serine	HO—CH_2—
Thr	Threonine	CH_3—CH(OH)—
Trp	Tryptophan	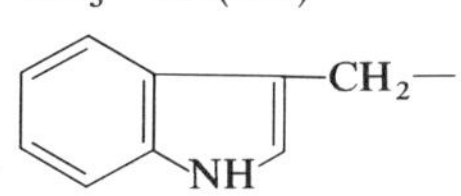
Tyr	Tyrosine	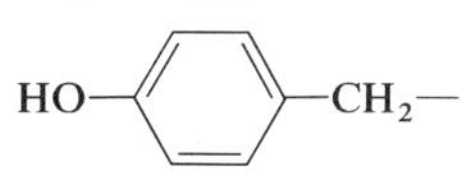
Val	Valine	$(CH_3)_2CH$—

APPENDIX II Physical and Mathematical Constants

Quantity	Symbol	Value: SI	Value: CGS emu
Physical Constants			
Permeability of vacuum	μ_0	$4\pi \times 10^{-7}$ kg m s^{-2} A^{-2}	1
Speed of light in vacuum	c	2.9979×10^{8} m s^{-1}	2.9979×10^{10} cm s^{-1}
Elementary charge (absolute value of the electron charge)	e	1.6022×10^{-19} A s	1.6022×10^{-20} abcoulombs
Planck constant	h	6.6262×10^{-34} J s	6.6262×10^{-27} erg s
	$\hbar = h/2\pi$	1.0546×10^{-34} J s rad^{-1}	1.0546×10^{-27} erg s rad^{-1}
Avogadro constant	N_A	6.0220×10^{23} mol^{-1}	
Electron rest mass	m_e	0.9110×10^{-30} kg	0.9110×10^{-27} g
Proton rest mass	m_p	1.6726×10^{-27} kg	1.6726×10^{-24} g
Electron g factor	g_e	2.0023	
Bohr magneton	μ_B	9.2741×10^{-24} J T^{-1}	9.2741×10^{-21} erg G^{-1}
Electron-to-proton magnetic moments ratio	μ_e/μ_p	658.21	
Proton magnetogyric ratio	γ_p	2.6752×10^{8} rad s^{-1} T^{-1}	2.6752×10^{4} rad s^{-1} G^{-1}
Boltzmann constant	k	1.3807×10^{-23} J K^{-1}	1.3807×10^{-16} erg K^{-1}
Hyperfine coupling constant of the hydrogen atom	$(A_c/h)_H$	1.4204×10^{9} Hz	
	$(A_c/\hbar)_H$	8.9247×10^{9} rad s^{-1}	
Mathematical Constants			
	e	2.7183	
	π	3.1416	

Appendix III Conversion Factors

Quantity	to convert from	to	multiply by
Length	angstroms (A)	meters (m)	1×10^{-10}
Pressure	atmospheres (atm)	pascals (Pa) ($kg\ m^{-1}\ s^{-2}$)	1.01325×10^{5}
Mass	atomic mass units (amu)	kilograms (kg)	1.6606×10^{-27}
Energy	calories (cal)	joules (J)	4.1840
	electron volts (eV)	joules (J)	1.6022×10^{-19}
	kilowatt-hours (kWh)	joules (J)	3.6×10^{6}
Viscosity	centipoises	$kg\ m^{-1}\ s^{-1}$	1×10^{-3}
Angles	degrees (deg)	radians (rad)	0.017453
Volume	liters (1)	cubic meters (m^3)	1×10^{-3}
Frequency	radians/seconds (rad s^{-1})	cycles/seconds (cps) (Hz)	0.15915
	wavenumber (cm^{-1})	energy (J)	1.9865×10^{-23}
		frequency (s^{-1}) (Hz)	2.9979×10^{10}
		frequency (rad s^{-1})	1.8837×10^{11}
	frequency (s^{-1}) (Hz)	energy (J)	6.6262×10^{-34}
	temperature (K)	energy (J)	1.3807×10^{-23}
		wave numbers (cm^{-1})	0.69467

Appendix IV Some Physical Quantities Used in the Book and Their SI Units

Physical Quantity	Symbol	Unit	Fundamental Units
Electric charge (quantity of electricity)	Q	C	A s
Electric current	I	A	A
Electric potential	V	V	$kg\ m^2\ s^{-3}\ A^{-1}$ ($=J\ A^{-1}\ s^{-1}$)
Energy	E	J	$kg\ m^2\ s^{-2}$
Frequency	ν	Hz	s^{-1}
Frequency (angular velocity)	ω	rad s^{-1}	rad s^{-1}
Magnetic field strength	H	$A\ m^{-1}$	$A\ m^{-1}$
Magnetic induction (flux density)	B	T	$kg\ A^{-1}\ s^{-2}$
Magnetic moment	μ	$J\ T^{-1}$	$A\ m^2$
Magnetic susceptibility	χ	m^3	m^3
Magnetization	M	$J\ T^{-1}\ m^{-3}$	$A\ m^{-1}$
Magnetogyric ratio	γ	rad $s^{-1}\ T^{-1}$	rad A s kg^{-1}
Molar magnetic susceptibility	χ_M	$m^3\ mol^{-1}$	$m^3\ mol^{-1}$
Power	P	W	$kg\ m^2\ s^{-3}$ ($=J\ s^{-1}$)
Viscosity (dynamic)	η	$kg\ m^{-1}\ s^{-1}$	$kg\ m^{-1}\ s^{-1}$

Author Index*

* Key:

Chapter and [reference number]

Alsaadi B.M. **7**[34]: 181, 200, 202, 221, *225;* **10**[20]: 278, *294*

Text pages on which reference is cited

Pages on which reference is listed

Subject Index